GENETICS, ETHICS, AND EDUCATION

Advances in human genetics and genomics are beginning to move outside the traditional realm of medicine and into the classroom. How will educational officials react when asked to incorporate personalized genomic information into the educational program?

This volume bridges the divide between science, education, and ethics around the emergent integration of genomics and education. By pairing comprehensive analysis of the issues with primers on the underlying science, the authors put all relevant parties on a level field to facilitate thorough consideration and educated discussion regarding how to move forward in this new era, as well as how best to support the future of education and the future of all students. The volume is unique in bringing together not only scholarly experts but also parents and laypersons. In doing so, it gives voice and understanding to a broad spectrum of disciplines that have a stake in the future of education.

Susan Bouregy served for thirteen years as the Director of the Yale University Human Subjects Committee and currently serves as Chief HIPAA Privacy Officer at Yale University.

Elena L. Grigorenko is a Distinguished Professor of Psychology at the University of Houston and is also affiliated with the Baylor College of Medicine, Yale University, Moscow State University for Psychology and Education, and St. Petersburg State University.

Stephen R. Latham is the Director of Yale University's Interdisciplinary Center for Bioethics and is also faculty chair of its Human Subjects Committee.

Mei Tan is a graduate research assistant in the Department of Psychology and Texas Institute of Measurement, Evaluation and Statistics at the University of Houston.

CURRENT PERSPECTIVES IN SOCIAL
AND BEHAVIORAL SCIENCES

Current Perspectives in Social and Behavioral Sciences provides thought-provoking introductions to key topics, invaluable to both the student and the scholar. Edited by world's leading academics, each volume contains specially commissioned essays by international contributors, which present cutting-edge research on the subject and suggest new paths of inquiry for the reader. This series is designed not only to offer a comprehensive overview of the chosen topics, but to display and provoke lively and controversial debate.

Published titles

Mindfulness and Performance edited by Amy L. Baltzell

Reflections on the Learning Sciences edited by Michael A. Evans, Martin J. Packer, and R. Keith Sawyer

Creativity and Reason in Cognitive Development, 2nd edition edited by James Kaufman and John Baer

Nurturing Creativity in the Classroom, 2nd edition edited by Ronald A. Beghetto and James C. Kaufman

Research and Theory on Workplace Aggression edited by Nathan A. Bowling and M. Sandy Hershcovis

Global Perspectives on Teacher Motivation edited by Helen M. G. Watt, Paul W. Richardson, and Kari Smith

Forthcoming title

Culture, Mind, and Brain: Emerging Concepts, Methods, Applications edited by Laurence J. Kirmayer, Marie-Francoise Chesselet, Shinobu Kitayama, Carol M. Worthman, and Constance A. Cummings

Genetics, Ethics, and Education

Edited by

Susan Bouregy
Yale University, Connecticut

Elena L. Grigorenko
Houston University and Baylor College of Medicine, Texas

Stephen R. Latham
Yale University, Connecticut

Mei Tan
Houston University, Texas

CAMBRIDGE
UNIVERSITY PRESS

CAMBRIDGE
UNIVERSITY PRESS

University Printing House, Cambridge CB2 8BS, United Kingdom

One Liberty Plaza, 20th Floor, New York, NY 10006, USA

477 Williamstown Road, Port Melbourne, VIC 3207, Australia

4843/24, 2nd Floor, Ansari Road, Daryaganj, Delhi – 110002, India

79 Anson Road, #06-04/06, Singapore 079906

Cambridge University Press is part of the University of Cambridge.

It furthers the University's mission by disseminating knowledge in the pursuit of education, learning, and research at the highest international levels of excellence.

www.cambridge.org
Information on this title: www.cambridge.org/9781107118713
DOI: 10.1017/9781316340301

First published 2017

Printed in the United Kingdom by Clays, St Ives plc

A catalogue record for this publication is available from the British Library.

ISBN 978-1-107-11871-3 Hardback
ISBN 978-1-107-54487-1 Paperback

CONTENTS

vii

FIGURES

ix

TABLES

CONTRIBUTORS

ASBURY, KATHRYN
Psychology in Education Research Centre, Department of Education, Derwent College, University of York, United Kingdom

BARNES, MARK
Ropes & Gray LLP, Boston, MA

BARROSO, CONNIE
Educational Psychology and Learning Systems Department, Florida State University, Tallahassee, FL

BECKMANN, KATHERINE
Senior Policy Advisor for Early Childhood Health and Development, Administration for Children and Families, US Department of Health and Human Services, Washington, DC

BOUREGY, KRISTA
Department of Education and Faith-Justice Institute, Saint Joseph's University, Philadelphia, PA

BOUREGY, SUSAN
HIPAA Privacy Office, Yale University, New Haven, CT

CONTRERAS, JORGE L.
S.J. Quinney College of Law, The University of Utah, Salt Lake City, UT

COWEN, CAROLYN D.
248 Highland Street, Marshfield, MA

DESHMUKH, VIKRANT G.
The University of Utah, Salt Lake City, UT

FISHER, CELIA B.
Marie Ward Doty University Chair in Ethics, Director Center for Ethics Education, Fordham University, Bronx, NY

FISHER, MARISA H.
Department of Counseling, Educational Psychology, and Special Education, Michigan State University, East Lansing, MI

GRIGORENKO, ELENA L.
Department of Psychology and Texas Institute of Measurement, Evaluation, and Statistics, University of Houston, TX; and Baylor College of Medicine, Departments of Pediatrics and Molecular and Human Genetics, Houston, TX

HARDEN, K. PAIGE
Department of Psychology and Population Research Center, University of Texas at Austin, TX

HART, SARA A.
Department of Psychology & Florida Center for Reading Research, Florida State University, Tallahassee, FL

HODAPP, ROBERT M.
Vanderbilt Kennedy Center & Department of Special Education, Peabody College, Vanderbilt University, Nashville, TN

KAPHINGST, KIMBERLY A.
Department of Communication, The University of Utah, Salt Lake City, UT

KORNILOV, SERGEY A.
Texas Institute for Measurement, Evaluation, and Statistics, University of Houston, TX; Department of Molecular and Human Genetics, Baylor College of Medicine, Houston, TX

KRAPOHL, EVA
MRC Social, Genetic and Developmental Psychiatry Centre, Institute of Psychiatry, Psychology and Neuroscience, King's College London, London, UK

LATHAM, STEPHEN R.
Interdisciplinary Center for Bioethics, Yale University, New Haven, CT

LITTLE, CALLIE
Department of Psychology, Florida State University, Tallahassee, FL

MANDELMAN, SAMUEL D.
Department of Neurological Surgery, Weill Cornell Medical College, New York, NY

O'DONNELL, KIERAN
Postdoctoral Research Fellow, Douglas Mental Health University Institute, Montreal, Canada

PELOQUIN, DAVID
Ropes & Gray LLP, Boston, MA

PETRILL, STEPHEN A.
The Ohio State University, Columbus, OH

RANDI, JUDI
University of New Haven, West Haven, CT

RIMFELD, KAILI
MRC Social, Genetic and Developmental Psychiatry Centre, Institute of Psychiatry, Psychology and Neuroscience, King's College, London, UK

SCHENKER, VICTORIA J.
The Ohio State University, Center on Education and Training for Employment, Columbus, OH

TAN, MEI
University of Houston, Department of Psychology and Texas Institute of Measurement, Evaluation, and Statistics, Houston, TX

TUCKER-DROB, ELLIOT M.
Department of Psychology and Population Research Center, University of Texas at Austin, TX

ABBREVIATIONS

A	adenine
AATD	alpha-1 antitrypsin deficiency
ACE	allelic variance (A), with contributions from shared (C) and unshared (E) environments
ACLA	American Clinical Laboratory Association
ACOG	American Congress of Obstetricians and Gynecologists
ACT	American College Testing
AD	atherosclerosis
ADHD	attention-deficit hyperactivity disorder
ADI-R	Autism Diagnostic Interview – Revised
ADOS	Autism Diagnostic Observation Schedule
APA	American Psychiatric Association
APOE4	alipoprotein E4
ART	alternative reproductive technique
ASD	autism spectrum disorder
ASRM	American Society for Reproductive Medicine
C	cytosine
CADDRE	California Center for Autism and Developmental Disabilities Research and Epidemiology
CAP	Colorado Adoption Project
CDC	Centers for Disease Control and Prevention
CDCV	common disease/common variant
CDRV	common disease/rare variant
CLIA	Clinical Laboratory Improvement Amendments of 1988
CMS	Centers for Medicare and Medicaid Services
CNP	copy number polymorphism
CNV	copy number variants

COMT	catechol-O-methyltransferase gene
CRISPR	clustered regularly interspaced palindromic repeats
CVD	cardiovascular disease
DDS	Department of Developmental Services
DIPBEC	Diagnosis of Psychotic Behavior in Children
DNA	deoxyribonucleic acid
DOE	Department of Energy
DOHaD	Developmental Origins of Health and Disease
DSM-IV	*Diagnostic and Statistical Manual of Mental Disorders,* 4th Edition
DSM-IV-TR	*Diagnostic and Statistical Manual of Mental Disorders,* 4th Edition, Text Revision
DSM-V	*Diagnostic and Statistical Manual of Mental Disorders,* 5th Edition
DTC	direct-to-consumer
DTCGT	direct-to-consumer genetic/genomic testing
DZ	dizygotic
EBN	evidence-based nutrition
EEG	electroencephalography
EGDS	Early Growth and Development Study
EPD	experience-producing drive
EPO	European Patent Office
ERPs	event-related potentials
ESTs	expressed sequence tags
EU	European Union
FAPE	Free Appropriate Public Education
FDA	Food and Drug Administration
FDCA	Food, Drug, and Cosmetic Act
FERPA	Family Educational Rights and Privacy Act
FH	familial hypercholesterolemia
fMRI	functional magnetic resonance imaging
FTC	Federal Trade Commission
G	guanine
GAO	US Government Accountability Office
GAT	genetic ancestry testing
GCTA	genome-wide complex trait analysis
GINA	Genetic Information Nondiscrimination Act
GPA	grade point average
GT	genetic testing
GWAS	genome-wide association studies

GxE	gene × environment interactions
HGP	Human Genome Project
HIPAA	Health Insurance Portability and Accountability Act
HPA	hypothalamic pituitary adrenal axis
HUGO	Human Genome Organization
IDEA	Individuals with Disabilities Education Act
IDEIA	Individuals with Disabilities Education Improvement Act
IEP	Individualized Educational Program/Plan
IQ	intelligence quotient
IRB	Institutional Review Board
IVF	in vitro fertilization
LD	learning disability
LD	linkage disequilibrium
LDT	laboratory developed test
MAF	minor allele frequency
MRI	magnetic resonance imaging
MRT	mitochondrial replacement techniques
MZ	monozygotic
NIEHS	US National Institutes of Environmental Health Sciences
NIH	US National Institutes of Health
NGS	next-generation DNA sequencing
OGOD	one gene one disorder
PCR	polymerase chain reaction
PGD	preimplantation genetic diagnosis
PGS	personal genome service
PKU	phenylketonuria
PMA	premarket approval process
PTSD	posttraumatic stress disorder
RAC	Recombinant DNA Advisory Committee
RD	reading disability
REAL-G	Rapid Estimate of Adult Literacy in Genetics
*r*GE	genotype–environment correlations
RRR	relative recurrence risk
RTI	response to intervention
SAT	Scholastic Aptitude Test
SB	Stanford-Binet Intelligence Scales
SES	socioeconomic status
SLD	specific learning disorder

SNP	single nucleotide polymorphism
SNV	single nucleotide variant
SRCD	Society for Research in Child Development
T	thymine
WAIS-R	Wechsler Adult Intelligence Scale – Revised
WES	whole-exome sequencing
WGS	whole-genome sequencing

Introduction

SUSAN BOUREGY, ELENA L. GRIGORENKO,
MEI TAN, AND STEPHEN R. LATHAM

It is widely accepted that two major scientific initiatives in the late twentieth/early twenty-first century, The Human Genome Project (HGP) and The Decade of the Brain, have resulted in tremendous progress in our understanding of how the genome and the brain work. Related fundamental discoveries have additionally reshaped classical biology and created new scientific fields, most notably the growth of neuroscience as a specific discipline. The former has bourgeoned, expanding the horizons of genetics and introducing a family of "-omic" disciplines (e.g., genomics, proteomics, epigenomics, transcriptomics), of which genomics (i.e., a genetics subdiscipline utilizing recombinant DNA, DNA sequencing, and bioinformatics methods to sequence, assemble, and analyze the function and structure of genomes – the complete set of DNA within a single cell of an organism), in particular, is featured almost daily both in the scientific literature and mass media. The latter, neuroscience, though, has been challenged to generate translational applications for critical areas of practice, in particular, education and neuroeducation (Ansari, De Smedt, & Grabner, 2012; Carew & Magsamen, 2010; Devonshire & Dommett, 2010; Fischer, Goswami, Geake, & The Task Force on the Future of Educational Neuroscience, 2010; Grigorenko, 2015; Hardiman, Rinne, Gregory, & Yarmolinskaya, 2012), and educogenomics (Grigorenko, 2007a, 2007c, 2010). Specifically, we are describing the need to define the real-life impact of genetic/genomics and neuroscience phenomena *on* educational practice rather than educating students *about* these sciences.

Two actions are critically important for the successful translation of research into everyday human practice: (1) the bolstering of public knowledge and comprehension, and (2) the critical appraisal of fundamental discoveries and their connections to practice. Such a translation into the field of education (hereafter, K–12 education) – notably underappreciated and understudied today – relies heavily on the views and beliefs of the general public, in particular, parents and educators, regarding the relevance and importance of translational applications of brain and genome research for education. Whereas there are limited informative "views-and-attitudes" studies (i.e., focus groups and surveys) on the role of the brain sciences in education (Howard-Jones & Fenton, 2012; Serpati & Loughan, 2012), there are no such studies on the role of the genomic sciences in education.

The literature on the integration of genetics/genomics and education is scarce compared to the literatures incorporating genetic/genomic knowledge with other sciences. It is represented primarily by writings on the heritable influences and molecular bases for individual differences in ability/achievement. Three ongoing developments substantiate the integration of education and genetics/genomics within the classroom: the ongoing mapping of high heritability estimates for ability/achievement onto testable genetic/genomic factors; the proliferation of direct-to-consumer genetic/genomic testing (DTCGT); and the spread of genetic/genomic literacy.

Academic ability/achievement is heritable, i.e., their development and manifestation are influenced by the genome, although these influences are exerted differently in different environments (Taylor, Roehrig, Soden Hensler, Connor, & Schatschneider, 2010). As the mapping of various facets of genetic/genomic and environmental control becomes more precise, there is growing interest in finding the most productive combinations of "predispositions" (i.e., characteristics of the genome) and "conditions" (i.e., characteristics of the environome) to maximize educational attainment, lifespan outcomes, and returns to schooling. Currently, this interest resides primarily within special interest groups dedicated to particular disorders impacting achievement (Collier, 2012; Greenbaum, 2012) and select families (Madsen, 2010; Maher, 2011), but scholars have long predicted that genetic test results will eventually become a driving force for the individualization of education (Nelkin & Tancredi, 1991). Educators and researchers need to understand this force and its pros and cons.

DTCGT, offered by companies such as 23andMe, deCODE, Navigenics, Pathway Genomics, and Athleticode, among others, allows families to obtain information about ancestry, carrier status and traits ranging from disease risk and drug response to behavior and propensities for various common diseases and disorders (Gollust, Hull, & Wilfond, 2002; Gurwitz & Bregman-Eschet, 2009; Kaye, 2008; McGuire & Burke, 2008; Wright, Hall, & Zimmern, 2011). Technology continually increases the affordability/attainability of such testing. Just as psychological testing has become central to schooling and educational decision making in the twentieth century, genetic/genomic testing will gain comparable significance in the twenty-first century.

The HGP and associated technology and information leap, as exemplified in such large-scale projects as The HapMap, ENCODE, the 1000 Genomes Project, 100,000 Genomes Project, and the Human Epigenome Project, triggered an outburst of data, resulting in unparalleled access to genetic/genomic information at multiple levels, from personal to systemic, giving it a prominent role in life decision making (Gymrek, McGuire, Golan, Halperin, & Erlich, 2013; Kung & Gelbart, 2012; Maher, 2011; Rodriguez, Brooks, Greenber, & Green, 2013). The concept of personal genetics/genomics, through professional utilization (e.g., in medicine and forensics) and public consumerism, has entered public life and, inevitably, will soon be as important as the concept of personal finances, contributing to present and future family and personal lifestyle decision making. As with personal finances or hygiene, a certain level of genetic/genomic literacy will be required to interpret and accept the notion that behavior, educational attainment, and other "features" of contemporary humans are influenced by the genome. How is that level defined? As the field's understanding of the genome is still a "work in progress," the initial key perhaps is in debunking common misconceptions (Bowling et al., 2008; Henderson & Maguire, 2000; Hook, DiMagno, & Tefferi, 2004; Lanie et al., 2004; Mills Shaw, Van Horne, Zhang, & Boughman, 2008) such as ideas of determinism, singularity of causation, and irreversibility of effects. Concepts that are crucial to becoming "genome-literate" include understanding family background, genetic risk and pleiotropic effects, and the co-action of the genome and environome in shaping traits and conditions. Yet, it is unclear how genetic/genomic literacy can be achieved. As a recent report indicates, these ideas are still inadequately covered both in school textbooks (Dougherty, Pleasants, Solow, Wong, & Zhang, 2011) and in professional courses (i.e., for professionals in healthcare [Feero & Green, 2011; Guttmacher, Porteous, & McInerney, 2007; Korf, 2011], social work [Kingsberry, Mickel, Wartel, & Holmes, 2011], and insurance [Korf, 2011]) and are loaded with unresolved ethical questions (Fisher & Harrington McCarthy, 2013).

In 2008 and 2011/2012, the National Cancer Institute Health Information National Trends Survey (2013) gauged public awareness of DTCGT in the US population, finding a significant increase of 7.6 percent. However, this awareness is not equally distributed throughout the population. Those in the age bracket of fifty to seventy-four years were significantly more aware of DTCGT than eighteen- to forty-nine-year-olds and individuals seventy-five years and older. This may be related to the finding that those with a prior cancer diagnosis (quite prevalent in this age group) were also more aware than those without a previous history of cancer. In addition, awareness increased with level of education, and those in urban settings were more aware than those in rural locations (Health Information National Trends Survey, 2013). While specific numbers and demographic details on the actual consumers of DTCGT are not readily available, researchers have generally characterized them according to their motivations for seeking out such services. They belong to three comprehensive categories: first, identity-seeking individuals engaging in GT to explore ancestry and ethnicity or to determine paternity; second, patients undergoing testing that has been ordered by a physician to check on the potentiality of disease; and, third, novelty seekers searching for new ways to improve their lifestyles (Su, 2013). These motivations are likely to expand in scope and complexity as genetic/genomic research continues to reveal more links between phenotype and genotype, including educational phenotype, and as awareness of DTCGT grows. Individuals who have undergone DTCGT themselves or used these services to learn more about their children, and who believe that their or their children's genetic/genomic profiles merit a modified educational approach, will be thrust into the existing mechanisms that govern the ways that school districts grant accommodations.

Currently, all states' special educational practices for children are guided by the Individuals with Disabilities Education Act (IDEA), first established in 1986 and reauthorized in 2004 (now known as Individuals with Disabilities Education Improvement Act – IDEIA). IDEIA is based on the principle that all children, including those with disabilities, are entitled to a free, appropriate public education that can meet their unique needs. Under this legislation, a child whom a school professional believes may have a disability is entitled to all relevant evaluations; the creation of an Individualized Education Plan (IEP), which outlines specifically what is needed to reach the educational goals set by parents, educational providers, and the child him- or herself; and mandates placement in as unrestricted an environment as possible, i.e., a typical classroom, if viable. In addition, if a parent feels that an IEP is not appropriate or that the child is not receiving the warranted services, it is the parent's right under the IDEIA to challenge the educational system and engage in due process. Schools are thus charged to provide sound curriculum-based instruction to children with disabilities, and to work closely with family members to provide the most appropriate educational supports to the child to maximize the child's potential to participate productively in society. These evaluations, whether the literature contains specific recommendations or not, may soon – and, in some cases, already do – include genome-related data.

Educators are ill-prepared to face parents who, armed with genetic/genomic data indicating a probability that their child is disabled or gifted, attempt to use those data to gain access for their children to enhanced education resources. There are anecdotal reports of such attempts already being made – and more will surely come. School systems need to decide what scientific benchmarks will justify their choice to regard any genetic/genomic data as a more reliable or useful indicator of a child's need for special intervention than currently utilized data from the educational process itself or related cognitive, academic, and behavioral assessments. Policies and procedures are needed for such determinations, lest ad hoc systems unfairly shower public resources upon squeaky-wheel parents. School systems, in turn, need to develop standards for the security and privacy of genetic/genomic data; standards governing the disclosure of such data to family members who may share traits controlled by the genome, rendering student data applicable not only to the student but also to close family members; and standards governing disclosure of "incidental findings" – health risks or genealogical information detected in genome-related data originally supplied for educational rather than diagnostic purposes.

If genetic/genomic data do end up playing a role in the allocation of school resources, this will raise difficult issues of justice and access. Poorer and less educated families may not be aware of the advantages to which GT may give them access. They may not know how to get such testing done, and, if they know, they may not be able to afford it. School systems will then have to decide whether fairness demands public support of genetic/genomic testing or whether, on the other hand, public schools should refuse to consider such test results because their constituents have unequal access to it. Whereas fair access to educational resources would appear to suggest a policy mandating the genome screening of all children, doing so would raise a number of concerns. Screening the genome is not the same as screening academic performance itself. In most cases, the presence or absence of a given allele (or other structural or functional variant) is unlikely to be perfectly correlated with student achievement, but will exhibit a complex and probabilistic pattern of additive and multiplicative effects, implicating other alleles/genes/variants, as well as environmental influences. Environmental influences that impact achievement (everything from family and peer contexts – from nutrition to study habits to parental role modeling) will moderate the impact in any given individual student. The subtleties of heritability and the amount of uncertainty generated by these subtleties can be difficult to grasp for students, parents and teachers alike, necessitating a plan to educate teachers and administrators in the basics of genetic and genomic sciences, so that a child's genetic/genomic profile does not result in pigeonholing or in the unintentional creation of a self-fulfilling prophecy about that child's potential.

The accumulation of genetic/genomic data is impossible to stop. The dramatic changes triggered by the HGP have already reshaped the lay of the land (Hoppe, 2013), such that massive amounts of relevant data (and services) are not only available but are readily accessible to a sizable group of consumers, whether firms or individuals. Yet, while discussions of the ethical uses of genetic/genomic data in medicine, forensics, and economics are in full force, discussion pertaining to these issues in education has been tentative at best. Such discussion is clearly needed as the incorporation of genetic information into the education sphere seems inevitable, and we as a society should prepare ourselves to respond in a scientifically, ethically, and fiscally responsible manner.

The HGP's impact on biotechnology and medicine has been monumental. However, the HGP also introduced privacy and social issues that have led to federal and private monitoring of the use of genetic/genomic information by individuals and institutions. Ethical issues related to the HGP are expected to become even more complex as the knowledge is applied to human behavior and penetrates multiple societal systems, including education (Buchanan, 2011; Grigorenko, 2007a, 2007b). The future of the utilization of HGP knowledge across and within these multiple systems depends on society's readiness to incorporate the HGP's scientific advances and deliver them to these systems' customers in accordance with ethical principles and the highest standards of practice (Buchanan, Brock, Daniels, & Wikler, 2000; Hook et al., 2004). Where ethicists have discussed the impact of genetics/genomics on education, they have concentrated primarily on ethical principles governing the possible use of future genetic learning enhancements; these discussions are part of a much larger bioethics literature on biomedical enhancement of humans (Harris, 2007; Savulescu & Bostrom, 2009). But, apart from one prescient article in 1991 (Nelkin & Tancredi, 1991), ethicists have ignored the more pressing ethical problems that GT results could pose for our educational system in the very near future – long before genetic/genomic mind enhancement becomes possible. Very soon, as GT results become increasingly available to parents and pediatricians (Hensley Alford et al., 2011), school officials will have to learn how to differentiate traits from conditions, and to make corresponding decisions about institutional accommodations for those with learning-related conditions. Schools will have to develop policies concerning issues of parental choice and student assent for interventions; right to know vs. obligation to share when it pertains to transmitting genetic risk to subsequent generations (Dickens, Pei, & Taylor, 1996); right to protect vs. right to breach confidentiality for the sake of the protection of relatives (Andrews, 1992); right to make a decision to test minors (Clayton, 1997; Hanson & Thomson, 2000; Howard, Avard, & Borry, 2011; Lucassen & Montgomery, 2010; McConkie-Rosell & Spiridigliozzi, 2004; Parker, 2010); the choices made in the aftermath of GT (Hook et al., 2004; Middleton, Hewison, & Mueller, 1998); potential for harm through stigmatization and discrimination (Kegley, 1996) or self-limitation; permissibility of certain types of genetic treatment (Hook et al., 2004); justice of unequal access to genetic/genomic information; and global and local issues of public protection.

Another special area of concern pertaining to genetic/genomic data addresses issues of informed consent, privacy, and confidentiality. Consenting individuals (whether DNA donors themselves or on behalf of their children) might not realize how much information they disclose by agreeing to subject their DNA to certain analytical techniques (Greenbaum, Sboner, Mu, & Gerstein, 2011; Gymrek et al., 2013; Rodriguez et al., 2013). This vulnerability arises from the very character of genetic/genomic data and their content (of which we still understand only a portion), size (which is massive, requiring specialized computing facilities and skills; in many cases, once generated and processed, the data cannot be "taken back"), and nature (possibly disclosing data across multiple generations of relatives – for an illustration, see Jim Watson's case study; Davies, 2010). Thus, current and future usage of genetic/genomic data presents a nontrivial issue, where the boundary between access and protection remains elusive. The most stunning lack, however, concerns the omniabsence of any conversation about the utilization of genetic/genomic tests and the potential need for regulation analogous to The Genetic Information Nondiscrimination Act of 2008 (GINA, Pub.L. 110–233, 122 Stat. 881, enacted May 21, 2008) for educational purposes.

In summary, principles and standards for the utilization of genetic and genomic data, while rapidly developing in medicine, have not even begun to be discussed in education. It is the shortage of such discourse that moved us to put this volume together. The volume is broadly focused on two objectives: (1) to delineate the relevance of genetics/genomics to child development, in general, and education, in particular; and (2) to outline applied and ethical issues concerning the integration of education and

genetics/genomics and to consider the legal, regulatory, and public perception issues specific to that integration.

The volume opens with two introductory chapters to equip the reader
with understanding of the relevant concepts and contexts. Mei Tan briefly
reviews highlights of the field of quantitative genetics, focusing specifically on the concept of heritability. Sergey Kornilov presents basic concepts
in the field of molecular genetics and genomics, preparing the reader to
understand the specific technical details of the discussions that follow.

The relevance of genetics/genomics to education is discussed in the following seven chapters. Kathryn Asbury, Kaili Rimfeld, and Eva Krapohl
briefly review research into the heritability of academic achievement,
particularly stressing the findings from investigations on what they call
the "dynamic" relationship between genes and experience. They straight-
forwardly pose the question of whether it is, or ever will be, possible to
personalize education along genetic lines, contributing to the discourse
by discussing the relevance, added value, and ethics of the utilization of
information about a child's genetic/genomic vs. environmental information in that child's education. Katherine Beckmann and Kieran O'Donnell
further develop the environmental line of reasoning briefly reviewing the
main actors in the acute stress response system, before discussing a proposed framework to describe the maladaptive effects of chronic stress. They
discuss how the emerging field of clinical epigenetics may contribute to the
field's understanding of how early-life experiences influence biology across
the lifespan, and the ethical considerations for this new field of research and
implications of recent findings for early care and education program and
policy development. Elena Grigorenko and Samuel Mandelman return the
volume's discourse to the discussion of co-contribution of genes and environments to child development and education, focusing on what is known
about the etiology of individual differences in general and specific cognitive
abilities. To follow, Elliot Tucker-Drob and Paige Harden focus on non-
cognitive skills and describe a transactional framework for understanding
how individual differences in such skills relate to cognitive development
and academic achievement. Then the volume's discourse turns, from the
discussion of abilities, to the discussion of disabilities. Callie Little, Connie
Barroso, and Sara Hart focus on learning disabilities and, within this discussion, argue that the personalized medicine approach, applied through
what they refer to as "precision education," might provide the best educational care for individuals with such disabilities. Robert Hodapp and Marisa
Fisher center their contribution on intellectual disabilities. In particular,
among other relevant comments, they state and restate the advantages

of a genetically informed approach to teaching students with intellectual disabilities, although many special education researchers and teachers continue to disregard genetic information. The final chapter of this section serves as a bridge to the next section and is aimed at its second objective. Victoria Schenker and Stephen Petrill examine the ethical implications of the role of genetics and environment on education from the perspective of behavioral genetics. They provide examples of behavioral genetic studies to examine some of the promises and barriers to using genetic information in educational settings. Specifically, they focus on three issues: (1) difficulty in translating genetic studies into educational practice; (2) misconceptions concerning how genetic effects operate in individuals versus populations; and (3) the misapplication and misinterpretation of genetic information. In their discussion of these three issues, they set up the context for the discourse that follows in the next seven chapters.

The second section of the book starts with Kimberly Kaphingst's chapter focusing on the issue of genomic literacy in general, with a particular emphasis on health decision making. The chapter describes various definitions of genetic and genomic literacy and then presents prior research regarding knowledge about genetics and genomics and the effects of literacy and numeracy skills on responses to genetic and genomic information. Priority areas for research on genomic literacy and educational practice are described with the supposition that enhancing and creating genomic literacy in the context of educational attainment and schooling will be, perhaps, even harder than in medicine (where it is essential!). In their contribution, David Peloquin and Mark Barnes provide an introduction to some of the relevant laws and regulations at the intersection of genomics and education. Specifically, they focus on issues pertaining to (1) the heritability of achievement and (2) DTCGT, discussing primarily federal law. In addressing heritability for achievement, they explore how the use of GT in the education system (1) raises similar constitutional concerns to those raised by school drug screening and newborn screening laws, (2) implicates informed consent laws, and (3) interacts with federal laws governing special education. They then address how GT used in the education system might be regulated by the Food and Drug Administration and the Centers for Medicare and Medicaid Services. Despite their federal focus, the authors remind readers that it is important to keep in mind that the public education system in the United States is funded primarily at the state and local levels, and thus any attempt to use GT in the public school system on a wide scale would likely need to grapple with myriad, disparate state laws. Celia Fisher enriches the volume's discourse further, illuminating ethical challenges arising at the junction of genomics and behavioral sciences, in general, and education, in particular, as well as steps that can be taken to ensure the responsible conduct of research involving GT. She addresses the need to incorporate principles of genetic literacy into informed consent practices and the unique ethical issues that arise for guardian permission and child assent procedures in cross-sectional and longitudinal studies and research involving data depositories and secondary analyses. She also talks about the tension between ensuring adequate privacy protections and the risks and benefits of disclosing research-derived personal genetic information to individual participants and their family members. Finally, she discusses ethical challenges of disseminating the results of susceptibility and intervention responsivity studies, with particular attention to the potential impact on marginalized populations. Next, Jorge Contreras and Vikrant Deshmukh bring into the discussion those segments of the commercial genomics industry that offer products and services to consumers, either directly or through intermediaries such as physicians, genetic counselors, or testing laboratories, a sector that we collectively refer to as "personal genomics" (Khoury et al., 2009). Specifically, their focus is on those products and services that provide genetic/genomic information to consumers, as opposed to drugs, vaccines or treatment regimens that may have been discovered using genomic information, or the administration of which may be influenced by a recipient's genomic characteristics. But even limited thus, the field, as the reader will discover, is complex and multifaceted. This complexity is reflected further in the contribution from Susan and Krista Bouregy who provide insight into the difficulties of a foreseeable penetration of genetic/genomic information into the education system to influence educational decision making. This chapter highlights select legal and ethical issues discussed in this section of the volume and transitions the discourse to the last two contributions, reflecting public perception of the relevant issues – one from an educator (i.e., the chapter from Judi Randi) and one from a parent (i.e., the chapter from Carolyn Cowen). Both of these chapters revise and interpret, from a lay person's viewpoint, a number of issues discussed throughout the volume.

Working on this volume has been extremely interesting and stimulating. We sincerely hope that getting familiar with these contributions will have the same effect on our readers. We are looking forward to a broad discussion of the related issues in both scientific and popular media outlets. We are confident that this discussion will unfold, and unfold intensely; it is only a matter of time.

ACKNOWLEDGMENT

We are grateful to the Spencer Foundation for supporting the preparation of this volume.

REFERENCES

Andrews, L. B. (1992). Torts and the double helix: Malpractice liability for failure to warn of genetic risks. *Houston Law Review, 29*, 149–184.

Ansari, D., De Smedt, B., & Grabner, R. H. (2012). Neuroeducation: A critical overview of an emerging field. *Neuroethics, 5*, 105–117. doi:10.1007/s12152-011-9119-3

Bowling, B. V., Huether, C. A., Wang, L., Myers, M. F., Markle, G. C., Dean, G. E., et al. (2008). Genetic literacy of undergraduate non-science majors and the impact of introductoy biology and genetics courses. *BioScience, 58*, 654–660.

Buchanan, A. (2011). Cognitive enhancement and education. *Theory and Research in Education, 2*, 145–162.

Buchanan, A., Brock, D., Daniels, N., & Wikler, D. (2000). *From chance to choice: Genetics and justice*. New York, NY: Cambridge.

Burton, A. (2009). Should patient confidentiality come second to the prevention of disease in others? *The Lancet Oncology, 10*, 210–211.

Carew, T. J., & Magsamen, S. H. (2010). Neuroscience and education: An ideal partnership for producing evidence-based solutions to guide 21st century learning. *Neuron, 67*, 685–688. doi:10.1016/j.neuron.2010.08.028

Clayton, E. W. (1997). Genetic testing in children. *Journal of Medicine & Philosophy, 22*, 233–251.

Collier, R. (2012). Genetic tests for athletic ability: Science or snake oil? *Canadian Medical Association Journal, 184*, E43–E44.

Davies, K. (2010). *The $1,000 genome: The revolution in DNA sequencing and the new era of personalized medicine*. New York, NY: Free Press.

Devonshire, I. M., & Dommett, E. J. (2010). Neuroscience: Viable applications in education? *The Neuroscientist, 16*, 349–356. doi:10.1177/1073858410370900

Dickens, B. M., Pei, N., & Taylor, K. M. (1996). Legal and ethical issues in genetic testing and counseling for susceptibility to breast, ovarian and colon cancer. *CMAJ Canadian Medical Association Journal, 154*, 813–818.

Dougherty, M. J., Pleasants, C., Solow, L., Wong, A., & Zhang, H. (2011). A comprehensive analysis of high school genetics standards: Are states keeping pace with modern genetics? *CBE Life Sciences Education, 10*, 318–327. doi:10.1187/cbe.10-09-0122

Feero, W., & Green, E. D. (2011). Genomics education for health care professionals in the 21st century. *JAMA, 306*, 989–990. doi:10.1001/jama.2011.1245

Fischer, K. W., Goswami, U., Geake, J., & the Task Force on the Future of Educational Neuroscience. (2010). The future of educational neuroscience. *Mind, Brain, and Education, 4*, 68–80. doi:10.1111/j.1751-228X.2010.01086.x

Fisher, C. B., & Harrington McCarthy, E. L. (2013). Ethics in prevention science involving genetic testing. *Prevention Science, 14*, 310–318. doi:10.1007/s11121-012-0318-x

Gollust, S. E., Hull, S. C., & Wilfond, B. S. (2002). Limitations of direct-to-consumer advertising for clinical genetic testing. *JAMA, 288,* 1762–1767.

Greenbaum, D. (2012). Introducing personal genomics to college athletes: Potentials and pitfalls. *The American Journal of Bioethics, 12,* 45–47. doi:10.1080/15265161.2012.656811

Greenbaum, D., Sboner, A., Mu, X. J., & Gerstein, M. (2011). Genomics and privacy: Implications of the new reality of closed data for the field. *PLoS Computational Biology, 7,* e1002278.

Grigorenko, E. L. (2004). Genetic bases of developmental dyslexia: A capsule review of heritability estimates. *Enfance, 3,* 273–287.

(2005). A conservative meta-analysis of linkage and linkage-association studies of developmental dyslexia. *Scientific Studies of Reading, 9,* 285–316.

(2007a). Bridging genomics and education. Retrieved from www.tcrecord.org/Content.asp?ContentId=13909.

(2007b). How can genomics inform education? *Mind, Brain, and Education, 1,* 20–27.

(2010). Bringing genomic sciences to the classroom. *Science.* Retrieved from www.sciencemag.org/cgi/eletters/328/5977/5512#13440

(2015). Genomic sciences for developmentalists: A merge of science and practice. *New Directions in Child and Adolescent Development, 147.*

Gurwitz, D., & Bregman-Eschet, Y. (2009). Personal genomics services: Whose genomes? *European Journal of Human Genetics, 17,* 883–889.

Guttmacher, A. E., Porteous, M. E., & McInerney, J. D. (2007). Educating healthcare professionals about genetics and genomics. *Nature Reviews Genetics, 8,* 151–157.

Gymrek, M., McGuire, A. L., Golan, D., Halperin, E., & Erlich, Y. (2013). Identifying personal genomes by surname inference. *Science, 339,* 321–324. doi:10.1126/science.1229566

Hanson, J. W., & Thomson, E. J. (2000). Genetic testing in children: Ethical and social points to consider. *Pediatric Annals, 29,* 285–291.

Hardiman, M., Rinne, L., Gregory, E., & Yarmolinskaya, J. (2012). Neuroethics, neuroeducation, and classroom teaching: Where the brain sciences meet pedagogy. *Neuroethics*(2), 135–143. doi:10.1007/s12152-011-9116-6

Harris, J. (2007). *Enhancing evolution: The ethical case for making better people.* Princeton, NJ: Princeton University Press.

Henderson, B. J., & Maguire, B. T. (2000). Three lay mental models of disease inheritance. *Social Science & Medicine, 50,* 293–301.

Hensley Alford, S., McBride, C. M., Reid, R. J., Larson, E. B., Baxevanis, A. D., & Brody, L. C. (2011). Participation in genetic testing research varies by social group. *Public Health Genomics, 14,* 85–93.

Hook, C. C., DiMagno, E. P., & Tefferi, A. (2004). Primer on medical genomics, Part XIII: Ethical and regulatory issues. *Mayo Clinic Proceedings, 79,* 645–650.

Hoppe, N. (2013). From omics to etics to policy and ethics: Regulating evolution. *Frontiers in Genetics, 4,* 1–2.

Howard-Jones, P. A., & Fenton, K. D. (2012). The need for interdisciplinary dialogue in developing ethical approaches to neuroeducational research. *Neuroethics, 5,* 119–134. doi:10.1007/s12152-011-9101-0

Howard, H. C., Avard, D., & Borry, P. (2011). Are the kids really all right? Direct-to-consumer genetic testing in children: Are company policies clashing with professional norms? *European Journal of Human Genetics, 19,* 1122–1126. doi:10.1038/ejhg.2011.94

Jin, J. (2000). An evaluation of the Ethical, Legal and Social Implications program of the U.S. Human Genome Project. *Princeton Journal of Bioethics, 3,* 35–50.

Kaye, J. (2008). The regulation of direct-to-consumer genetic tests. *Human Molecular Genetics, 17*(R2), 15.

Kegley, J. A. (1996). Using genetic information: The individual and the community. *Medicine and Law, 15,* 377–389.

Khoury, M. J., McBride, C. M., Schully, S. D., Ioannidis, J. P. A., Feero, W. G., Janssens, A. C. J. W., et al. (2009). The Scientific Foundation for Personal Genomics: Recommendations from a National Institutes of Health-Centers for Disease Control and Prevention Multidisciplinary Workshop. *Genetics in Medicine, 11*(8), 559–567.

Kingsberry, S. Q., Mickel, E., Wartel, S. G., & Holmes, V. (2011). An education model for integrating genetics and genomics into social work practice. *Social Work in Public Health, 26,* 392–404. doi:10.1080/10911350902990924

Korf, B. R. (2011). Genetics and genomics education: The next generation. *Genetics in Medicine, 13,* 201–202.

Kung, J. T., & Gelbart, M. E. (2012). Getting a head start: The importance of personal genetics education in high schools. *Yale Journal of Biology & Medicine, 85,* 87–92.

Lacroix, M., Nycum, G., Godard, B., & Knoppers, B. M. (2008). Should physicians warn patients' relatives of genetic risks? *CMAJ Canadian Medical Association Journal, 178,* 593–595.

Lanie, A. D., Jayaratne, T. E., Sheldon, J. P., Kardia, S. L., Anderson, E. S., Feldbaum, M., et al. (2004). Exploring the public understanding of basic genetic concepts. *Journal of Genetic Counseling, 13,* 305–320.

Lucassen, A., & Montgomery, J. (2010). Predictive genetic testing in children: Where are we now? An overview and a UK perspective. *Familial Cancer, 9,* 3–7.

Madsen, A. (2010, October 1). Peeking at "the cards you were dealt." *Quest, 17.*

Maher, B. S. (2011). Human genetics: Genomes on prescription. *Nature, 478,* 22–24. doi:10.1038/478022a

McConkie-Rosell, A., & Spiridigliozzi, G. A. (2004). "Family matters": A conceptual framework for genetic testing in children. *Journal of Genetic Counseling, 13,* 9–29.

McGuire, A. L., & Burke, W. (2008). An unwelcome side effect of direct-to-consumer personal genome testing: Raiding the medical commons. *JAMA, 300,* 2669–2671.

Meslin, E. M., Thomson, E. J., & Boyer, J. T. (1997). The Ethical, Legal, and Social Implications Research Program at the National Human Genome Research Institute. *Kennedy Institute of Ethics Journal, 7,* 291–298.

Middleton, A., Hewison, J., & Mueller, R. F. (1998). Attitudes of deaf adults toward genetic testing for hereditary deafness. *American Journal of Human Genetics, 63,* 1175–1180.

Mills Shaw, K. R., Van Horne, K., Zhang, H., & Boughman, J. (2008). Essay contest reveals misconceptions of high school students in genetics content. *Genetics*, *178*, 1157–1168.

The National Cancer Institute. (2013). *HINTS briefs: Public awareness of direct-to-consumer genetic tests*. Washington, DC: The National Cancer Institute.

Nelkin, D., & Tancredi, L. (1991). Classify and control: Genetic information in the schools. *American Journal of Law & Medicine, 17*, 51.

Parker, M. (2010). Genetic testing in children and young people. *Familial Cancer*, *9*, 15–18.

Rodriguez, L. L., Brooks, L. D., Greenber, J. H., & Green, E. D. (2013). The complexities of genomic identifiability. *Science, 339*, 275–276. doi:10.1126/science .1234593

Savulescu, J., & Bostrom, N. (Eds.). (2009). *Human enhancement*. New York, NY: Oxford University Press.

Serpati, L., & Loughan, A. R. (2012). Teacher perceptions of neuroeducation: A mixed methods survey of teachers in the United States. *Mind, Brain & Education, 6*, 174–176. doi:10.1111/j.1751-228X.2012.01153.x

Skiba, T., Landi, N., Wagner, R., & Grigorenko, E. L. (2011). In search of the perfect phenotype: An analysis of linkage and association studies of reading and reading-related processes. *Behavior Genetics, 41*, 6–30.

Su, P. (2013). Direct-to-consumer genetic testing: A comprehensive view. *Yale Journal of Biology and Medicine, 86*, 359–365.

Taylor, J., Roehrig, A. D., Soden Hensler, B., Connor, C. M., & Schatschneider, C. (2010). Teacher quality moderates the genetic effects on early reading. *Science, 328*, 512–514.

Watson, J. D. (1990). The Human Genome Project: past, present, and future. *Science, 248*, 44–49.

Wright, C. F., Hall, A., & Zimmern, R. L. (2011). Regulating direct-to-consumer genetic tests: What is all the fuss about? *Genetics in Medicine, 13*(4), 295–300.

1

What Is Heritability and Why Does It Matter?

MEI TAN

In the world of genetics, the concept of genetic identity is fairly straightforward: how similar or close to identical are the genotypes (or genomes) of different species, or different individuals? This is something that can be estimated (Guo, 1996), and the relevant calculations are used to determine genetic relatedness or comparative genetic "identicalness" or identity, the commonly used term (e.g., Carlsson, McDowell, Carlsson, & Graves, 2007; Gharghani et al., 2009; Phelps & Allendorf, 1983). In the workaday world of most individuals, however, the concept of a biological or genetic identity is expressed neither mathematically nor objectively. That is, the notion of a genetic identity for most people is shaped by a much more basic drive, as framed by Eric Cohen,[1] who said, "The first reason for engaging in modern genetics is simply man's desire to know himself, a desire that nearly all of us share, if not in equal degrees" (Cohen, 2005, p. 31). Yet the nature and implications of genetic knowledge are not so easy to grasp.

The conception of how our genetic information contributes to our personal identities, for most of us, is not fully realized. We may know that our DNA came from our parents and that it reflects our relatedness to them. We may know that our DNA has contributed to our development as unique individuals and endowed us with particular traits, like hair color, eye color, the shape of our nose or chin. The true nature of this inheritance, still, is a bit fuzzy for most of us. How about the way we laugh? Our sense of humor? To what degree is our health inherited, or our intelligence, or how we perform in school? When we know what we have inherited, what do we know about ourselves? The high subjectivity with which genetic information is

The author would like to thank the Spencer Foundation for its generous support of this chapter.

[1] From his inaugural lecture of the Genomics Forum at the University of Edinburgh in Scotland, 2004.

regarded is illustrated in Klitzman's (2009) interviews with sixty-four individuals who had or were known to be genetically at risk for Huntington disease, breast cancer, or alpha-1 antitrypsin deficiency (a common pulmonary condition caused by the inherited lack of the alpha-1 antitrypsin enzyme; Sharp, 1971). The struggles of these individuals to incorporate this information into their sense of self (e.g., as "vulnerable," "diseased," "healthy") revealed how genetic information is viewed highly subjectively, and so affects individuals' sense of identity in varied ways.

The intent of this chapter is to interrogate the field to see how and why issues of ethics have followed genetics closely as the field has developed in its ability to inform what we know about our health, our heritage, and our futures. The author, a student of genetics-related psychology rather than of genetics itself, will attempt to answer the questions: What is heritability, and when and why does it matter? And what does it have to do with an individual's sense of identity?

Because the answers to questions of genetic identity may have important consequences for people – for example, by affecting how they are perceived and understood by themselves and by others – this chapter begins by presenting some fundamental ideas about identity that come with the knowledge of a biological inheritance. In the second part of the chapter, the methods used to understand the inheritance of complex traits will be described. The concepts of concordance and heritability will then be exemplified in some detail to explicate what they mean, how they are estimated, and how they are used to help us understand what our genetic inheritance is.

JUST ONE QUESTION: WHO AM I?

There comes a point (or perhaps several) in our lives when the question, "Who am I?" seems to be the most urgent, most important question to answer. In fact it has many facets and may never be fully answerable. Facets of identity are things such as what we think, what we like, how we look, who our friends are. But there are also aspects of identity that are broad and deep, such as our cultural heritage, our ancestors, and parentage. This latter point – who our parents are – is particularly weighty. We need not understand the intricacies of genetics to feel that who we are and who we become is largely influenced by the genes we have inherited. Thus, who our parents are – the source of our genes – looms large as an important aspect of identity, and so genes are inextricably entangled in the ethics of how we are identified.

One compelling example that illustrates the importance of one's genetic inheritance to one's conception of self is the search by adoptees for their biological parents. This is a common phenomenon that at one time was hypothesized to be an indication of some form of psychopathology – the result of a traumatic adoption revelation, problematic adoptive family relationships, poor self concept and personal adjustment, or stressful life events (Sobol & Cardiff, 1983; Triseliotis, 1973). While early policy and practice in adoption was based on a mission to "rescue" a child from adverse circumstances and to sever the connection with the family of origin to protect the child, there has been a gradual realization in the fields of social work and psychology that there are significant and important issues of identity that eventually arise for children who have been separated from their biological families, no matter what the circumstances (Matthews, 2014; Winter & Cohen, 2005; Wrobel, Grotevant, & McRoy, 2004). There is now a rich and extensive literature on the situations of those who have been affected by the adoption process and the efforts of adoptees to locate and communicate with their biological families (a process known as "searching") (e.g., Brodzinsky & Palacios, 2005; Haimes & Timms, 1985; Howe, Feast, & Coster, 2000; Matthews, 2014; Triseliotis, Feast, & Kyle, 2005).

More prevalent now is the view that, rather than a manifestation of emotional and mental troubles, the need to search is a natural and normal process (Matthews, 2014; Winter & Cohen, 2005; Wrobel et al., 2004), emergent generally in adolescence and young adulthood and that may last a lifetime (Baden & Wiley, 2007; Grotevant, 1997). Its basic source appears to be an inherent understanding that biological parentage and the DNA received from parents (one's genetic identity) is an essential aspect of one's construction of a sense of self, along with one's social relationship with one's community (i.e., what is my role in my community, according to myself and others?); one's interpretation of formative experiences (i.e., what does this event mean about me?); and one's sense of development over time (i.e., what are my possible futures?) (Winter & Cohen, 2005), all of which aspects may be more complex in an adoptive situation (Grotevant, Dunbar, Kohler, & Esau, 2000; Matthews, 2014). Along similar lines, afforded by the latest technologies in alternative reproductive techniques, is the desire to discover one's donor parent/s (Cushing, 2010; Haimes, 1988, 1993), and the subsequent debate over whether offspring from donated eggs or sperm (third-party reproduction) may be able to legally access information about their biological "parents," as donor anonymity and privacy continue to be protected (Braverman, 2010, 2013). Such children may in fact want to know from their parents why certain eggs or sperm were selected. (For more on the complexities of these issues, see Cowan, this volume.)

The results of both of these pursuits appear equally difficult to interpret. Suppose an adoptee discovers that his or her biological parent is a world renowned genius in physics? Or a criminal currently in jail? How might this impact one's sense of identity? Similarly, suppose an individual should discover that her DNA contains implications of risk for a certain debilitating disease, or connects him to an ethnicity about which he has mixed feelings?

Another phenomenon that has capitalized, in a sense, on people's desire to know their genetic identities, is the direct-to-consumer (DTC) genotyping industry (e.g., 23andMe, Decode; see Peloquin and Barnes, this volume). Access to this service allows individuals willing and able to pay the ability to learn about their identity (e.g., ancestry or paternity) and/or disease risk (e.g., for certain types of cancer), to potentially make better lifestyle choices (Su, 2013). In a systematic review of user perspectives on DTC genetic testing (Goldsmith, Jackson, O'Connor, & Skirton, 2012), Goldsmith and colleagues searched for primary research papers published between 2001 and 2011 that focused on the knowledge and use of DTC genetic tests to members of the general population, particularly user attitudes, and perceptions of benefits and risks. In a set of seventeen reviewed papers, they found that the main reason for engaging in DTC genetic testing given by individuals was curiosity about their genetic make-up, some just generally, but most to learn about genetic risk for certain health conditions, such as heart disease. Their overall intention was that the information would influence lifestyle choices or stimulate family discussion on health issues. Concerns that arose about genetic DTC pertained to issues of privacy and who would have access to the information, worry that the tests might yield "unwanted information" that they would not be ready to hear or might affect them negatively, and the reliability of the tests. These findings were reiterated in a brief overview of the DTC testing industry composed by Su (2013). However, people's motivations to engage in such testing and how much they fully comprehend the implications of the results are far from consistent, clear, or understood.

A systematic review and critique of the literature on predictors of who might decide to get a genetic test revealed conflicting findings (Sweeny, Ghane, Legg, Huynh, & Andrews, 2014). Looking at both subjective (e.g., perceived risks and perceived benefits of testing, disease-specific worries) and objective (e.g., family history, general health, education) predictors of testing, the authors found most predictors to be inconsistent across studies. To those who do decide to undergo genetic testing, the full implications of the information they gain from such testing may not be clear. A study on public knowledge and attitudes toward genetic testing in a sample of 300 individuals from the general public in the area of Durham, North Carolina, showed among other findings that while many seemed aware of the medical consequences or uses of genetic testing, significantly fewer were aware of some of the social implications of genetic testing – such as how findings might affect other family members, one's work situation, or one's insurance (Haga et al., 2013). Thus, although DTC provides access to one's genetic information – whether for personal, familial, or health reasons – a great deal of education is still needed to support individuals' understanding of the meaning and uses of such information, its benefits and potential harms, and why people might access such tests. In sum, the point we have tried to make thus far is that one's biological origins/genetics/DNA are intimately connected to one's sense of identity in important ways that are difficult to understand and may be equally difficult to express. This is why issues concerning genetic information can be highly sensitive and must be examined for their moral and ethical implications.

What follows in the second half of this chapter is an explication of the methods most commonly used to initially explore the genetic similarities between blood relatives. Studies using these methods provide the first insights that can be gained into the etiology of human traits, and indicate how further information may be pursued. The statistical concepts of concordance and heritability quantify these insights so that they can be used as indicators of probability and risk.

THE ETIOLOGY OF COMPLEX HUMAN TRAITS

Long before the revolutionary outcomes of the Human Genome Project, which mapped the nucleotide sequences contained in a "collective" human DNA sample and was officially completed in 2003, and the advent of consumers' interest in personal genetics, the first inquiries into familial human "likenesses" (or genetics, before the word was coined) were driven by early scientists and philosophers who observed that certain human behaviors or traits, such as intelligence, and physical characteristics, such as eye color, seemed to run in families. They could discern that there were discrete traits, i.e., present versus not present (also referred to as categorical, qualitative, or diagnostic traits) and continuous traits, i.e., occurring in a range of degrees that could be found to be normally distributed across a population (also referred to as quantitative) traits (Vinkhuyzen, Wray, Yang, Goddard, & Visschler, 2013). Yet these very early scientists were baffled by how such traits were passed from generation to generation. Pythagoras argued that they were passed along via male semen. Two hundred years later, Aristotle challenged this view by arguing not only that both male and female parents must contribute to the development of their offspring, but that what they passed on had to be in the form of both information and the material to be shaped by that information (Mukherjee, 2016). While a bit closer to the truth, it would not be until several centuries later that more plausible models explaining "likeness" would be proposed. Even then, their full import could not be grasped. The mechanisms of inheritance suggested in the mid-nineteenth century by the work of Charles Darwin (i.e., natural selection) and Gregor Mendel, whose breeding of discrete traits in pea plants illustrated some simple rules for genetic inheritance, were not capitalized on in their time. As with any complex picture, these initial disparate pieces were not enough to see the whole.

It was not until the turn of the twentieth century that scientists realized that many traits, particularly those that were continuous, did not follow the simple rules mapped out earlier by Mendel. In a breakthrough in 1918, Ronald Fisher proposed that continuous traits that are normally distributed within a population could be accounted for by many genes operating according to simple Mendelian mechanisms at once, with additional influence from the environment (Griffiths, Wessler, Carroll, & Doebley, 2015). Such traits, resulting from both genetic and environmental factors, became known as "complex" traits, and the field of quantitative genetics began to emerge. *Quantitative genetic or behavioral genetic studies* (like the more recently developed molecular-genetic approaches) focus not on individuals but the observable traits (discrete and continuous) they carry and what might explain the particular distributions of these traits within a *group* of people (Falconer & Mackay, 1996; Griffiths et al., 2015; Vinkhuyzen et al., 2013). More complex questions could now be addressed: what human traits or behaviors may be inherited or influenced by genes? And for those traits and behaviors that are inherited, how much of their manifestation may be attributed to genes, and how much to the individual's exposure to different contexts, experiences, and environmental elements?

Quantitative genetic studies examine variation (i.e., similarities and differences) in populations – i.e., large discrete groups of people – in an attempt to capture the balance of genetic versus environmental contributions to the manifestation of specific traits and behaviors. These studies capitalize upon the natural occurrences of various configurations of families (nuclear and extended) and pairs of relatives (blood and adoptive relatives; siblings,

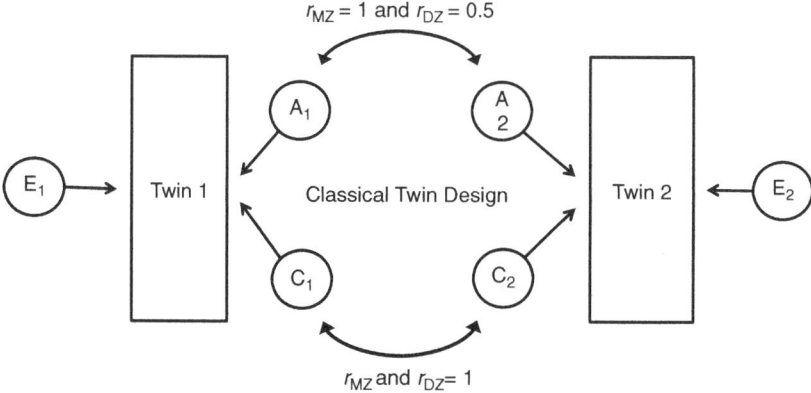

FIGURE 1.1 Classical twin design. A = additive genetic influences; C = shared environmental influences; E = unshared environmental influences. r_{MZ} = correlation between monozygotic twins; r_{DZ} = correlation between dizygotic twins.

including twins; parents and children; cousins, and so forth). These particular configurations of biologically related and unrelated people provide rough indicators of shared genetic material and shared environment such that their relative contribution to the trait-related phenotypic variation in a particular population may be quantitatively estimated. Briefly, observed phenotypic variance (P) of a particular trait in a population may be parsed into variance attributable to the combined influence of genes (A), variance attributable to a common or shared environment (C), and the residual variance due to unique (unshared) environmental factors (E) and/or error. This relationship, commonly denoted as $P = A + C + E$, represents one of the simplest models of heritability (see Figure 1.1), based on several assumptions, and will be discussed in more depth later.

Study designs that capitalize on known familial and genetic relationships, exemplified below, led to the development of the statistical concepts of concordance and heritability. Concordance is the probability that an individual will inherit the same discrete trait as another individual (such as a brother inheriting the same genetic disorder as his sister) within a particular population. Heritability is the proportion of trait variance within a population (such as weight) that can be accounted for by genetic variations within that population (Plomin, DeFries, Knopik, & Neiderhiser, 2013). Concordance and heritability are estimates used to quantify genetic risks for or genetic contributions to complex traits. They will be discussed in greater detail later in this chapter.

MEASURING (DIS)SIMILARITIES

Quantitative Genetic Designs

Although first conceived by Sir Francis Galton in the late nineteenth century, the first ostensible so-called twin and adoption studies (Merriman, 1924; Theis, 1924, respectively) were reported in the early 1920s; both investigated the etiology of intelligence and related indicators (Munsinger, 1975; Plomin & Spinath, 2004). In the first, Merriman (1924) addressed three research questions concerning twins, including whether intellectual data would support the biological belief that there are two genetically distinct types of twins, fraternal (dizygotic, DZ) and duplicate (identical or monozygotic, MZ). Based on twin and sibling performance on four indicators of intelligence, results conclusively supported a clear difference in the degrees of similarity between fraternal and identical twins, with identical twins showing greater similarity to each other than both fraternal twins and non-twin siblings. Specifically, on the Stanford-Binet Test of Intelligence (Terman, 1916), the Army Beta Test (Yerkes, 1921), the National Intelligence Test (Yerkes, Haggerty, Terman, Thorndike, & Whipple, 1920), and teacher ratings of students' intellectual traits (on a scale of 1–5), twin pairs of increasing likeness achieved more similar scores on all indicators. This is indicated by Pearson correlations, r, which reflect the closeness of twins' scores, with higher r scores indicating higher degrees of score similarity (see Table 1.1; based on Merriman, 1924, table 15). Using a different approach to the same general question, Theis (1924) collected data on adopted children's potential to achieve in formal education (as a rough proxy for intelligence) and related their scores to the social-educational-occupational status of both their adoptive parents and their biological parents. A higher degree of correspondence was found between the children's mental capabilities and their biological parents' social background (see Table 1.2, based on Munsinger 1975, table 3), suggesting a stronger influence of heredity over environment. Lack of information, however, about age of adoption and other details rendered the results inconclusive (Munsinger, 1975).

Despite their weaknesses, both of these studies provided the foundation for thousands of future studies on the contributions of heredity and environment to individual differences in human characteristics in different populations.

Twin Study Designs. The twin studies carried out today are based on the fact that twins, depending on how they were conceived (i.e., via two

TABLE 1.1 *Pearson correlations of between-twin pairs of increasing likeness in performance on four indicators of intelligence*

	All twin pairs		All same-sex twin pairs		"Similar" same-sex twin pairs[a]	
	N	r	N	r	N	r
Stanford-Binet	105	.782	67	.867	22	.986
Army Beta Test	76	.841	45	.908	18	.887
National Intelligence Test	143	.891	92	.925	12	.987
Teacher Ratings	90	.512	53	.654	12	.940

[a] Same sex twin pairs reported to be "similar" enough to possibly be mistaken for each other.
Based on table 15 in Merriman (1924).
N = sample size; r = Pearson correlation.

TABLE 1.2 *Biological social background and capability of adopted children to achieve in higher education*

Biological family backgrounds	Poor	Mixed	Good
% children capable of higher education	63%	67%	92%

Based on table 3 in Munsinger (1975).
The designations for family background, "poor," "mixed," and "good" were devised and applied by Theis.

independent zygotes or one zygote that later divided), may have almost identical or almost non-identical genetic material. Identical twins are conceived when a single fertilized egg (or a zygote resulting from one egg fertilized by one sperm) splits into two zygotes, either on its way to the uterus or after implantation in the uterus. In the first instance, the two resulting fetuses will have their own sacs (chorions) within the placenta; in the second they will share a sac; yet in both cases these twins – termed "monozygotic" (MZ) – will have almost identical DNA. Fraternal twins – "dizygotic" (DZ) – result when two eggs are fertilized by two different sperm and both zygotes successfully implanted in the uterus. Like other non-twin siblings, DZ twins are only, on average, about 50 percent genetically similar (i.e., the degree of similarity for each pair of twins can be between 0 percent and 100 percent); they may be of the same or opposite sex. Most classical twin studies generally access same-sex DZ twins only, as MZ twins are always of the same sex. However, there are statistical methods now available to work with all possible types of children conceived (semi)simultaneously (i.e., twins, triplets, quadruplets, and so forth) of any combination of sex. Thus, modern twin studies tend to include all such types of multiple births, and often their families as well; these combinations of relatives may increase the statistical power of quantitative genetic studies.

Knowing relative degrees of genetic similarity allows us to form expectations about twin similarity in traits that may be inherited. Those twins who are MZ would be expected to be more similar to each other than DZ twins. For example, given a large sample of MZ twins and an equal sample of same-sex DZ twins, if they are all tested with the same intelligence test, one would expect the scores within MZ pairs of twins to be closer to each other than the scores within DZ pairs of twins, and in fact this observation has been borne out, as in the Merriman study above, in many subsequent studies (Deary, Johnson, & Houlihan, 2009). These degrees of similarity and difference between different types of twins can be measured and quantified as estimates of *concordance* and *heritability*, as defined earlier.

Now, some important caveats must be acknowledged with respect to twin studies. First, results depend upon the accurate assignment of zygosity for the twin pairs. Identical versus fraternal twins can be identified by DNA testing, of course (Plomin et al., 2013). Alternatively, similarities on physical traits that are known to be heritable and affected by many genes, such as eye color, hair color, and hair texture, may also be fairly reliable indicators of zygosity; differences in even just one of such traits are thought likely to only occur in fraternal pairs of twins. Many studies make use of parent or self-report zygosity questionnaires (e.g., Cohen, Dibble, Grawe, & Pollin, 1973; Hill Goldsmith, 1991; Nichols & Bilbro, 1966) to distinguish MZ from DZ twins, resolving ambiguous cases using DNA testing (e.g., Olson et al., 2013). Yet, because different twin studies may use different methods to determine zygosity, comparisons or meta-analyses of twin studies need to take these differences in method into account, as some methods carry higher reliability (i.e., DNA testing) than others (McCartney, Harris, & Bernieri, 1990), and, therefore, the corresponding concordance and heritability estimates may contain different amounts of error.

The second caveat is the assumption that for both types of twins, identical and fraternal, the contributions of the shared home environment to twins' degrees of similarity are roughly the same. That is, that MZ twins' similarity is not more influenced by shared environment than the similarity of DZ twins. This is called the "equal environments assumption." If this could not be assumed to be true, the contributions of twins' shared genetic material to their similarity could not be estimated with any meaningful degree of certainty. The assumption has been tested empirically and, for most traits, appears to hold true (Plomin et al., 2013), yet it warrants attention in all twin studies to be sure that it is not violated.

Third, it is now known that MZ twins, even those with the highest degree of pre-natal shared environment (a shared chorion) do not have exactly identical DNA (either structurally or functionally), and it is unclear as yet the extent to which pre-natal de novo mutations and post-natal epigenetic changes in each twin may create genetic differences between two MZ twins (Casselman, 2008; Skipper, 2008).

And finally, the issue of generalizability poses possible confounding variables for twin studies. Can twins be considered representative of the general population? Studies of twin births, brain development, language development, cognitive and motor development have examined this question, along with studies of their personalities. In general, and in the long term, twins do not appear to be significantly different from non-twin individuals (Plomin et al., 2013); however, again, this is an important issue that bears consideration in all twin studies.

Adoption Study Designs. Situations of child adoption result in specific configurations of relatives that can be similarly informative to study. For example, if two siblings (same biological mother and father) are adopted into the same family at an early age, and if that adoptive family already has small children, one might obtain concordance and heritability estimates for certain traits in light of the specific assumptions about genetic and environmental (dis)similarities between these siblings. More specifically, similarities between the two genetically related children relative to their adoptive siblings can be estimated. The expectation would be that for traits with a high genetic influence, the adopted biological sibling pair will be more like each other than either of them will be like their adoptive siblings. Similarly, the biological siblings of the adopting family will be more like each other than either of them will be like their adopted siblings. In cases where the biological parents are known, the similarities between the adopted children, their biological parents and their adoptive parents can be compared (as in the Theis study). For those traits and behaviors that are influenced more by the environment, adopted and adoptive siblings and parents will be more similar than different. A variant of this situation is when two biological siblings (with ~50 percent genetic material in common) are adopted by different families and thus raised in different environments. Again, relative contributions of genetic and environmental influence on certain measurable traits can be estimated based on what is known about the genetic material and environments shared by the individuals involved (Plomin et al., 2013).

Like twin studies, certain caveats must be acknowledged in the consideration of adoption study designs. First, age at adoption must be considered, as the period of time spent with the biological family may muddy the division between genetic and environmental effects. That is, the adopted child's similarity to his or her biological family may be due to both genetic and environmental effects if delivery to the adoptive family was delayed. Optimal designs include only those children adopted immediately after birth (Rice & Borecki, 2001). Second, the possible occurrence of selective placement must be addressed, as women or couples offering children for adoption and the corresponding adoptive families can tend to fall within particular and, importantly, rather different, demographic and socio-economic strata (Coulter, 2015; Vandivere, Malm, & McKlindon, 2011). This possibility leads directly to the question, as in twin studies, of the generalizability of the data generated by adoption studies.

Adoption designs have yielded and continue to yield interesting and useful information concerning the concordance and heritability estimates of human traits and behaviors. Two examples of large-scale adoption studies designed to further elucidate the genetic and environmental contributions to human behavioral development are the Colorado Adoption Project (CAP; Plomin & DeFries, 1983), started in 1975, and the Early Growth and Development Study (EGDS; Leve et al., 2013), started in 2003. Both are longitudinal, prospective studies with a focus on how environment and genetic inputs interact to guide early childhood development. Both involve biological parent, child, and adoptive parent triads. The CAP, involving more than 200 adoptive families (i.e., families where only environments are shared), also recruited matched control families (non-adoptive, i.e., families where both genes and environments are shared) for comparison. Comprehensive sets of data were collected from both sets of parents on several variables, including demographics, cognitive abilities, personality and temperament, mood, drug use, talents, and common health and behavior indicators (Plomin & DeFries, 1983, 1985). The children were similarly assessed on dimensions of infant and child development. Two common scales were used to characterize the home environment, and videos of mother–child interaction were taken to assess maternal behavior. The resulting data were used to estimate the variance of the family environment, to examine specific environmental effects without the interference of genetic effects, to identify some environmental factors that affect only some individuals but not others (genotype–environment interaction), and to investigate how much children may create their own environments (genotype–environment correlations) (Plomin & DeFries, 1983). The EGDS extends the work of the CAP by collecting a similar range of behavioral data from 561 adoptive triads, along with measures of parenting and marital relations, and samples of DNA and salivary cortisol. The four aims of the study involve not only understanding emergent child behaviors without the effect of shared genes, but to isolate genetic and environmental influences on behavior while examining genetic influences on the social environment, and how gene–environment interactions may affect child outcomes (Leve et al., 2013). Thus, carefully designed adoption studies continue to be carried out to better understand the interactions between genes and environments and how they affect a child's development.

Combination Studies. It should be noted that there have also been studies that combine twin and adoption study designs by following the development of twins adopted into separate homes at birth. One example of such a study is the Minnesota Study of Twins Reared Apart, which was established in 1979 (Bouchard, Lykken, McGue, Segal, & Tellegen, 1990). It followed more than 100 sets of twins reared apart in Australia, Canada, China, New Zealand, Sweden, and West Germany, as well as the United States and the United Kingdom. Results yielded evidence of high heritability for most of the studied psychological traits, demonstrating that on indicators of personality, temperament, occupational and leisure interests, and social attitudes, MZ twins reared apart were as similar to each other as MZ twins reared together (Bouchard, 2005; Bouchard et al., 1990). These echo findings from several studies, both before and after the Bouchard study, examining many disparate traits, such as type A personality characteristics (Pedersen et al., 1989), susceptibility to dental problems (Boraas, Messer, & Till, 1988), and body mass index (Stunkard, Harris, Pedersen, & McClearn, 1990).

In summary, twin and adoption studies are designed to parse genetic and environmental contributions to the manifestation and development of human traits and behaviors by assessing individuals' performance, and examining (dis)similarities in performance with regard to expectations based on known genetic relationships. These types of studies have been and continue to be informative in investigations on the etiologies of complex human traits. What is important to know about twin and adoption studies is what information they can and cannot provide. While such studies are unable to provide any specific information about the actual genes involved, nor the mechanisms and systemic interactions that comprise these genes' respective spheres of influence, studying these naturalistic family configurations can provide initial estimates and insights into possible balances of genetic and environmental influence on human development. Thus, quantitative genetic studies are entry points into studies of the etiology of complex human traits.

Characterizing in greater detail the information that the quantitative studies carried out with the designs described above actually yield will bring us back full circle to the sensitive nature of this information, its importance for our fuller understanding of the many human traits left to explore, and how studies in heritability may contribute to future research.

ESTIMATING HERITABILITY

The estimates of concordance and heritability derived from family, twin, and adoption studies are statistical, mathematically derived estimates. Both of these estimates describe the probable sources of phenotypic variability *within a population* by positing the proportion of variation in that population that is due to genetic variation. Four aspects of heritability are key to understanding the concept:

- First, concordance and heritability pertain to populations, or large groups of people, and therefore do not apply to individuals or the probability of any trait manifesting in any individual;
- Second, they are designed to explain variation of a trait within a population; therefore, the trait in question must show variability within that population (such as variation in hair color rather than the existence of hair itself, which generally applies to close to 100 percent of the individuals in any given human population);
- Third, these estimates may differ for any given population at different points in time, i.e., an estimate does not remain a constant but is specific to a specific population at a given time;
- And, fourth, heritability encompasses the notion of environmentality – the proportion of variance that is due to or explained by environmental variation; the heritability and environmentality estimates for any given trait always add up to 1 (Baker, 2004).

The Example of Height

These points about heritability can be illustrated briefly using the classic example of height (Galton, 1886; Plomin et al., 2013; Vinkhuyzen et al., 2013). First, we might note that the heritability of height has been found to be between ~70–90 percent (Plomin et al., 2013; Vinkhuyzen et al., 2013), which means that much of the variation in height (i.e., 70–90 percent) observed within a population is generally due to *genetic* differences among the individuals of that population. The balance of that variation – ~10–30 percent, according to the definition of heritability – is accounted for by environmental factors. However, if we hone in on a population located in a region experiencing drought or famine, it might be observed that the environmental (nutritional) conditions exert a strong negative influence on height for some individuals, subsequently accounting for more than 10 percent of the height variation found in the population. Thus, the heritability estimates of height are altered according to the specific characteristics of the time and place in question. In other words, in these circumstances, if the environment appears to account for more than 10 percent of the variation in this population's height – say, if it accounts for 15 percent now – the heritability of height, or the amount of its variability in the population that can be attributed to genetic influence, is now less than 90 percent and is closer to 85 percent. As the influence of environmental factors increase under certain conditions, the influence of the genome decreases. A closer look at how concordance and heritability are parsed within behavioral genetic studies delineates further the complexity of these estimates, and how they are dependent upon many factors, such as elements of the study design, definitions applied, and assumptions made about populations and disease prevalence.

The Example of Autism

To illustrate these points, we now consider the example of autism spectrum disorder (ASD), a complex neurodevelopmental disorder that disrupts the normal acquisition of social, communication, and cognitive skills. It manifests in individuals to a high degree of heterogeneity in the abovementioned skill domains, hence its characterization as a spectrum (Blake, Hoyme, & Crotwell, 2013; Chen, Peñagarikano, Belgard, Swarup, & Geschwind, 2015; Hallmayer et al., 2011). The transmission of ASD is believed to be polygenic, involving multiple genes. Different papers have projected the numbers of genes involved to range from several dozen (Geschwind, 2011) to as many as 200–1,000 (Chen et al., 2015). Similarly, the estimated heritability of ASD has varied widely between studies. Here, we will briefly describe a 2011 study that attempted to rigorously address and resolve this variation through a highly transparent and reasoned process, aimed to estimate more accurately the concordance and heritability of ASD.

In 2011, Hallmayer and his colleagues noted that ASD has been considered one of the most heritable of neurodevelopmental disorders, and that this was largely due to notable differences, from 92 percent to 0 percent, found in the concordance rates between MZ and DZ twins, respectively (Hallmayer et al., 2011). That is, in a sample of twins, given a MZ pair of twins, one of whom has been diagnosed with ASD, there is a much higher chance that the other MZ twin has ASD (MZ concordance) than would be the case if the pair were dizygotic (DZ concordance). Although many studies have aimed to estimate the genetic contributions to ASD, using different methods and finding variable concordance rates, three studies in particular reported extremely high MZ concordance rates (92 percent; Bailey et al., 1995; 36 percent; Folstein & Rutter, 1977; 91 percent; Steffenburg et al., 1989) for various aspects of autism, with low corresponding DZ (0 percent) concordance rates. The resulting MZ–DZ differences in concordance led to an estimated heritability of ASD of about 90 percent (Hallmayer et al., 2011). While acknowledging that even some more recent studies have produced similarly high concordance rates for MZ twins (e.g., MZ 88 percent, DZ 31 percent; Rosenberg et al., 2009; MZ 95 percent, DZ 31 percent; Taniai, Nishiyama, Miyachi, Imaeda, & Sumi, 2008) for ASD (Ronald & Hoekstra, 2011), Hallmayer and colleagues challenged these surprisingly high results. Their main argument was that methods of diagnoses, and their standards and rigor, had changed over time. In the study by Steffenburg and colleagues (1989), caregivers were interviewed using a researcher-designed tool. Children were examined and interviewed to elicit the relevant information on known autism symptomatology; three published tools were used as aids, the Autism Behavior Checklist (Krug, Arick, & Almond, 1980), the Lotter (Lotter, 1966), and the Diagnosis of Psychotic Behavior in Children (Haracopos & Kelstrup, 1975, 1978). Folstein and Rutter (1977) selected their sample based on detailed case review of referred or solicited individuals. Other studies used case vignettes (Taniai et al., 2008) and telephone interviews (Lichtenstein, Carlstrom, Rastam, Gillberg, & Anckarsater, 2010). In yet other inquiries yielding a broad range of heritability estimates (e.g., MZ 87 percent, DZ 12 percent, for males; Constantino & Todd, 2005; MZ 74 percent; Skuse, Mandy, & Scourfield, 2005) measures of single core features of autism, such as social responsiveness, had been used as a proxy for holistic diagnosis (Ronald & Hoekstra, 2011). Hence, the California Autism Twins Study, carried out by Hallmayer and colleagues, sought to correct these estimates by accessing carefully defined ASD probands from a population-based sample, and verifying their diagnoses using current, standard measures. They ascertained sex-specific concordance rates for both narrow and broad definitions of ASD to estimate genetic and environmental contributions to ASD susceptibility (Hallmayer et al., 2011).

Their methods exemplify a careful, rigorous approach to concordance and heritability estimates, with an intention to yield the most accurate results.

First, potentially affected twin pairs were identified using the records of California's Department of Developmental Services (DDS), which coordinates services for individuals with autism, mental retardation, and other developmental disabilities. Links to California's live births records identified twins and obtained their demographic data. Twin pairs had to meet specific eligibility criteria concerning age, state of residence, mental ability, absence of confounding conditions, and parents with language capabilities that would allow them to participate in the study fully. Initial qualifying diagnoses included possible ASD based on recorded information indicating a suspected ASD, *Diagnostic and Statistical Manual of Mental Disorders, 4th Edition* (*DSM-IV*) diagnosis of ASD, mental retardation of unknown etiology, or other developmental disability. Trained staff at the California Center for Autism and Developmental Disabilities Research and Epidemiology then characterized these client records as containing (or not) a diagnosis of ASD, criteria used for diagnosis, specific autistic behaviors, and other relevant clinical characteristics. Expert clinical review was then carried out to verify that children met the criteria for a qualifying diagnosis. Only twins born in the years 1987– 2004 were included, as the diagnosis of autism can be most reliably established if a child is at least four years or older. In this way, a broad capture from a statewide representative sample of previously diagnosed or possible undiagnosed cases of ASD twins were targeted for invitation into the study.

Second, the zygosity of those twin pairs whose families consented to participate in the study was then confirmed through genetic testing. That is, specific loci on particular chromosomes were targeted (nine short tandem repeat loci and the X/Y amelogenin locus). Twin pairs not matching at least one of these loci were considered dizygotic (DZ); twin pairs matching at all loci were considered monozygotic (MZ).

Third, all participating twins were assessed using the Autism Diagnostic Interview – Revised (ADI-R, a structured parent interview; Lord, Rutter, & Le Couteur, 1994) and the Autism Diagnostic Observation Schedule (ADOS, a structured play and interview scale administered to children and adolescents; Lord et al., 2000). At the time of the study, these were the most widely used and well-validated research measures for diagnosing narrowly defined autism and its more broadly defined spectrum, ASD. Verbal and non-verbal IQ were ascertained using the Stanford-Binet Intelligence Scales (SB), 5th edition (Roid & Barram, 2004), or the Mullen Scales of Early Learning Skills (Shank, 2011), for those children whose achievement levels fell below those ascertained by the SB. Based on their ADI-R and ADOS scores, each twin received a narrow (strict autism) or broad (ASD) diagnosis. Individuals diagnosed in the narrow category had to meet the standard criteria for autism on both the ADI-R and the ADOS. Individuals in the broader ASD category included those in the narrow category, along with individuals who met specific published criteria for ASD combining information from the ADI-R, the ADOS, and the ADOS revised algorithm for diagnosis (Hallmayer et al., 2011; Risi et al., 2006).

The sample was finalized when diagnoses of the twin pairs ascertained via the DDS files and records were confirmed by the California Autism Twins Study procedure described earlier. Probands were those individuals who were identified via the DDS initial qualifying procedures and by subsequent diagnosis by the study. Individuals in the strict autism analyses had to be diagnosed as such by the study, even if they were identified as diagnosed through the DDS records. Twin pairs in the strict autism analysis had to include at least one twin who had met the study criteria for strict autism (Hallmayer et al., 2011).

Hallmayer and his group then calculated probandwise concordance, rather than pairwise concordance. Pairwise concordance is calculated as the proportion of all twin pairs with two affected individuals out of a total number of twin pairs, when the pairs of twins are recruited according to whether at least one member is affected. Probands are individuals who have been independently ascertained. Therefore, probandwise concordance represents the risk of an affected proband co-twin also being diagnosed. That is, it looks at risk to individuals rather than to pairs, and it has been argued to be the stronger method of concordance determination (Hallmayer et al., 2011; McGue, 1992). Given:

$$N = \text{total number of twins in the sample}$$
$$R = \text{number of pairs with two affected twins}$$
$$t = \text{number of pairs with two probands}$$
$$S = \text{number of pairs with one proband}$$
$$U = \text{number of pairs with one proband}$$
$$R + U = N, \text{ and } t + S = R$$

Pairwise concordance would be given by R/N, which compares the number of twin pairs containing two affected individuals to the total number of twins in the study. Probandwise concordance compares numbers of individuals (co-twins of probands) and is calculated as $(R + t)/(N + t)$. This reflects, effectively, the ratio of the number of proband co-twins ($S + t + t$) to the total number of individual co-twins in the sample.

TABLE 1.3 *Results from Hallmayer and colleagues (2011), concordance estimates*

Probandwise concordance	Strict autism		ASD	
MZ male twins	40 pairs	58% (95% CI, 42%–74%)	45 pairs	77% (95% CI, 65%–86%)
MZ female twins	7 pairs	60% (95% CI, 28%–90%)	9 pairs	50% (95% CI, 16%–84%)
DZ male twins	31 pairs	21% (95% CI, 9%–43%)	45 pairs	31% (95% CI, 16%–46%)
DZ female twins	10 pairs	27% (95% CI, 9%–69%)	13 pairs	36% (95% CI, 11%–60%)
DZ female co-twins of male probands	$n = 54$	3.7% (95% CI, 0.5%–13%)	$n = 76$	5.3% (95% CI, 1.5%–13.0%)
DZ male co-twins of female probands	$n = 2$	50% (95% CI, 1%–99%)	$n = 6$	50.0% (95% CI, 11.8%–88.2%)

MZ = monozygotic; DZ = dizygotic; n = sample size; CI = confidence interval.

After systematically checking for any evidence of demographic, clinical, or selection bias in the selected sample (by statistically comparing the selected and unselected groups from the original DDS qualifying sample on a number of factors), the concordance rates obtained above were applied in a model-fitting exercise to determine the liability of inheriting autism, both in the strict and broad senses. Here, liability is assumed to be a continuous, normally distributed, unobserved trait consisting of numerous genetic and environmental factors of small effect combined. Liability values above a certain threshold indicate the manifestation of the disease. The threshold is based on disease prevalence. For this study, a prevalence of 1 percent for males and 0.3 percent for females for ASD was assumed; and 0.5 percent for males and 0.15 percent for females for strict autism. The classical model described previously, which assumes that phenotypic variance can be explained by allelic variance (A), with contributions from shared (C) and unshared (E) environments, was then applied in its mathematical form, $A + C + E = 1$, where A = the proportion of liability variance due to genetic factors; C = the proportion of liability variance due to shared environmental factors; and E = the proportion of liability variance due to random or non-shared environmental factors. The liability was estimated separately for the two sexes using the concordance data generated in the first part of the study. Their results are summarized in Table 1.3.

In all of the models listed above, both the genetic and shared environmental components were significant, for both strict and broad autism, and it was concluded that the heritability and shared environment components for males and females could be considered equal. For the entire sample, then, the best fitting model for the narrow definition of autism estimated a heritability of 37 percent (95% CI, 8%–84%), and a shared environment effect of 55 percent (95% CI, 9%–81%). For the broader definition, ASD, the estimated heritability was 38 percent (95% CI, 14%–67%), and the shared environmental component, 58 percent (95% CI, 30%–80%). Unlike the conclusions of previous heritability studies on autism, these numbers suggest that for both strict autism and ASD, genetic heritability is moderate, and a large proportion of the variance in liability is indeed due to shared environmental factors (Hallmayer et al., 2011).

WHAT DOES IT ALL MEAN?

There are a few important lessons to gain from this example. First, estimates of concordance and heritability for complex traits are generated via lengthy and sophisticated scientific efforts that may differ in the measurement instruments used, ascertainment of the sample, and the data analytic methods and models ultimately applied. A recent case in point illustrates how these differences come about. With respect to the diagnosis of ASD, the release of the new *Diagnostic and Statistical Manual for Mental Health Disorders* (DSM-V; APA, 2013) in 2013 raised concerns because its description and structuring of the criteria for the diagnosis of autism had changed significantly from the previous version (DSM-IV-TR; APA, 2000). It was widely feared that such a change would result in a loss of sensitivity in the diagnostic process, thereby decreasing the diagnosis of actual cases of ASD (Carrington, 2016; Christiansz, Gray, Taffe, & Tonge, 2016; McPartland, Reichow, & Volkmar, 2013). Indeed, several reviews and meta-analyses have been carried out comparing the percentage of people diagnosed using the DSM-IV-TR or DSM-IV criteria against those who would qualify for an ASD diagnosis under the DSM-V (Bennett & Goodall, 2016; Kulage, Smaldone, & Cohn, 2014; Smith, Reichow, & Volkmar, 2015; Sturmey & Dalfern, 2014). The findings show an ostensible reduction in the number of individuals who would be diagnosed using the DSM-V versus DSM-IV. However, it has been suggested that the earlier criteria had perhaps led to an overdiagnosis of the disorder (Christiansz et al., 2016), and that different instrumentation or methods of diagnosis that cover the new criteria more appropriately would result in acceptable levels of both sensitivity and specificity of the diagnosis as outlined in DSM-V (Carrington, 2016). Thus, criteria and methods for diagnosis continue to evolve, and in the process may be challenged, rejected, or adapted to. These lead to variations in outcomes.

Second, these heritability estimates may change or differ, even if methods and approaches do not, due to changes in prevalence estimates from one time period to another, or from one population to another. In fact, the reported prevalence of autism has increased dramatically over the last forty years, twenty to thirty times over, according to the US Centers for Disease Control and Prevention (CDC, 2012). It has been estimated that in the United States alone, from the years 1993–2003, prevalence of autism in schools increased 800 percent, and rates have continued to rise over the decades (Kroncke, 2016). Today, prevalence is reported at one in sixty-eight (CDC, 2014). Yet such changes in prevalence, we are reminded, need to be regarded with caution (Mandell & Lecavalier, 2014) as they may be highly influenced by diagnostic methods and/or the realization of previous misdiagnoses (Blumberg et al., 2013). Like methods of diagnosis and their outcomes, prevalence can be highly variable.

Third, over and above the computational details, concordance and heritability estimates are built upon a series of calculated probabilities that incorporate multiple assumptions along the way. For example, Hallmayer and colleagues acknowledge that the ACE model assumes, among other implicit assumptions, that no interaction exists between genes and environment. Similarly, the model also assumes that the shared environment for MZ twins and DZ twins is the same, which may in fact not be the case (the "equal environments assumption" mentioned earlier). The authors offer the caveat that their estimate of A may in fact be an overestimate due to these and other such assumptions (Hallmayer et al., 2011). However, such caveats beg the question: In comparing these different methods to generate estimates of concordance and heritability, how is one to weigh these assumptions?

It is interesting to note that, despite the seeming soundness of the procedure conducted by Hallmayer and colleagues, the relative contributions of genetics and environment to autism continue to be reported in vastly different proportions. One recent review points to a "large accumulation of evidence" that indicates a "strong genetic component of ASD" (Chen et al., 2015, p. 113). It even cites Hallmayer's work as supportive of this argument, even though Hallmayer's intention was to emphasize the opposite view. A recent meta-analysis that includes the Hallmayer study concludes that the combined data of all of the twin studies of ASD to date reveal heritability estimates of 64–91 percent, and suggest that previously reported significant shared environmental effects may be due to the overinclusion in such studies of concordant DZ twins (Tick, Bolton, Happé, Rutter, & Rijsdijk, 2016). On the other side of this debate, a large, countrywide heritability study in Sweden (Sandin et al., 2014) found both a moderate heritability estimate of .50 (95% CI, 0.45–0.56) and a non-shared environmental component of .50 (95% CI, 0.44–0.55). To take advantage of access to extended family members, the researchers measured relative recurrence risk, the risk that a relative of a child with autism will be also be diagnosed, compared to the risk of a relative of a child without autism. This measurement was applied across all siblings and cousins of affected individuals. A later study and subsequent commentaries call attention not only to possible family (age of parents at time of birth) and other environmental effects, but to the larger issue of less studied gene–environment interactions that have yet to be explored in any substantial way with respect to ASD (Charman & Chakrabarti, 2016; Sandin et al., 2016; Schendel, Gronborg, & Parner, 2014).

These commentaries remind us that despite the usefulness of the general indications given by heritability studies, theoretical models have certain limitations and it may be useful to re-examine autism both historically and in its social context to gain new insights into the relationship between its genetic and environmental mechanisms. One such recent attempt worth touching upon explores the sociology of autism, identity, and behavior, and how these things might change the genetic make-up of a diagnosed population over time (Navon & Eyal, 2016). Using a concept called "looping," conceived by philosopher of science, Ian Hacking (1995, 2006), Navon and Eyal investigate how the "geneticization" of autism may have established conceptions of biological identity and biological community that have ultimately changed diagnostic practice, and also increased autism's genetic heterogeneity over time. The basic idea is that, as people are labeled and classified – in one way or another – their behavior as individuals change as they adapt their sense of self and their behavior to these bestowed identities. This so-called "looping" happened, Navon and Eyal propose, when knowledge of a genetic etiology for autism became known, changing how individuals with autism and their parents saw themselves and were treated by others. This change caused shifts in practices of medical classification (i.e., diagnosis), broadening prevalence and the genetic heterogeneity of the ASD community over a series of feedback loops happening over time (Navon & Eyal, 2016).

Whether or not one is convinced by these proposals and arguments, autism presents a particularly apt example of such possible "looping." Its social dimensions, initially thought in 1912 to represent a childhood form of schizophrenia, were first described in detail by psychiatrist and physician Leo Kanner (1943), who attributed the condition to emotionally cold, detached parents. Psychologist Bruno Bettelheim later popularized the term "refrigerator mother" (Bettelheim, 1959), which cast hard blame on the family environment, and this perception lasted for decades, into the late 1960s (Blake et al., 2013); it may even persist today as vestigial remnants in the minds of many. What happened when society lifted that blame? It seems conceivable that parents, if not the children themselves, certainly felt and behaved differently over time. Even as the search for other environmental causes of autism continues (de Zeeuw et al., na; Matelski & Van de Water, 2016; Tick et al., 2016), knowing the biological sources of the disorder ostensibly freed many parents to become more confident advocates for their children, and more psychologically secure caregivers (Silverman, 2012).

LOOPING BACK: UNDERSTANDING THE GENETICS OF HUMAN BEHAVIOR, UNDERSTANDING OURSELVES

Heritability, as we have seen in the previous sections, can be estimated for physical and behavioral traits, for plants, animals, and humans. However, its estimates seem to stir up more controversy and angst when they touch upon particular behavioral human traits, i.e., those that may be more difficult to measure reliably and may be associated with sensitive societal issues, e.g., intelligence, personality, and neurodevelopmental disorders, such as depression and schizophrenia (Sesardic, 2005). The objective of the field of human behavioral genetics is to understand etiologies of behavior, differences among populations, and human development within the contexts of different environments. It is important to realize that beyond the noble aims of objective scientific inquiries, etiologies – whatever they concern – matter deeply because they hold a powerful place in humans' conceptions of themselves and others.

Quantitative genetic studies seek to appraise, preliminarily, the role of genetic sources of variation in individual differences in human psychological processes and behavior. The information we learn from such studies comes with great promise and, also, numerous cautionary notes, holding possibilities for a better understanding of ourselves, and for predicting and shaping our future. It also comes, as broached in the chapters of this volume, with all of the complexity and responsibility of a Pandora's box. The controversy of the so-called nature–nurture debate, which arose with Francis Galton's investigations into the heritability of genius and sparked heated discussions on what aspects of human behavior are endowed from nature (i.e., ostensibly unchangeable) and which can be nurtured or altered through the environment (Press, Chapman, & Parens, 2006), though dissipated, lingers on. Take a relatively recent headline that read, "Genetic background of extreme violent behavior" (Tiihonen et al., 2014), published in *Molecular Psychiatry* in October 2014. Intended or not, such a headline may be misinterpreted by passing readers as an indication of a direct causal relationship between the variation in the *MAOA* and *CDH13* genes and the variation in violent behavior. While such a conclusion would be erroneous, the all too easily inferred message of such a headline could be harmful to anyone with a relative who may have these variants or who has displayed extreme and violent behavior, by affecting the way they see themselves or how others see them and treat them.

Similarly, in the field of education, there is sensitivity to genetic aspects of learning as schooling takes place in a formal setting where genetic indications may generate unfounded positive or negative expectations for performance, or labels where none are yet warranted. Yet it is important to continue to gather and apply the information afforded to us by our ever-developing understanding of genetic mechanisms and our advances in technology. Issues of ethics and genetics, as we will see, must remain close companions as we learn more about the relationships between DNA and behavior and illness, and struggle to come to grips with their implications for our sense of identity.

REFERENCES

APA. (2000). *Diagnostic and statistical manual of mental disorders, 4th ed. (DSM-4)*. Washington, DC: American Psychiatric Publishing.

(2013). *Diagnostic and statistical manual of mental disorders, 5th ed. (DSM-5)*. Washington, DC: American Psychiatric Publishing.

Baden, A. L., & Wiley, M. O. L. (2007). Counseling adopted persons in adulthood integrating practice and research. *The Counseling Psychologist, 35*(6), 868–901.

Bailey, A., Le Couteur, A., Gottesman, I. I., Bolton, P., Simonoff, E., Yuzda, E., et al. (1995). Autism as a strongly genetic disorder: Evidence from a British twin study. *Psychological Medicine, 25*(01), 63–77.

Baker, C. (2004). *Behavioral genetics: An introduction to how genes and environments interact through development to shape differences in mood, personality, and intelligence*. Washington, DC: The American Association for the Advancement of Science.

Bennett, M., & Goodall, E. (2016). A meta-analysis of DSM-5 autism diagnoses in relation to DSM-IV and DSM-IV-TR. *Review Journal of Autism and Developmental Disorders, 3*, 119–124.

Bettelheim, B. (1959). Joey: A mechanical boy. *Scientific American, 200,* 116–127.

Blake, J., Hoyme, H. E., & Crotwell, P. L. (2013). A brief history of autism, the autism/vaccine hypothesis and a review of the genetic basis of autism spectrum disorders. *SD Med, 15,* 58–65.

Blumberg, S. J., Bramlett, M. D., Kogan, M. D., Schieve, L. A., Jones, J. R., & Lu, M. C. (2013). Changes in prevalence of parent-reported autism spectrum disorder in school-aged US children: 2007 to 2011–2012. *National Health Statistics Reports, 65*(20), 1–7.

Boraas, J., Messer, L., & Till, M. (1988). A genetic contribution to dental caries, occlusion, and morphology as demonstrated by twins reared apart. *Journal of Dental Research, 67*(9), 1150–1155.

Bouchard, T. J. (2005). Identical twins reared apart. *eLS.* doi:10.1038/npg.els.0005156

Bouchard, T. J., Lykken, D. T., McGue, M., Segal, N. L., & Tellegen, A. (1990). Sources of human psychological differences: The Minnesota study of twins reared apart. *Science, 250*(4978), 223–228.

Braverman, A. M. (2010). How the internet is reshaping assisted reproduction: From donor offspring registries to direct-to-consumer genetic testing. *Minn. JL Sci. & Tech., 11,* 477.

(2013). Defining, understanding, and managing the complex psychological aspects of third-party reproduction. In M. V. Sauer (Ed.), *Principles of oocyte and embryo donation* (pp. 185–193). London: Springer.

Brodzinsky, D., & Palacios, J. (2005). *Psychological issues in adoption: Research and practice*: Westport, CT: Praeger.

Carlsson, J., McDowell, J. R., Carlsson, J. E. L., & Graves, J. E. (2007). Genetic identity of YOY bluefin tuna from the eastern and western Atlantic spawning areas. *Journal of Heredity, 98*(1), 23–28. doi:10.1093/jhered/esl046

Carrington, S. J. (2016). Implications of *ICD* and *DSM* on screening, diagnosis, and monitoring. In J. L. Matson (Ed.), *Handbook of assessments and diagnosis of autism spectrum disorder.* Switzerland: Springer International Publishing.

Casselman, A. (2008). Identical twins' genes are not identical. *Scientific American.*

CDC. (2012). Prevalence of autism spectrum disorders: Autism and Developmental Disabilities Monitoring Network, 14 sites, United States, 2008. *Morbidity and Mortality* Weekly Report. Surveillance Summaries, *61*(3).

(2014). Prevalence of autism spectrum disorder among children aged 8 years: Autism and Developmental Disabilities Monitoring Network, 11 sites, United States, 2010. *Morbidity and Mortality Weekly Report. Surveillance Summaries, 63*(2), 1.

Charman, T., & Chakrabarti, B. (2016). Not just genes – Reclaiming a role for environmental influences on aetiology and outcome in autism. A commentary on Mandy and Lai. *Journal of Child Psychology and Psychiatry, 57*(3), 293–295.

Chen, J. A., Peñagarikano, O., Belgard, T. G., Swarup, V., & Geschwind, D. H. (2015). The emerging picture of autism spectrum disorder: genetics and pathology. *Annual Review of Pathology: Mechanisms of Disease, 10,* 111–144.

Christiansz, J. A., Gray, K. M., Taffe, J., & Tonge, B. J. (2016). Autism spectrum disorder in the DSM-5: Diagnostic sensitivity and specificity in early childhood. *Journal of Autism & Developmental Disorders, 46,* 2054–2063. doi:10.1007/s10803-016-2734-4

Cohen, D. J., Dibble, E., Grawe, J. M., & Pollin, W. (1973). Separating identical from fraternal twins. *Archives of General Psychiatry, 29,* 465–469.

Cohen, E. (2005). The real meaning of genetics. *The New Atlantis, Summer* (9), 29–41.

Constantino, J. N., & Todd, R. D. (2005). Intergenerational transmission of sub-threshold autistic traits in the general population. *Biological Psychiatry, 57,* 655–660.

Coulter, C. M. (2015). Needs of families post-international adoption. *Master of Social Work Clinical Research Papers.* Paper 433.

Cushing, A. L. (2010). "I just want more information about who I am": The search experience of sperm-donor offspring, searching for information about their donors and genetic heritage. *Information Research, 15*(2), 1.

de Zeeuw, E. L., van Beijsterveldt, C. E. M., Hoekstra, R. A., Bartels, M., & Boomsma, D. I. The etiology of autistic traits in preschoolers: a population-based twin study. *Journal of Child Psychology and Psychiatry,* n/a-n/a. doi:10.1111/jcpp.12741

Deary, I. J., Johnson, W., & Houlihan, L. M. (2009). Genetic foundations of human intelligence. *Human Genetics, 126,* 125–232.

Falconer, D. S., & Mackay, T. F. C. (1996). *Introduction to quantitative genetics* (4th ed.). Harlow, UK: Longmans Green.

Folstein, S., & Rutter, M. (1977). Infantile autism: A genetic study of 21 twin pairs. *Journal of Child Psychology and Psychiatry, 18*(4), 297–321.

Galton, F. R. S. (1886). Regression toward mediocrity in hereditary stature. *Journal of the Anthropological Institute of Great Britain and Irelenad, 15,* 17.

Geschwind, D. H. (2011). Genetics of autism spectrum disorders. *Trends in Cognitive Sciences, 15*(9), 409–416.

Gharghani, A., Zamani, Z., Talaie, A., Oraguzie, N. C., Fatahi, R., Hajnajari, H., et al. (2009). Genetic identity and relationships of Iranian apple (Malus× domestica Borkh.) cultivars and landraces, wild Malus species and representative old apple cultivars based on simple sequence repeat (SSR) marker analysis. *Genetic Resources and Crop Evolution, 56*(6), 829–842.

Goldsmith, L., Jackson, L., O'Connor, A., & Skirton, H. (2012). Direct-to-consumer genomic testing: Systematic review of the literature on user perspectives. *European Journal of Human Genetics,* 1–6.

Griffiths, A. J. F., Wessler, S. R., Carroll, S. B., & Doebley, J. (2015). *Introduction to genetic analysis.* New York: W. H. Freeman and Company.

Grotevant, H. D. (1997). Coming to terms with adoption. *Adoption Quarterly, 1*(1), 3–27. doi:10.1300/J145v01n01_02

Grotevant, H. D., Dunbar, N., Kohler, J. K., & Esau, A. M. L. (2000). Adoptive identity: How contexts within and beyond the family shape developmental pathways. *Family Relations, 49*(4), 379–387. doi:10.2307/585833

Guo, S.-W. (1996). Variation in genetic identity among relatives. *Human Heredity, 46*(2), 61–70.

Hacking, I. (1995). The looping effects of human kinds. *Causal Cognition: A Multidisciplinary Debate,* 351–394.

    (2006). Genetics, biosocial groups and the future of identity. *Daedalus, 135*(4), 81–95.

Haga, S. B., Barry, W. T., Mills, R., Ginsburg, G. S., Svetkey, L., Sullivan, J., et al. (2013). Public knowledge and attitudes toward genetics and genetic testing. *Genetic Testing and Molecular Biomarkers, 17*(4), 327–335. doi:10.1089/gtmb.2012.0350

Haimes, E. (1988). "Secrecy": What can artificial reproduction learn from adoption? *International Journal of Law, Policy and the Family, 2*(1), 46–61.

(1993). Ethics and society: Do clinicians benefit from gamete donor anonymity? *Human Reproduction, 8*(9), 1518–1520.

Haimes, E., & Timms, N. (1985). *Adoption, identity and social policy: The search for distant relatives*: Aldershot, UK: Gower.

Hallmayer, J., Cleveland, S., Torres, A., Phillips, J., Cohen, B., Torigoe, T., et al. (2011). Genetic heritability and shared environmental factors among twin pairs with autism. *Archives of General Psychiatry, 68*(11), 1095–1102.

Haracopos, D., & Kelstrup, A. (1975). *DIPAB observationsskema [DIPAB observation scheme]*. Herning, Denmark: Special-Pædagogisk Forlag A/S.

(1978). Psychotic behaviour in children under the institutions for the mentally retarded in Denmark. *Journal of Autism & Childhood Schizophrenia, 8*, 1–12.

Hill Goldsmith, H. (1991). A zygosity questionnaire for young twins: A research note. *Behavior Genetics, 21*(3), 257–269. doi:10.1007/BF01065819

Howe, D., Feast, J., & Coster, D. (2000). *Adoption, search & reunion: The long term experience of adopted adults*: London: Virago Press.

Kanner, L. (1943). Autistic disturbances of affective contact. *Nervous Child, 2*, 217–250.

Klitzman, R. (2009). "Am I my genes?" Questions of identity among individuals confronting genetic disease. *Genetics in Medicine, 11*(12), 880–889.

Kroncke, A. P. (2016). What is autism? History and foundations. In A. P. Kroncke, M. Willard, & H. Huckabee (Eds.), *Assessment of autism spectrum disorder: Critical issues in clinical, forensic, and school settings* (pp. 3–9). Switzerland: Springer International.

Krug, D. A., Arick, J., & Almond, P. (1980). Autism behavior checklist for identifying severely handicapped individuals with high levels of autistic behaviour. *Journal of Child Psychology and Psychiatry, 21*, 221–229.

Kulage, K. M., Smaldone, A. M., & Cohn, E. G. (2014). How will DSM-5 affect autism diagnosis? A systematic literature review and meta-analysis. *Journal of Autism & Developmental Disorders, 44*(8), 1918–1932.

Leve, L. D., Neiderhiser, J. M., Shaw, D. S., Ganiban, J., Natsuaki, M. N., & Reiss, D. (2013). The early growth and development study: A prospective adoption study from birth through middle childhood. *Twin Research and Human Genetics, 16*(1), 412–423. doi:10.1017/thg.2012.126

Lichtenstein, P., Carlstrom, E., Rastam, M., Gillberg, C., & Anckarsater, H. (2010). The genetics of autism spectrum disorders and related neuropsychiatric disorders in childhood. *American Journal of Psychiatry, 167*(11), 1357–1363.

Lord, C., Risi, S., Lambrecht, L., Cook, E. H. J., Leventhal, B. L., DiLavore, P. C., et al. (2000). The Autism Diagnostic Observation Schedule – Generic: A standard measure of social and communication deficits associated with the spectrum of autism. *Journal of Autism & Developmental Disorders, 30*(3), 205–233.

Lord, C., Rutter, M., & Le Couteur, A. (1994). Autism Diagnostic Interview – Revised: A revised version of a diagnostic interview for caregivers of individuals with possible pervasive developmental disorders. *Journal of Autism & Developmental Disorders, 24*(5), 659–685.

Lotter, V. (1966). Epidemiology of autistic conditions in young children I. Prevalence. *Social Psychiatry, 1*, 163–173.

Mandell, D., & Lecavalier, L. (2014). Should we believe the Centers for Disease Control and Prevention's autism spectrum disorder prevalence estimates? *Autism, 18*(5), 482–484. doi:10.1177/1362361314538131

Matelski, L., & Van de Water, J. (2016). Risk factors in autism: Thinking outside the brain. *Journal of Autoimmunity, 67*, 1–7. doi:http://dx.doi.org/10.1016/j.jaut.2015.11.003

Matthews, J. A. (2014). Family context and searching among internationally adopted adolescents (Unpublished master's thesis). Tufts University, Medford, MA.

McCartney, K., Harris, M. J., & Bernieri, F. (1990). Growing up and growing apart: A developmental meta-analysis of twin studies. *Psychological Bulletin, 107*, 226–237.

McGue, M. (1992). When assessing twin concordance, use the probandwise not the pairwise. *Schizophrenia Bulletin, 18*(2), 171–176.

McPartland, J. C., Reichow, B., & Volkmar, F. R. (2013). Sensitivity and specificity of proposed DSM-5 diagnostic criteria for autism spectrum disorder. *Journal of the American Academy of Child & Adolescent Psychiatry, 51*(4), 1236–1242.

Merriman, C. (1924). The intellectual resemblance of twins. *Psychological Monographs, 33*(5), i-57.

Mukherjee, S. (2016). *The gene: An intimate history.* New York: Scribner.

Munsinger, H. (1975). The adopted child's IQ: A critical review. *Psychological Bulletin, 82*(5), 623–659.

Navon, D., & Eyal, G. (2016). Looping genomes: Diagnostic change and the genetic makeup of the autism population. *American Journal of Sociology, 121*(5), 1416–1471. doi:10.1086/684201

Nichols, R. C., & Bilbro, W. C. (1966). The diagnosis of twin zygosity. *Acta Genetica, 16*, 265–275.

Olson, R. K., Hulslander, J., Christopher, M., Keenan, J. M., Wadsworth, S. J., Willcutt, E. G., et al. (2013). Genetic and environmental influences on writing and their relations to language and reading. *Annals of Dyslexia, 63*, 25–43.

Pedersen, N. L., Lichtenstein, P., Plomin, R., DeFaire, U., McClearn, G., & Matthews, K. A. (1989). Genetic and environmental influences for type A-like measures and related traits: A study of twins reared apart and twins reared together. *Psychosomatic Medicine, 51*(4), 428–440.

Phelps, S. R., & Allendorf, F. W. (1983). Genetic identity of pallid and shovelnose sturgeon (*Scaphirhynchus albus* and *S. platorynchus*). *Copeia*, 696–700.

Plomin, R., & DeFries, J. C. (1983). The Colorado Adoption Project. *Child Development*, 276–289.

(1985). Origins of individual differences in infancy: The Colorado Adoption Project. *Science, 230*, 1369–1371.

Plomin, R., DeFries, J. C., Knopik, V. S., & Neiderhiser, J. M. (2013). *Behavioral genetics.* New York, NY: Worth.

Plomin, R., & Spinath, F. M. (2004). Intelligence: Genetics, genes, and genomics. *Journal of Personality and Social Psychology, 86*(1), 112–129.

Press, N., Chapman, A. R., & Parens, E. (2006). Introduction. In E. Paren, A. R. Chapman, & N. Press (Eds.), *Wrestling with behavioral genetics: Science, ethics, and public conversation* (pp. xiii–xxxv). Baltimore: The Johns Hopkins University Press.

Rice, T. K., & Borecki, I. B. (2001). Familial resemblance and heritability. *Advances in Genetics*, 42, 35–44.

Risi, S., Lord, C., Gotham, K., Corsello, C., Chrysler, C., Szatmari, P., et al. (2006). Combining information from multiple sources in the diagnosis of autism spectrum disorders. *Journal of the American Academy of Child & Adolescent Psychiatry*, 45(9), 1094–1103.

Roid, G., & Barram, R. (2004). *Essentials of Stanford–Binet Intelligence Scales (SB5) assessment*. Hoboken, NJ: John Wiley & Sons.

Ronald, A., & Hoekstra, R. A. (2011). Autism spectrum disorders and autistic traits: A decade of new twin studies. *American Journal of Medical Genetics*, 156B(3), 255–274.

Rosenberg, R. E., Law, J. K., Yenokyan, G., McGready, J., Kaufman, W. E., & Law, P. A. (2009). Characteristics and concordance of autism spectrum disorders among 277 twin pairs. *Archives of Pediatric Adolescent Medicine*, 163, 907–914.

Sandin, S., Lichtenstein, P., Kuja-Halkola, R., Larsson, H., Hultman, C. M., & Reichenberg, A. (2014). The familial risk of autism. *JAMA*, 311(17), 1770–1777.

Sandin, S., Schendel, D., Magnusson, P., Hultman, C., Surén, P., Susser, E., et al. (2016). Autism risk associated with parental age and with increasing difference in age between the parents. *Molecular Psychiatry*, 21, 693–700.

Schendel, D. E., Gronborg, T. K., & Parner, E. T. (2014). The genetic and environmental contributions to autism: Looking beyond twins. *JAMA*, 311(17), 1738–1739.

Sesardic, N. (2005). *Making sense of heritability*. Cambridge, UK: Cambridge University Press.

Shank, L. (2011). Mullen scales of early learning. In J. S. Kreutzer, J. DeLuca, & B. Caplan (Eds.), *Encyclopedia of clinical neuropsychology* (pp. 1669–1671). New York, NY: Springer.

Sharp, H. L. (1971). Alpha-1-antitrypsin deficiency. *Hospital Practice*, 6(5), 83–96.

Silverman, C. (2012). *Understanding autism: Parents, doctors, and the history of a disorder*. Princeton, NJ: Princeton University Press.

Skipper, M. (2008). Human genetics: Not-so-identical twins. *Nat Rev Genet*, 9(4), 250–251.

Skuse, D. H., Mandy, W. P. L., & Scourfield, J. (2005). Measuring autistic traits: Heritability, reliability and validity of the Social and Communication Disorders Checklist. *The British Journal of Psychiatry*, 187(6), 568–572. doi:10.1192/bjp.187.6.568

Smith, I. C., Reichow, B., & Volkmar, F. R. (2015). The effects of DSM-5 criteria on numer of individuals diagnosed with autism spectrum disorder: A systematic review. *Journal of Autism & Developmental Disorders*, 45(8), 2541–2552.

Sobol, M. P., & Cardiff, J. (1983). A sociopsychological investigation of adult adoptees' search for birth parents. *Family Relations*, 32(4), 477–483. doi:10.2307/583686

Steffenburg, S., Gillberg, C., Hellgren, L., Andersson, L., Gillberg, I. C., Jakobsson, G., et al. (1989). A twin study of autism in Denmark, Finland, Iceland, Norway and Sweden. *Journal of Child Psychology and Psychiatry*, 30(3), 405–416.

Stunkard, A. J., Harris, J. R., Pedersen, N. L., & McClearn, G. E. (1990). The body-mass index of twins who have been reared apart. *New England Journal of Medicine, 322*(21), 1483–1487. doi:10.1056/NEJM199005243222102

Sturmey, P., & Dalfern, S. (2014). The effects of DSM-5 autism diagnostic criteria on number of individuals diagnosed with autism spectrum disorders: A systematic review. *Review Journal of Autism and Developmental Disorders, 1*(4), 249–252.

Su, P. (2013). Direct-to-consumer genetic testing: A comprehensive view. *The Yale Journal of Biology and Medicine, 86*(3), 359–365.

Sweeny, K., Ghane, A., Legg, A., Huynh, H., & Andrews, S. (2014). Predictors of genetic testing decisions: A systematic review and critique of the literature. *Journal of Genetic Counseling, 23*(3), 263–288. doi:10.1007/s10897-014-9712-9

Taniai, H., Nishiyama, T., Miyachi, T., Imaeda, M., & Sumi, S. (2008). Genetic influences on the broad spectrum of autism: A study of proband-ascertained twins. *American Journal of Medical Genetics, 147B*(6), 844–849.

Terman, L. M. (1916). *The measurement of intelligence: An explanation and complete guide for the use of the Stanford revision and extension of the Binet-Simon Intelligence Scale.* Boston, MA: Houghton-Mifflin.

Theis, S. V. S. (1924). *How foster children turn out (Publication No. 165).* New York: State Charities Aid Association.

Tick, B., Bolton, P., Happé, F., Rutter, M., & Rijsdijk, F. (2016). Heritability of autism spectrum disorders: A meta-analysis of twin studies. *Journal of Child Psychology and Psychiatry, 57*(5), 585–595. doi:10.1111/jcpp.12499

Tiihonen, J., Rautiainen, M. R., Ollila, H. M., Repo-Tiihonen, E., Virkkunen, M., Palotie, A., et al. (2014). Genetic background of extreme violent behavior. *Mol Psychiatry.* doi:10.1038/mp.2014.130

Triseliotis, J. (1973). *In search of origins: The experiences of adopted people.* London: Routledge and Kegan Paul.

Triseliotis, J., Feast, J., & Kyle, F. (2005). *The adoption triangle revisited.* London: BAAF.

Vandivere, S., Malm, K., & McKlindon, A. (2011). Adoption USA: Summary and highlights of a chartbook on the national survey of adoptive parents. In E. A. Rosman, C. E. Johnson, & N. M. Callahan (Eds.), *Adoption factbook V* (pp. 115–124). Alexandria, VA: National Council for Adoption.

Vinkhuyzen, A. A., Wray, N. R., Yang, J., Goddard, M. E., & Visschler, P. M. (2013). Estimation and partitioning of heritability in human populations using whole genome analysis methods. *Annual Review of Genetics, 47,* 75–95.

Winter, K., & Cohen, O. (2005). Identity issues for looked after children with no knowledge of their origins: Implications for research and practice. *Adoption & Fostering, 29*(2), 44–52. doi:10.1177/030857590502900206

Wrobel, G. M., Grotevant, H. D., & McRoy, R. G. (2004). Adolescent search for birth-parents: Who moves forward? *Journal of Adolescent Research, 19*(1), 132–151. doi:10.1177/0743558403258125

Yerkes, R. M. (1921). *Psychological examining in the United States army.* Washington, DC: Surgeon General's Office.

Yerkes, R. M., Haggerty, M. E., Terman, L. M., Thorndike, E. L., & Whipple, G. M. (1920). *National intelligence tests.* Yonkers-on-Hudson, NY: World Book Co.

2

Molecular Genetics and Genomics

SERGEY A. KORNILOV

MOLECULAR GENETICS AND GENOMICS: A PRIMER

The current chapter aims to serve as a basic primer or introduction to the core concepts of molecular genetics relevant to study designs used to evaluate the role of one or more types of genetic variation in the development and manifestation of complex traits and common disorders, including neurodevelopmental and learning disorders. Thus, the chapter will first provide a brief overview of the structure of DNA and its role as the main substrate of genetic information. It will then discuss major types of genetic variation that have gained prominence in genetic epidemiological studies and present the basic logic behind genetic linkage and association study designs. It will conclude with a discussion of the emerging role of DNA sequencing in elucidating the genetic architecture of complex traits.

DNA and the Human Genome

Most human cells contain a full set of information of the genetic make-up of the individual stored in the cell nucleus in chromosomes – threadlike structures made of DNA and other proteins. The main carrier of genetic information is DNA – a two-stranded molecule that resembles a twisted ladder and forms a double helix. DNA is comprised of two sugar phosphate backbones, connected by rungs composed of four organic DNA bases: purines (guanine [G] and adenine [A]) and pyrimidines (thymine [T] and cytosine [C]). The basic structural unit of a single strand of DNA is called a nucleotide and consists of one sugar and one phosphate group, and one base. The two complementary strands of the DNA molecule are connected by a hydrogen bond between the opposing bases. Adenine always pairs with thymine (A–T), and cytosine always pairs with guanine (C–G) – the specificity of these two pairing rules is driven by the dimensional properties of the four bases (i.e., purines are larger than pyrimidines) that is critical for DNA stability and has been referred to as the complementary base-pairing rule. An important implication of the complementary base-pairing rule is that if one knows the sequence of bases in one strand of the DNA molecule, the other strand can be reconstructed using the rule; this process lies at the heart of DNA replication. DNA replication serves the purpose of producing an identical copy of the original template DNA molecule. Such replication occurs when the template molecule is "unzipped" and separated into two separate single strands. For both, the opposite strand is reconstructed using the rule-driven hydrogen bonding of free nucleotides that results in the synthesis of two identical DNA molecules.

In humans, the DNA is tightly packed into a set of twenty-three chromosomes. Each chromosome has a centromere and two arms, a short one (p arm) and a long one (q arm), that end with telomeres. Each arm of a chromosome can be visually split into regions called cytogenetic bands. The bands correspond to the patterns of light and dark areas of chromosomal regions that can be detected under a microscope after staining and are labeled consecutively starting from the centromere out toward the telomeres. Cytogenetic bands (e.g., 18q11) can be further subdivided into subbands (e.g., 18q11.1). Twenty-two chromosomes are non–sex-determining autosomes, consecutively labeled from 1 to 22 approximately according to their length: chromosome 1 is the longest with a length of ~249 million base pairs (bp); 21 is the shortest with ~47 million bp. The remaining sex chromosome is either an X (156 million bp) or a Y (57 million bp), depending on the sex of the individual. Most human cells contain a double (i.e., diploid) set of genetic information ($23 \times 2 = 46$ chromosomes), with half of the material inherited from the mother, and half from the father. Human egg and sperm cells, on the other hand, only contain a single set of twenty-three chromosomes and are, therefore, termed haploid.

All of the genetic information that is necessary to build the full organism is coded in the linear sequence of DNA nucleotides. Triplets of nucleotides – called codons – specify the amino acids to be used to build the proteins encoded by the genes. Codons encode twenty different amino acids that are themselves used as building blocks during protein biosynthesis (e.g., the AAA sequence codes for the amino acid called Lysine; CCC codes for Proline). Although the three-letter code can be used to generate sixty-four different codons, there are only twenty amino acids that need to be encoded; correspondingly, the majority of the amino acids can be encoded by more than one codon.

The central dogma of molecular biology postulates that the flow of genetic information in biological systems follows a DNA → RNA (ribonucleic acid) → protein pathway. The first step in this pathway is called transcription and involves the synthesis of a single-stranded messenger RNA (mRNA) molecule that carries the information out of the cell nucleus into the plasma where protein synthesis takes place. The RNA molecule is a less stable single-stranded molecule that resembles DNA but contains uracil (U) as the complementary base to adenine. Transcription is enabled by a specific enzyme, RNA polymerase, which uses the DNA sequence as a template to produce mRNA (which in turn is the molecular template that will be used in actual protein synthesis during the second phase called translation). Translation takes place in ribosomes – cellular structures that link together amino acids in the order specified by the mRNA to produce polypeptide chain molecules that are then folded and modified into actual functional proteins.

The complete sequence of DNA is called the genome. The traditional view of the human genome stipulates that the majority of the DNA sequence does not undergo transcription and translation, with long stretches of this intergenic DNA separating DNA regions that code for amino acids and functional proteins – genes. A gene can be formally defined as "a locatable region of genomic sequence, corresponding to a unit of inheritance, which is associated with regulatory regions, transcribed regions, and/or other functional significant regions" (Pearson, 2006, p. 401). The latest human genome build (i.e., a publicly released assembly of reference DNA sequences in a chromosomal order that is a mosaic representative of the human genome) included 19,817 coding genes, accounting for only about 1 percent of the total human genome length. However, according to the current state-of-the-art knowledge, ~80 percent of the genome might have some other (i.e., non–protein-coding) biochemical function (The ENCODE Project Consortium, 2012). In other words, the distinction between genic and intergenic DNA regions is complicated by the existence of previously ignored regulatory elements in the DNA sequence and the presence of transcribed but non–protein-coding RNAs and pseudogenes – functional relatives of genes that resemble them but have lost the ability to code for a functional protein; pseudogenes can nonetheless regulate the expression of other protein-coding genes (Gerstein et al., 2007).

Transmission of Genetic Information

Genetic information is transmitted from parent to offspring through (1) the transfer of this information into sex cells (gametes) during meiosis, the process of cell division during which one diploid cell gives rise to four "daughter" haploid sex cells; and (2) the subsequent fusion of the haploid parental gametes to form a diploid zygote during conception. At the very beginning of this process, a diploid progenitor cell contains a double set of genetic material with each of the chromosomes in a homologous pair coming from one of the parents. Homologous chromosomes are similar but are not necessarily fully identical as one is coming from the mother, and the other one – the father. When the DNA is duplicated and the cell contains four sets of the genetic material, homologous chromosomes form connected pairs that can exchange portions of genetic material during cross-over. After two rounds of meiotic division, four haploid gametes are formed. Importantly, cross-over results in the recombination of the maternal and paternal genetic variants (alleles) on the same chromosome and increases genetic variability. The probability of a cross-over occurring in-between two genomic segments at specific locations (loci) in the genome depends on the distance between these segments – i.e., segments that are located close to each other are less likely to be "separated" during cross-over than segments located further apart from each other (and, correspondingly, they are more likely to be inherited together). This results in the deviation of the observed frequencies of sequences of alleles called haplotypes from their expected frequencies in the population, the phenomenon referred to as linkage disequilibrium (LD). The notion of LD captures the non-random association between genetic loci and points to the correlated nature of the human genome at the population level. DNA segments are considered linked if they are jointly transmitted from parent to offspring more often than under independent inheritance: segments that are located close together on the same chromosome and undergo recombination in less than 50 percent of meiotic events in the population are considered to be linked; segments that are located on different ends of the same chromosome or even on different chromosomes and undergo recombination in approximately 50 percent of meiotic events are considered to be inherited independently.

GENETIC VARIATION

The child can inherit same or different alleles (alleles can be thought of as variant forms of genes) from their mother and father at the same position or locus in the genome. In general, although most humans can be said to have the same set of genes, they usually (with the exception of monozygotic twins) do not share the same exact sequence of DNA nucleotides. In diploid cells, the combination of alleles at a particular locus defines the genotype at that locus (e.g., if an individual has inherited the C allele from the mother and the A allele from the father, they are said to have a CA genotype at that locus). If the two alleles are the same on both chromosomes, an individual is said to be homozygous at this locus, whereas different alleles form a heterozygous genotype.

Variation in the human genome is a consequence of different types of mutation events that alter the base sequence composition of the DNA. These include large events that occur during cell division and lead to deletions, insertions, duplications, translocations, and inversions of relatively large chromosomal segments, as well as smaller events that typically occur during DNA replication. Genetic variants can be private (i.e., limited to a particular family or individual) or present in a larger population at a particular frequency. A particular location in the genome is considered polymorphic (i.e., exhibiting variation in the population) if the mutant or derived (typically less frequent, also called "minor") allele is present in the referent population (e.g., individuals of European ancestry) at the frequency of at least 1 percent, while the referent wild-type or original (more frequent, also called "major") allele is found with the complementary frequency. The corresponding genetic variation in the DNA sequence at that locus is called a polymorphism.

Single Nucleotide Polymorphisms

Multiple types of variation are found in the human genome, ranging in length from single nucleotide polymorphisms (SNPs, also referred to as single nucleotide variants, SNV, i.e., A vs. G) that involve substitutions of a single base in the genetic code, to short tandem repeats (repetitions of a pattern of one or more nucleotides adjacent to one another otherwise known as micro- and mini-satellites, e.g., AAA AAA vs. AAA AAA AAA), to short insertions and deletions (indels, e.g., A-G-A vs. A- -A) that generally range in size from 1 bp to 1 kb (1 kb = 1,000 bp), to structural and copy number variants (CNVs; deletions, duplications, translocations, inversions, and insertions of genetic material that are over 1 kb long), and larger events such as chromosomal aberrations and aneuploidies (deviations from the expected number of chromosomes, e.g., as seen in Down syndrome characterized by presence of all or part of a third copy of chromosome 21; MIM#190685).

SNPs are single base pair substitution polymorphisms that usually have two alleles; this type of genetic variation represents around 90 percent of the total variation in the primary DNA sequence, with SNPs occurring approximately every 1,000 bp in the human genome. SNPs are more frequently found in intergenic compared to genic DNA. Moreover, among those that are located within coding regions, less than half are considered synonymous (i.e., not resulting in any change of the identity of the encoded amino acid), with the rest considered to be non-synonymous changes. The latter result in in alterations of protein structure (missense SNPs) or a truncated/ non-functional protein. For example, rs4680 (reference sequence 4680, an ID assigned to this known polymorphism in the dbSNP database–www. ncbi.nlm.nih.gov/SNP/) is a SNP located on chromosome 22 at position 19,963,748 within the *COMT* (catechol-O-methyltranferase) gene; the two alleles of this SNP differentiate between the COMT proteins that contain valine (G allele) or methionine (A allele) at the 158th codon. Carrying the G allele results in the higher enzymatic activity of the COMT enzyme that degrade neurotransmitters in neuronal synapses, especially in the prefrontal brain regions. However, the functional significance of the majority of SNPs in the human genome is unknown and requires either experimental or analytical examination of their possible pathogenicity based on the evaluation of their impact on the 3D structure of the altered protein, its stability, and physiochemical similarity to the non-mutant (wild-type) protein. SNPs that are located outside of the coding regions might nevertheless also have important regulatory and transcription-altering functions. In fact, the majority of SNPs that have been found to be associated with complex human disorders and traits are located outside of the protein-coding regions of the DNA.

Other Structural and Copy Number Variants

The human genome contains several types of structural genetic variants. Small insertions and deletions (indels) represent the smallest class of structural genetic variants and generally range in size from 1 bp to 1 kb; the human genome contains over 1,000,000 indels combined over different populations (The 1000 Genomes Project Consortium, 2010). Only approximately a third of indels are found in the coding regions of the genome. Given the triplet nature of the genetic code, indels that affect base pairs in numbers that are not multiples of three can be particularly damaging due to their ability to change the grouping of codons (called the "reading frame") and significantly alter the protein structure. For example, loss-of-function "frameshift" mutations in the *FOXP1* gene have been recently identified as one of the possible causes of sporadic (i.e., caused by de novo rather than inherited mutations) intellectual disability (Sollis et al., 2015).

Duplications or deletions of DNA segments over 1 kb long are called CNVs. Some types of structural variants do not involve changes in the amount of genetic material (e.g., inversions of the DNA sequence). Most of the structural variants can be considered low frequency (0.50 percent to 1–5 percent minor allele frequency, or MAF) or rare (<0.50 percent MAF) variants. Those found at higher frequencies are sometimes called copy number polymorphisms (CNP). The human genome is estimated to have fewer CNVs and CNPs (e.g., ~20,000) than indels and SNPs. However, given the size of the structural variants, the cumulative sum of base pairs affected by them can actually be larger than the sum of the base pairs affected by SNPs and indels. Thus, these variants represent another major source of genetic variation in the human genome.

Like SNPs, structural variants differ with respect to their functional impact. CNVs, especially deletions that lead to loss of protein function, are biased away (i.e., found at a frequency lower than expected) from ultra-conserved genomic regions that are known to be under strong selection (Cooper, Nickerson, & Eichler, 2007), which might be explained by their ability to affect gene expression. Thus, CNVs that increase or decrease the number of copies of a particular gene might affect that gene's transcript abundance (i.e., the amount of gene transcription products). Structural variants can disrupt transcription by creating breakpoints within transcripts or disturb the spatial distribution of chromosomes in the cell nucleus. Copy number and structural variants found in regulatory regions might also disrupt gene expression – sometimes for genes that are several millions of base pairs away, depending on their location and the type of regulatory element they express – and even result in more severe effects like chromosomal fragility. Recent studies indicate that CNVs are associated with a host of neurodevelopmental conditions including developmental delays, intellectual disability, developmental disorders of language, and autism spectrum disorders (Coe et al., 2014; Girirajan et al., 2011).

MAIN TYPES OF STUDIES OF THE MOLECULAR GENETIC BASES OF COMPLEX HUMAN TRAITS AND DISORDERS

The main goal of molecular genetic epidemiological studies is to reveal the genetic architecture of a particular trait by identifying genomic regions, genes, and/or genetic variants that play a role in the development and manifestation of the trait in a particular population, thus linking the genetic variation in the population with variation in a particular phenotype – usually a behavioral or a neurobiological trait or a clinical diagnosis. In clinical settings, the main goal is usually to identify the specific causal genetic factors (usually highly damaging single nucleotide and copy number variants that affect the structure, function, or the expression profile of a specific protein) behind a particular phenotype (e.g., intellectual disability) in a single individual or family. The choice of molecular genetic methods utilized to interrogate the genome largely depends on study design and the specifics of the trait under investigation.

Mendelian vs. Common Disorders and Traits

Simple or Mendelian traits are characterized by low prevalence (<1 percent), Mendelian patterns of inheritance (autosomal dominant, autosomal recessive, and X-linked), and high penetrance (i.e., the probability of observing the trait given the presence of the mutation is high). These traits and disorders are usually monogenic, i.e., caused by pathogenic mutations in a single gene, and typically present with a distinctive clinical phenotype. To date, over 3,500 (out of 7,000 estimated) such traits have been identified (Boycott, Vanstone, Bulman, & MacKenzie, 2013). Among them are such developmental disorders as Williams syndrome (MIM#194050; caused by a deletion of a 1.5–1.8 Mb [1 Mb = 1,000,000 bp] region of chromosomes 7q11.23, containing approximately 28 genes) and Charcot-Marie-Tooth syndrome (MIM#606482; caused by mutations in the *DNM2* gene located on chromosome 19p13.2).

In contrast, complex traits (e.g., height and intelligence) and common disorders (e.g., developmental disorders of speech and language) are usually polygenic, have higher prevalence (>1 percent), and are characterized by complex non-Mendelian patterns of vertical transmission and genetic architecture; for the majority of these traits, the specifics of this architecture (i.e., the location and pathogenicity of causal variants in the genes contributing to the disorder) are unknown. Complex traits and common disorders are frequently thought to be caused by a multitude of genetic variants that are common (e.g., with a MAF of at least 5 percent in the population) but have relatively low penetrance and individually exert only small influences on the trait.

This assumption, called the common disease/common variant hypothesis, has been challenged by the recent realizations that most of the common variants identified as susceptibility loci for such traits account for only a minor proportion of their genetic variance and frequently do not have a clear functional significance. An alternative approach dubbed the common disease/rare variant hypothesis postulates that rare genetic variants with high penetrance are major contributors not only to Mendelian but also to complex traits (Schork, Murray, Frazer, & Topol, 2009). It is, however, possible that the middle-ground solution that takes into account both types of genetic variation would be more realistic if the field develops theoretical and mechanistic accounts of how common and rare variants operate together. For example, complex traits can be thought of as regulated by a large number of polymorphisms that modulate a set of biochemical traits, and individuals at the extremes of the continuous liability conferred by these polymorphisms exhibit disease states when additional risk is conferred by rare variants (Gibson, 2012).

Linkage and Association Studies

The discovery of the genes responsible for a large portion of Mendelian disorders in the past several decades has been driven by the development and application of linkage analysis – a method that aims to establish the locations of genomic segments shared by related individuals who are also phenotypically similar (i.e., are affected by the same disorder). Thus, linkage is established when individuals affected by a Mendelian disorders are found to harbor the same co-inherited haplotype that is not shared by the unaffected individuals within the same extended pedigree(s) and/or nuclear families. The resolution of linkage studies depends on the type of genetic variation used to probe different genomic locations and is generally relatively low, and they frequently identify from a few to a few hundred genes (but not specific genetic variants) predicted to be relevant for the trait and located within about 1 Mb from the locus that is displaying a statistical linkage signal (Botstein & Risch, 2003). Linkage analysis of Mendelian traits depends on several assumptions regarding the genetic model of the trait. When the trait is complex or quantitative in nature, however, it typically does not conform to a specific mode of inheritance. Adjustments to analytical procedures that take this into account have been developed and generally assume that affected related individuals should show an increased sharing of inherited haplotypes in the region close to the causal gene(s); i.e., phenotypically similar related individuals should have similar alleles for a particular genetic marker (a DNA sequence with a known location that exhibits variation and can thus be used to differentiate between individuals in a population) located close to the gene(s) contributing to the trait. The ability to carry out linkage analysis depends on the availability of phenotypes and genotypes from families and extended pedigrees.

The genetic epidemiological investigations of complex traits in the past several decades has largely moved from genetic linkage to genetic association designs, although the latter is experiencing a renaissance due to the increasing popularity and availability of precise high-throughout DNA sequencing. Linkage studies permit different alleles to be linked with a disorder or trait in different families, as long as the location of the genomic segment harboring these alleles shows evidence for linkage. In contrast, genetic association studies assume that the same allele of a particular genetic variant is associated with the trait in a similar fashion in the population.

One of the major approaches to testing the association between a genetic variant and a trait is to recruit two groups of individuals that differ with respect to a particular dichotomous trait (i.e., affected cases and unaffected controls) or a sample of individuals that exhibit significant variation in the quantitative trait. The DNA from cases and controls is then genotyped for a set of informative genetic markers. In this case, association testing examines whether the distribution of alleles is similar across affected and unaffected individuals (assuming that the over-representation of a particular allele among the affected individuals is indicative of the possible involvement of that variant or the gene in the trait etiology), and/or whether individuals who carry different alleles at a particular locus differ with respect to a quantitative trait (e.g., IQ or reading achievement scores). In both cases, the genetic effect on the trait is quantified in terms of odds ratios (i.e., increases the odds of being affected with a disorder associated with carrying a particular allele) or other effect sizes, such as regression coefficients (e.g., the magnitude of the change in standardized achievement scores associated with carrying the allele).

SNP genotyping and association analysis have become the leading tools for the discovery of genes associated with human diseases and traits due to their abundance in the human genome, low mutation rates, and simplicity of methods for SNP genotyping. Commercial genotyping platforms rely on prior knowledge about the reference sequence of the human genome and interrogate known sites of genetic variation (from dozens of genetic markers, selected based on a priori assumptions or the results of previous studies, to several million genetic markers!). Thus, genetic association studies frequently assume that although the causal variant might not be present in the set of studied genomic loci, the correlated nature of the human genome (driven by LD) would still permit detecting an indirect association. In this case, loci can be evaluated for association with a trait using "proxy" SNPs that are correlated with other SNPs in the region because of the shared history and, therefore, "tag" common SNP haplotypes, removing the necessity to genotype SNPs at all possible loci. However, it is certainly possible for an individual to carry private or rare mutations that are missed by genotyping platforms due to their low frequency (and novel mutations that have never been described and, correspondingly, could not have made it to the commercial SNP genotyping platforms).

Genome-wide association studies (GWAS) are a special case of genetic association studies and typically involve genotyping large samples of unrelated (although family-based studies have also been employed) individuals on hundreds of thousands to millions of SNP markers that are more or less evenly distributed across the genome at a particular density. Over 2,000 GWAS studies have been published since 2005, and produced an array of results for a variety of traits. However, although multiple loci have been identified for many complex traits, many were likely missed due to the low statistical power of GWAS studies, which require samples on the order of 2,000–3,000 cases and controls to detect significant associations; GWAS also require extensive and careful validation in replication studies, transforming them into two-stage investigations where association signals detected in the discovery cohort are validated in a replication cohort. The replicated findings are then typically interrogated further experimentally (e.g., by studying the molecular/biochemical impact the variant has in a particular tissue) or in animal models by engineering the mutation in a model organism (e.g., mouse).

In addition to SNPs, GWAS studies sometimes utilize CNV-specific genotyping techniques to obtain genotypes that indicate the number of copies of a particular segment(s) of DNA. Common CNVs can be analyzed in a relatively straightforward fashion in case–control CNV GWAS by comparing CNP allele (i.e., copy number) frequencies in cases and controls. Family-based designs offer an important advantage over case–control CNV association designs as they permit evaluating whether the CNV that showed an association with a trait is de novo (non-inherited) vs. inherited when parental DNA is available. The popularity of the de novo or spontaneous/sporadic framework for CNV analysis is illustrative of the recent attention to the high genetic mutation rate in humans. For example, on average a newborn is expected to have about seventy new single nucleotide mutations in their genome with most of these mutations seemingly being of paternal origin (Keightley, 2012; Lynch, 2010). Although structural variants have not been precisely evaluated for their mutational properties yet, some estimates suggest that their mutation rate is lower than that of single nucleotide variants, and decreases with size. New large CNVs over 100 kb long are estimated to occur once every forty-two births (Itsara et al., 2009). De novo CNVs are estimated to affect a significantly higher number of base pairs per birth (16–50 kb) cumulatively than single nucleotide variants (61 bp; Campbell & Eichler, 2013), and their deleterious potential is very high.

Gene Expression Studies

Developmental biological processes that involve cellular proliferation, differentiation/specialization, and plasticity heavily rely on the carefully orchestrated cascades of changes in gene expression patterns. For example, in the human brain, what genes are expressed and when they are expressed varies greatly by anatomical location and cell type; importantly, different locations and their corresponding cell types display expression patterns that are consistent across different individuals (Hawrylycz et al., 2012). At the same time, they are subject to systematic and identifiable changes throughout development (Kang et al., 2011; Naumova et al., 2013).

The full set of gene transcripts (transcribed from genomic DNA) in a cell at a particular stage of development is called a transcriptome, and gene expression analysis has received a wide variety of applications in recent studies, most notably in studies of differential gene expression in different types of cancers with the goal of obtaining their distinct molecular signatures, which could guide personalized treatment. Since transcription results in the synthesis of an mRNA molecule that is used as a template during translation, one can measure the concentration of RNAs specific to a particular (gene and) protein as an index of the level of expression of the corresponding gene in the studied tissue. The main objective of gene expression analyses is thus to establish patterns of differential gene expression between samples obtained from the same individuals (e.g., from different areas of the brain to obtain evidence for the differential expression of genes hypothesized to be particularly involved in a specific process with identifiable neuroanatomical and functional correlates) or from several groups of individuals (e.g., children with and without a particular neurodevelopmental disorder to obtain evidence for the differential expression of genes that have been proposed as candidate genes for that disorder). When such differential expressions are registered, genes that are found to be differentially expressed between the two samples are likely involved in one or more etiological pathways for the studied trait. For example, a recent study found atypical levels of expression of neurotrophins (genes that code for a family of proteins critical for neuronal development and function) in adolescents and adults with autism spectrum disorders, compared to their unaffected age peers (Segura et al., 2015).

Human expression studies that focus on complex neuropsychiatric and developmental phenotypes depend on the availability of relevant tissue samples. An ideal investigation might focus on evaluating differential gene expression in the brain. However, that would require justification for the selection of particular brain regions for the analysis of samples obtained post-mortem. In addition to ethical and practical concerns associated with obtaining these samples, such studies face the issue of post-mortem RNA instability. In the absence of post-mortem samples, it is possible to evaluate global differential gene expression patterns between two samples using other types of tissue that are more readily available, such as peripheral whole blood. However, whole-blood gene expression levels are not perfectly correlated with gene expression levels in the central nervous system (Sullivan, Fan, & Perou, 2006), consequently requiring extra care when interpreting results obtained from peripheral blood samples, especially with respect to genes that are not expressed in either brain or blood, or, in general, show differential expression between blood and brain.

DNA SEQUENCING

Genetic linkage and association studies, as well as clinical genetic testing, largely rely on the availability of commercial platforms that permit the interrogation of known genetic variants that have been characterized with respect to their location and allele frequencies for populations of different ancestries (and in some cases – pathogenicity). When an association is detected, the ultimate goal is to obtain the location and identity of the causal variant, which by itself is neither known precisely nor genotyped directly. Under these conditions, the causal mutation can be identified if the precise order of nucleotides in the DNA sequence is established for the region that is supposed to harbor it. The Human Genome Project, the largest public scientific collaboration project in biology that took thirteen years and $3 billion to complete, resulted in the publication of the working draft of the full sequence of human genome (International Human Genome Sequencing Consortium, 2001, 2004), which revolutionized genetic science in general and genomics in particular. Today, DNA sequencing is becoming the leading tool for both applied and fundamental genetic research and clinical practice, driven by technological advances in the field and associated decreases in cost, lowering the price to ~$1,500 per genome (compare to $100,000,000 in 2001 and ~$50,000 in 2010; van Dijk, Auger, Jaszczyszyn, & Thermes, 2014).

The Human Genome Project relied largely on the gold standard method of DNA sequencing based on Sanger's chemistry, developed in the 1970s (Sanger & Coulson, 1975; Sanger, Nicklen, & Coulson, 1977). Although Sanger sequencing remains the gold standard for targeted DNA sequencing, the application of this method in large-scale human projects (i.e., with a large number of individuals or multiple long DNA fragments) is limited by its relatively high cost and low throughput.

Modern sequencing platforms outperform Sanger sequencing by a factor of 100–1,000 while reducing the cost of sequencing to .10–4 percent of Sanger sequencing costs when sequencing long (over 1 Mb) regions (Kircher & Kelso, 2010). These methods are referred to as next-generation DNA sequencing (NGS), to reflect the massive shift from Sanger-based sequencing to new high-throughput approaches developed in the mid-2000s. Unlike Sanger sequencing, these approaches can be used to efficiently and quickly sequence the whole genome of an individual or large target regions of the genome. The existing NGS platforms achieve that by cutting the genomic DNA into smaller fragments and sequencing each of the fragments in parallel in the same reaction (massive parallel sequencing). Next-generation whole-genome sequencing (WGS) is becoming increasingly common in the discovery of mutations responsible for Mendelian diseases, as well as in the discovery of the genetic architecture of complex traits and disorders, and clinical diagnostics of childhood and neurodevelopmental disorders (Goldstein et al., 2013). In the first case, DNA sequencing can be performed using samples of unrelated individuals who represent cases/controls or extremes of the quantitative phenotype variation continuum; allele frequency can then be compared between the two cohorts at all sequenced loci. In family-based designs, samples of related individuals are used to both detect mutations within the shared regions of the genome in extended pedigrees and reveal de novo mutations in family trios.

One of the most popular applications of NGS in both research and clinical practice is whole-exome sequencing (WES). Developed as a cost-effective alternative to WGS, WES focuses only on the coding regions of the DNA. Given the cumulative abundance of rare coding variants of potentially high functional impact in the human genome and the exonic nature of most mutations underlying Mendelian disorders, WES has recently become the state-of-the-art tool for Mendelian disorder gene discovery (Bamshad et al., 2011). Next-generation sequencing studies generate a wealth of data (e.g., an individual genome is expected to have ~20,000 departures from the reference sequence!), and pose significant analytical and interpretational challenges. To overcome these, both clinical and research applications of WGS/WES rely on careful variant filtration, annotation, and prioritization to reduce the search space for the causal variant(s).

Next-generation sequencing revolutionized the field of genetics, and this technology offers the field of developmental psychopathology a new tool for identifying the genetic architecture of complex traits in a comprehensive fashion. In the largest WGS study of eighty-five families with two children affected with autism spectrum disorders, Yuen et al. (2015) found that in approximately 69 percent of the families the two siblings carried different rather than the same genetic risk variant, highlighting the genetic heterogeneity and complexity of autism spectrum disorders. The authors also found that the majority of the small pathogenic CNVs detected in their study would have been missed if other genotyping methods (i.e., CNV microarrays) had been used.

CONCLUSION

Next-generation whole-genome and whole-exome sequencing are becoming the gold standard tool for the investigation of Mendelian as well as complex traits and common disorders both in research and clinical practice. The complexity of the interpretation of results obtained through genetic association studies or even DNA sequencing at the individual family level poses unique challenges not only for specialists in the fields of genomics, but also pediatricians, parents, and children themselves, raising a set of important technical, as well as educational and ethical concerns. As mentioned in the introduction, the main goal of this chapter was thus to provide the reader of this volume with an initial set of conceptual pointers that would hopefully aid them with the thoughtful and informed digestion of the more specific issues explicated in other chapters.

In addition to this overview, several online resources have been developed in recent years to support genetics and genomics education. Recognizing the need for widespread genomics education, the National Human Genome Research Institute has developed a set of online materials geared toward providing overviews of key concepts and issues in human genetics and genomics to students, educators, patients, and health professionals in a consumer-friendly fashion. It also partnered with the National Library of Medicine to develop GeneEd (Genetics, Education, Discovery) – a web portal dedicated to providing information about human genetics to students and teachers in grades 9–12. The websites are accessible at www.genome.gov/Educators/ and http://geneed.nlm.nih.gov, respectively. Teaching resources are also available at www.genome.gov/10005911.

ACKNOWLEDGMENTS

The writing of this chapter was supported by grants R01 DC007665, R21 HD070594, and P50 HD052120 from the US National Institutes of Health, and grant № 14.Z50.31.0027 from the Government of the Russian Federation. Grantees undertaking such projects are encouraged to express freely their professional judgment. The chapter, therefore, does not necessarily reflect the position or policies of the abovementioned funding agencies, and no official endorsement should be inferred.

REFERENCES

Bamshad, M. J., Ng, S. B., Bigham, A. W., Tabor, H. K., Emond, M. J., Nickerson, D. A., et al. (2011). Exome sequencing as a tool for Mendelian disease gene discovery. *Nature Review Genetics, 12*(11), 745–755. doi:10.1038/nrg3031

Botstein, D., & Risch, N. (2003). Discovering genotypes underlying human phenotypes: Past successes for Mendelian disease, future approaches for complex disease. *Nature Genetics, 33*(suppl.), 228–237. doi:10.1038/ng1090

Boycott, K. M., Vanstone, M. R., Bulman, D. E., & MacKenzie, A. E. (2013). Rare-disease genetics in the era of next-generation sequencing: Discovery to translation. *Nature Reviews Genetics, 14*(10), 681–691. doi:10.1038/nrg3555

Campbell, C. D., & Eichler, E. E. (2013). Properties and rates of germline mutations in humans. *Trends in Genetics, 29*(10), 575–584. doi:10.1016/j.tig.2013.04.005

Coe, B. P., Witherspoon, K., Rosenfeld, J. A., van Bon, B. W., Vulto-van Silfhout, A. T., Bosco, P., et al. (2014). Refining analyses of copy number variation identifies specific genes associated with developmental delay. *Nature Genetics, 46*, 1063–1071. doi:10.1038/ng.3092

Cooper, G. M., Nickerson, D. A., & Eichler, E. E. (2007). Mutational and selective effects on copy-number variants in the human genome. *Nature Genetics, 39*(S), 22–29. doi:10.1038/ng2054

Gerstein, M. B., Bruce, C., Rozowsky, J. S., Zheng, D., Du, J., Korbel, J. O., et al. (2007). What is a gene, post-ENCODE? History and updated definition. *Genome Research, 17*(6), 669–681. doi:10.1101/gr.6339607

Gibson, G. (2012). Rare and common variants: Twenty arguments. *Nature Reviews Genetics, 13*(2), 135–145. doi:10.1038/nrg3118

Girirajan, S., Brkanac, Z., Coe, B. P., Baker, C., Vives, L., Vu, T. H., et al. (2011). Relative burden of large CNVs on a range of neurodevelopmental phenotypes. *PLoS Genetics, 7*(11), e1002334. doi:10.1371/journal.pgen.1002334

Goldstein, D. B., Allen, A., Keebler, J., Margulies, E. H., Petrou, S., Petrovski, S., et al. (2013). Sequencing studies in human genetics: Design and interpretation. *Nature Review Genetics, 14*(7), 460–470. doi:10.1038/nrg3455

Hawrylycz, M. J., Lein, E. S., Guillozet-Bongaarts, A. L., Shen, E. H., Ng, L., Miller, J. A., et al. (2012). An anatomically comprehensive atlas of the adult human brain transcriptome. *Nature*, 489, 391–399. doi:10.1038/nature11405

International Human Genome Sequencing Consortium. (2001). Initial sequencing and analysis of the human genome. *Nature*, 409(6822), 860–921. doi:10.1038/35057062

(2004). Finishing the euchromatic sequence of the human genome. *Nature*, 431(7011), 931–945. doi:10.1038/nature03001

Itsara, A., Cooper, G. M., Baker, C., Girirajan, S., Li, J., Absher, D., et al. (2009). Population analysis of large copy number variants and hotspots of human genetic disease. *The American Journal of Human Genetics*, 84(2), 148–161. doi:10.1016/j.ajhg.2008.12.014

Kang, H. J., Kawasawa, Y. I., Cheng, F., Zhu, Y., Xu, X., Li, M., et al. (2011). Spatio-temporal transcriptome of the human brain. *Nature*, 478(7370), 483–489. doi:10.1038/nature10523

Keightley, P. D. (2012). Rates and fitness consequences of new mutations in humans. *Genetics*, 190(2), 295–304. doi:10.1534/genetics.111.134668

Kircher, M., & Kelso, J. (2010). High-throughput DNA sequencing – Concepts and limitations. *Bioessays*, 32, 524–536. doi:10.1002/bies.200900181

Lynch, M. (2010). Rate, molecular spectrum, and consequences of human mutation. *Proceedings of the National Academy of Sciences*, 107(3), 961–968. doi:10.1073/pnas.0912629107

Naumova, O. Y., Lee, M., Rychkov, S. Y., Vlasova, N. V., & Grigorenko, E. L. (2013). Gene expression in the human brain: The current state of the study of specificity and spatiotemporal dynamics. *Child Development*, 84(1), 76–88. doi:10.1111/cdev.12014

Pearson, H. (2006). Genetics: What is a gene? *Nature*, 441, 398–401. doi:10.1038/441398a

Sanger, F., & Coulson, A. R. (1975). A rapid method for determining sequences in DNA by primed synthesis with DNA polymerase. *Journal of Molecular Biology*, 94(3), 441–448. doi:10.1016/0022-2836(75)90213-2

Sanger, F., Nicklen, S., & Coulson, A. R. (1977). DNA sequencing with chain-terminating inhibitors. *Proceedings of the National Academy of Sciences*, 74(12), 5463–5467.

Segura, M., Pedreno, C., Obiols, J., Taurines, R., Pamias, M., Grunblatt, E., et al. (2015). Neurotrophin blood-based gene expression and social cognition analysis in patients with autism spectrum disorder. *Neurogenetics*, 16(2), 123–131. doi:10.1007/s10048-014-0434-9

Schork, N. J., Murray, S. S., Frazer, K. A., & Topol, E. J. (2009). Common vs. rare allele hypotheses for complex diseases. *Current Opinion in Genetics & Development*, 19(3), 212–219. doi:10.1016/j.gde.2009.04.010

Sollis, S., Graham, S. A., Vino, A., Froehlich, H., Vreeburg, M., Dimitropoulou, D., et al. (2015). Identification and functional characterization of de novo FOXP1 variants provides novel insights into the etiology of neurodevelopmental disorder. *Human Molecular Genetics*. doi:10.1093/hmg/ddv495

Sullivan, P. F., Fan, C., & Perou, C. M. (2006). Evaluating the comparability of gene expression in blood and brain. *American Journal of Medical*

Genetics Part B: Neuropsychiatric Genetics, 141B(3), 261–268. doi:10.1002/ajmg.b.30272

The 1000 Genomes Project Consortium. (2010). A map of human genome variation from population-scale sequencing. *Nature, 467*(7319), 1061–1073. doi:10.1038/nature09534

The ENCODE Project Consortium. (2012). An integrated encyclopedia of DNA elements in the human genome. *Nature, 489*(7414), 57–74. doi:10.1038/nature11247

van Dijk, E. L., Auger, H., Jaszczyszyn, Y., & Thermes, C. (2014). Ten years of next-generation sequencing technology. *Trends in Genetics, 30*(9), 418–426. doi:10.1016/j.tig.2014.07.001

Yuen, R. K. C., Thiruvahindrapuram, B., Merico, D., Walker, S., Tammimies, K., Hoang, N., et al. (2015). Whole-genome sequencing of quartet families with autism spectrum disorder. *Nature Medicine, 21*, 185–191. doi:10.1038/nm.3792

3

Can (and Should) We Personalize
Education Along Genetic Lines?
Lessons from Behavioral Genetics

KATHRYN ASBURY, KAILI RIMFELD, AND EVA KRAPOHL

INTRODUCTION

Genes are a major player in education, if not always readily acknowledged as such (Deary et al., 2007; Johnson et al., 2009; Wadsworth et al., 1995). Studies have found that individual variation in academic abilities such as reading (Byrne et al., 2008; Harlaar et al., 2008; Hayiou-Thomas et al., 2010; Olson, 2007), writing (Oliver et al., 2007), mathematics (Kovas et al., 2007, 2009; Tosto et al., 2014), science (Haworth et al., 2008), and achievement in public examinations (Krapohl et al., 2014; Shakeshaft et al., 2013) can be explained, to a substantial extent, by genetic differences between children. Unsurprisingly, it turns out that children do not leave their DNA at the school gates (Asbury & Plomin, 2013).

In this chapter, we review research into the heritability of academic achievement, particularly research that has focused on the dynamic relationship between genes and experience. We ask whether it is, or ever will be, possible to personalize education along genetic lines. Specifically, will information about a child's DNA ever be relevant to the decisions we make about that child's education? We also ask whether, regardless of pragmatics, such genetic information *should* ever be taken into account in educational decision making.

Our genes influence everything we do, and there is no conceivable reason why learning should be any different. As well as influencing our appetites and aptitudes, our genes go a step beyond and influence the learning environments we experience (Plomin, 1994). Furthermore, relationships between achievement and environmental factors such as family chaos (Hanscombe et al., 2011), socio-economic status (SES; Krapohl & Plomin, 2016), or the perceived home and school environments (Krapohl et al., 2014) have been found to be mediated by genetic factors; DNA differences largely explain the associations between environment and achievement.

We know that genes have an impact on how well and how easily we learn, and consideration of what that really means seems important. One key issue is whether our growing understanding of the relationship between genetics and education has anything of actual practical value to offer to those who teach and learn in schools and, if so, how it should be used. In some ways, the question presented in this chapter's title is easy to address head-on. In response to the *"can"* part of the question, our answer is a simple *no* – we do not currently have the knowledge or expertise to personalize education along lines that are clearly genetic. It is worth bearing in mind though that this "no" is likely to become less certain in the near future. Already we can use a polygenic score to predict almost 10 percent of individual differences in exam grades at the end of compulsory education (Selzam et al., 2016). In answer to the "should" part of the question, we offer an honest *don't know* and open it up for discussion.

IS IT POSSIBLE TO PERSONALIZE EDUCATION ALONG GENETIC LINES?

Not only do we know that genes are linked to achievement, but we also know that the majority of human behavior, including achievement, is polygenic in origin. Human behavior is affected by many genes, each with a miniscule effect and embroiled in complex relationships with the environment and other genes and non-coding parts of the genome (Yang et al., 2011). Together, these genes can explain meaningful proportions of individual differences. It is important, therefore, to discuss what will happen when we reach a time when genetic variants associated with educational achievement can be reliably identified and combined with others into polygenic predictors of learning strengths and difficulties. Currently, if someone were to sequence the genomes of every child about to start school in any country in the world, we would not have a practical educational use for the data. Although we can now predict almost 10 percent of individual differences in educational achievement at the end of compulsory education (Selzam et al., 2016), prediction for achievement in earlier school years is much lower. There is a great deal of work still to be done. We believe we should use the time this work takes to establish a constructive dialog between researchers and policymakers to ensure that when we have meaningful genetic prediction for academic achievement at all ages, we also have a better idea of what it means and how to use it.

What we know is there is no single gene "for" reading, writing, mathematics, or science. The Generalist Genes Hypothesis (Kovas & Plomin, 2006;

Plomin & Kovas, 2005) predicts that many of the same variants (versions of a gene) will influence a wide range of human traits. Research has tended to find that genes are generalists and environments are specialists; that is, genes affect a wide range of cognitions or areas of learning, while experiences have more domain-specific, or specialist, effects. The polygenic nature of the heritability of behavior explains the relatively slow progress we have seen in identifying specific genes involved in educational achievement and intelligence, progress which is just now speeding up. This applies to most other traits in the life sciences as well. The process of understanding the relationship between genes and the environment has proved equally challenging. The weaving of an intricate web of genetic and environmental effects begins before a child is even born. Genetic effects can never be independent of the environment as they can only be expressed in an environmental context. Rutter (2006) offers the rather lovely example of a white flamingo. Famously pink, flamingos not exposed to the typical flamingo diet of shrimp, and plankton will always be white. Although a flamingo's ability to *be* pink is entirely explained by genes, the fact of actually becoming pink – fulfilling their flamingo potential if you like – is explained by diet, an environmental influence. The flamingo, therefore, is dependent on the complementary action of genes and experience to become the pink bird we know. If this is true for a simple trait such as a flamingo color, then the likelihood of it not being true for the far more complex traits linked to children's learning appetites and aptitudes has to be zero. Neither genetic nor environmental effects operate in isolation but instead are caught up in constant interplay with each other (Kendler, 2005).

The most basic reason why we cannot currently personalize education along genetic lines is that we simply do not have enough information – we do not know which groups of genes matter in relation to which traits. This is known as the "missing heritability problem" and is a challenge for all genetic research, not just that related to education (Maher, 2008; Plomin & Simpson, 2013). This means that while we know approximately how much variance in a trait or behavior can be explained by genes in a particular population at a particular point in history we do not yet know anywhere near enough about which genes are involved, or understand their mechanisms of influence sufficiently well to be able to personalize education along genetic lines. However, as discussed, we can now explain nearly 10 percent of the variance in exam performance at age sixteen, using measured genes combined in a polygenic score (Selzam et al., 2016). Times are changing and we may be witnessing a tipping point for the "missing heritability problem."

There is clear evidence of incremental scientific progress and we are optimistic about the chances of gradually addressing the deficits in current knowledge and making a real contribution to education. Progress has been spectacularly fast in genetics. Rutter (2006) points out that as recently as the mid-1950s we did not know how many chromosomes there are in the human genome, and there was discussion of Down syndrome being caused by stress (Valentine, 1986). New, powerful and affordable techniques are emerging all the time. The Human Genome Project was completed in 2003, exactly fifty years after the discovery of the structure and function of DNA (Watson & Crick, 1953). It required the effort of 2,000 researchers and cost $3 billion. And yet, just one decade later, we can sequence the genome of an individual in a few hours for less than $20,000 and the cost is decreasing all the time. In fact, one commercial company, *Illumina*, can now sequence an individual's DNA for $1,000 per genome albeit only in the context of extremely large studies. The natural implication of this is that, as whole-genome sequencing becomes faster and more affordable – which it will – more people and more institutions will have access to both the procedure and its results. We are likely to enter an era in which the entire sequence of the genome is known for many people. There is no doubt whatsoever that this raises new questions about who we are and how we live as individuals and as a species. Nor is there doubt that it will create new opportunities to study genetic effects on learning and genotype–environment interplay in education.

This era has not yet dawned but it would be wise to prepare for it. Meanwhile, scientists are already using DNA arrays (gene chips) to genotype sections of DNA that vary between individuals. A gene chip is a small slide dotted with short DNA sequences that can help us identify which genes are active in which cells by making it possible to look at thousands of single nucleotide polymorphisms (SNPs) at the same time in Genome-Wide Association Studies (GWAS). Most DNA is shared between human beings but behavioral geneticists are interested in the genetic variants that differ between individuals. Gene chips allow us to look at this in a new way. DNA arrays have been commercially available since 2000 and one of their particular strengths is that they can be customized, with possible implications for education. Customization of DNA arrays has led to the development of CardioChip, a microarray for variants related to cardiovascular function (Barrons et al., 2001), and ImmunoChip for immunological dysfunction (Cortes & Brown, 2011). We have a long way to go before a useful learning chip becomes a realistic proposition but it is fair to say that such a thing represents a genuine possibility – although any need for it may be eradicated by widespread sequencing. Although behavior may always be the best guide for educators (as the manifestation of both genetic and environmental influences) it is possible that in time polygenic prediction will pave the way for prediction and intervention before difficulties actually manifest in behavior.

Another methodological development in genetic research is that we have recently seen the emergence of the first new human quantitative genetic technique in a century – Genome-Wide Complex Trait Analysis (GCTA; Yang et al., 2010, 2011, 2013). GCTA uses hundreds of thousands of SNPs from thousands of individuals to calculate quantitative genetic parameters for any phenotype of interest (Plomin et al., 2013). The technique opens up the potential for a much wider community of researchers to employ behavioral genetic methods as it works with the DNA of unrelated individuals and does not rely on large samples of special populations such as twins, adoptive families, or step families, i.e., populations in which degree of familial relatedness is a salient and measurable factor. It does, however, rely on very large samples. These emerging techniques are all designed to speed up progress in identifying genetic risk and protective factors. It is fair to say that nobody was prepared for the level of complexity that would be involved in understanding how the genome works after the huge achievement of cracking the code in the first place. The expectation in the early days of this research was that single genes of large effect would be identified whereas the reality is that complex traits have been found to be influenced by multiple genes of small effect. However, researchers are working hard to find genes and to understand mechanisms of influence, and are devising new means of doing so all the time. We need to take this seriously.

Although we are not yet capable of using genetic information to inform educational decisions, there is a reasonable chance that this will change and that such a change might benefit society. Nor is it impossible that before that time arrives offerings may appear in the market even if their prediction of probable risk of learning strengths and difficulties is rather limited. This phenomenon can be observed with the current trend for neuroscience in education and with commercially available genetic tests for children's sporting ability. We already use genetic information to screen for diseases in a baby's first days of life (e.g., cystic fibrosis) and Francis Collins, Director of the US National Institutes of Health and former Director of the Human Genome Project has written that: "I am almost certain … that whole-genome sequencing will become part of newborn screening in the next few years … It is likely that within a few decades people will look back on our current circumstances with a sense of disbelief that we screened for so few conditions" (Collins, 2010, p. 50). It looks like Collins' prediction of wholesale whole-genome sequencing within just a few years of 2010 was a little wide of the mark. Although the NIH is currently funding four sites to trial whole-genome sequencing at birth, an undeniably exciting development, there is still a major leap to be taken from sequencing to prediction. However, Collins' "few decades" prediction remains realistic and its implications – practical, ethical, moral, legal, political, and educational – are immense. The science may currently be lagging behind a vision of the future in which genetic information is used to predict academic strengths and difficulties, as well as health and disease risk and protective factors, but the vision does not belong to the realm of science fiction. Not to consider this future and its implications would seem negligent. The responsible approach to take in this time of uncertainty is to imagine a future in which we do have enough information about genes and their relationships with experiences to be able to calculate probabilities of reading problems, numeracy difficulties, and particular strengths or talents. We should prepare ourselves for a future in which this information is available because scientists are working toward exactly this and we have no reason to assume they will not succeed in their endeavor. Not doing so is akin to planning a Mars landing with no plan for what to do once you get there.

SHOULD WE PERSONALIZE EDUCATION ALONG GENETIC LINES?

When we move from an acknowledgment that we cannot currently personalize education along genetic lines to consideration of the possibility that we may one day be able to do so the important question becomes whether we *should* or whether doing so will cause more harm than good. Detractors argue that such an approach could lead to discrimination, which is of course a possibility that merits serious consideration. Further understanding of how that could manifest, and how it might be avoided, is needed. At the same time, it is worth considering the amount of environmental discrimination we currently accept in society. For instance, it is not unusual to make educational decisions on the basis of a child's family income or social status. Although it is not difficult to see shadows lurking in a debate such as this one, and although it is important to address the risks that would be involved in using DNA in the educational decision-making process, there are also benefits to be considered (Asbury & Plomin, 2013; Panofsky, 2009; Parens, 2004).

In the ideal scenario, the availability of well-understood genetic information could help teachers and school leaders to optimize the educational environment in such a way that every child's personal learning needs might be fully met. Schools are already awash with data about children's achievements, progress, and social backgrounds, and it is important to note that relationships between environmental factors and ability have been found to be genetically mediated (e.g., Trzaskowski et al., 2014). Genetic information is potentially another piece in the puzzle that can help teachers identify pupils who may need extra support with reading, extension work in mathematics, or a total change of approach. It may even help governments to direct resources to children at risk of educational difficulties as a result of their genetic make-up, their environment, and the likely interplay between the two, long before they start school. Early intervention is one very plausible potential benefit of openly taking genetic influence into account in education. Genetic research could, if we think it through carefully in parallel with emerging scientific developments, be a force for good.

There are good precedents here with genetic and chromosomal learning difficulties such as Down syndrome, Prader-Willi syndrome, and Williams syndrome. These disorders are less complex than common behavioral traits in that they are influenced by single genes, deletions, or additional genetic material, all of which are relatively rare and have large effects. Spotting the genetic risk factors very early on, often prenatally, has led to the development of disorder-specific teaching and learning profiles, which can be adapted to the needs of individual children. If we know a child has an extra chromosome 21 (Down syndrome), for instance, we know that the probability of spoken language being difficult is much higher than for other children. We also know that strategies such as teaching signing, and even reading, from a very young age have a good chance of supporting this child in developing language and communication skills while easing some of the frustrations associated with not finding it easy to talk (Fidler, Hodapp, & Dykens, 2002; Rondal & Buckley, 2003). We can use our knowledge of probabilistic risk to make our best guess at what kind of learning environments and teaching strategies are likely to yield the best results, and can then tailor that best guess to fit the actual child. If we can, in time, recognize genetic profiles that can predict who will struggle to acquire number skills, reading or social communication skills, then we may be able to design equivalent tailored interventions and possibly deliver them before the problem ever becomes a problem.

The educational environment does not always cater very well for diversity. This has more to do with governmental influences on curricula and standards than it does with the ability or desire of individual teachers to understand and meet the needs of individual children. In a landscape dominated by high stakes testing, there is simply not enough time to nurture individual appetites and aptitudes unless they are directly relevant to the curriculum or the test (Kohl, 2009). The focus, therefore, is on means rather than individual differences and this emphasis is encouraged by powerful international comparison studies such as the Program for International Student Assessment (PISA), a study which tests the reading, mathematics and scientific literacy of fifteen-year-old pupils from around the world (OECD, 2014). There is huge pressure on governments to improve, or for high achievers to maintain, their average level of performance. This pressure filters down to schools, who are also encouraged by phenomena such as league tables to aim for a high average level of performance. In this climate a focus on individual differences is too often lost. This is an unfortunate trend, not least because the empirical data show that differences between schools only explain around 20 percent of the differences between children in terms of their academic achievement (Rasbash et al., 2010), and differences between countries even less.

Although we do not yet have genetic information that could be used to make strong predictions about children's learning appetites and aptitudes a case can be made that just talking openly about the moderate to high heritability of educational achievement can help to create conditions in which individual differences can be attended to, ideally by offering new support for the argument that we need to encourage systems that genuinely support personalized teaching and learning. The aim of education in many countries is to improve average attainment, that is, to move the bell curve along to the right. This yields better results in international comparison studies such as PISA and, therefore, plays well in the media. However, this approach is meaningless in the classroom and does not speak to the diverse needs of individual children. It is an overly simplistic approach to nurturing the strengths and weaknesses of a society's children. Children differ for biological as well as social reasons, and it seems important to take their great variability, and its etiology, into account and to support all children in fulfilling their individual potential by tailoring their education to their biologically and socially influenced profiles. A genetically sensitive education should be the most inclusive education there is, not a source of discrimination. Perhaps just talking publicly about individual differences and their genetic as well as social roots might help with a change of culture in this regard, as long as the message that genes are not deterministic is communicated very clearly. This might be enough, even without untangling the intricate web of genes and experiences. Acceptance of genetics as a major player in how well children learn might also promote a new research emphasis on using education to address individual differences, not just means. The public already acknowledges and accepts genetic influence on health and disease, including psychopathology, and we need to understand more about why genetic influence on learning seems so different to some people. One likely explanation is that behavioral genetic research into teaching and learning has not yet been successfully disentangled in the public eye from the race–IQ debate invoked by publications as recent as *The Bell Curve* (Hernstein & Murray, 2010) and *The Mismeasure of Man* (Gould, 1996). There are many who still see our view that genetic research can promote a positive approach to individual differences as idealistic and insufficient to negate the potential disadvantages.

One problem with the whole idea of personalizing education along genetic lines is that it may be just too simplistic. If we could say that general cognitive ability (*g*-factor or *g*) was 100 percent heritable – which it is not – and 100 percent predictive of school achievement – no again – then it would be less difficult to envision a future in which we hunt for the many genes that influence *g*, develop an empirical understanding of how those genes interplay with other physiological forces and with experiences in the womb, the family home, and the classroom. (Not that we are suggesting for one moment that such a vast program of research would be either straightforward or speedy.) However, it is not even *that* simple. A recent UK study gathered data from more than 6,000 pairs of twins on GCSE (General Certificate of Secondary Education) grades (Krapohl et al., 2014). GCSEs are standardized assessments taken by most pupils at age sixteen in the United Kingdom (a minority take alternative but equivalent qualifications). Pupils usually take approximately ten GCSEs of which English, Mathematics, and Science are compulsory subjects. By using a genetically sensitive design, the study was able to show that genes explained 62 percent (95% CI, .58–.67) of the variance in GCSE achievement. It also went further than this simple analysis in using a range of related measures to explain the heritability of achievement. Cognitive ability was the strongest predictor of GCSE achievement, but other measures such as self-efficacy and perceptions of the school environment were not far behind. Nine variables, including cognitive ability, were found to explain 75 percent of the genetic influence on pupils' GCSE grades. Excluding cognitive ability (*g*) eight predictors explained 50 percent of the heritability of GCSE results. When *g* was added, prediction increased to 75 percent. By itself, cognitive ability was able to explain 51 percent of the heritability of GCSE. So, it turns out that the genes influencing achievement also influence cognitive ability, personality, health, self-efficacy beliefs, the perceived school environment, well-being, behavior problems (parent and child report), and the perceived home environment. Furthermore, each of these domains was shown to be heritable in its own right (heritability estimates ranged from 35 percent to 58 percent) and relationships between them and GCSE achievement were shown to be mediated by genes. This study brings us a step closer to understanding the interplay between genes and experience and its effects on achievement, but there remains much to learn. It illustrates the point that children are active agents in their own education, not passive recipients, and that they select, modify, and create experiences that are correlated with their genetic propensities. More fine-grained analysis of this dataset is planned with a view to identifying more precise predictors of achievement and assessing their likely value as targets for intervention. It is not unreasonable to consider the likelihood that it will never be possible for us to turn these scientific findings into practical teaching and learning tools, especially as we consider the complex and dynamic nature of genotype–environment interplay. The question to ask is whether it is worth trying? If we could identify genes and understand pathways to influence sufficiently well to tailor educational opportunities to individual children would that be a good thing? We believe it would, and that there is enough evidence to justify maintaining optimism and continuing to try.

Another valid concern relates to the potentially negative effects of labeling. The Pygmalion effect and its negative corollary, the golem effect, in education are well known (Rosenthal & Jacobsen, 1968). Specifically, Rosenthal and Jacobsen (1968) found that if teachers were led to expect high performance from children, then the children's performance would be enhanced, an observer-expectancy effect. The researchers argued that biased expectancies led to self-fulfilling prophecies. It seems reasonable to consider whether providing teachers or parents with genetically based probabilities of children's strengths and difficulties might lead to biased expectancies in this way. If genetic information had a strong effect on teachers' and parents' expectations of children and young people, then that could very easily become a problem. Although genes are not deterministic, it is not inconceivable that labels might become so. One important way of preparing for a future in which genetic information may be taken into account in education would be to study the likely impact of such information on teacher expectations and to consider new ways of countering any undesired effects of this nature. There may be similarly deleterious effects on children's motivation if they feel, for instance, that they are fighting their own genetic make-up in trying to pass their mathematics test. Such possibilities need serious theoretical and empirical consideration. Being able to target genetically and environmentally vulnerable individuals and tailor their education to their specific needs may, if labeling does have these negative effects on expectations and motivation, backfire. We need to establish whether this is in fact a risk and, if so, how we might address it.

GENOTYPE–ENVIRONMENT INTERPLAY

Understanding the implications of gene–environment interplay research, and a systematic push to focus on interplay as well as a program of gene hunting can help with this line of inquiry. We need a strong focus on understanding genes in their environmental context and, equally, environments in the context of genes. Gene–environment interplay research typically falls under two main headings, genotype–environment correlations (known as *r*GE) and gene × environment interactions, known as GxE (Kendler & Eaves, 1986). The first phenomenon, *r*GE, represents a mediation model in which an individual's exposure to certain experiences depends to some extent on their genes. That is, genes mediate, or account for, *how* rather than *whether* two variables are related. The second, GxE, involves moderation in the sense that it represents genetic influences on differences in individuals' responses to the environment, that is, the strength of a relationship between variables in different contexts (Baron & Kenny, 1986). For instance, the effect of attending a failing school may affect different children differently depending on their genotype. We start by considering the implications of *r*GE research for the likelihood and desirability of personalizing education along genetic lines before moving on to the implications of GxE research.

GENOTYPE–ENVIRONMENT CORRELATION (*r*GE) RESEARCH

The behavioral genetic literature describes three types of *r*GE: passive, active, and evocative (Scarr & McCartney, 1983). (See additional discussion of these correlations in Tucker-Drob and Harden, this volume.) In a *passive r*GE a child will receive both their genes and their environment from their parents. For example, this may occur if a child inherits genes that are associated with struggling to learn to read (from one or both parents) and is also brought up by those parents in a house with few books and limited exposure to text. In an *evocative r*GE, children evoke certain responses from the environment in part based on their genetic propensities, for instance a child predisposed to having a strong, clear voice and confidence in public speaking may be picked for leading roles in school plays or assemblies, thereby enhancing their confidence, experience, and reputation at a good rate of knots. Finally, in an *active* rGE, children actively seek out environments correlated with their genes. They will choose to work hard or not, to join the chess club or not, to dance or to run or to play the banjo or to read horror stories … or not. These choices will be influenced by their genetically influenced appetites and aptitudes. Understanding the genetic influences on a child may, therefore, help, in concert with watching the child's own behavior, with exposing them to opportunities they may love and thrive in. If we lived in a world where all opportunities were available to all young people, then children would be free to find out what they love and are good at in their own time. We do not live in such a world. Children do not currently have equal opportunities to identify their appetites and aptitudes, their individual potential. It is our hope that gene–environment interplay research may in time help by using genetic and environmental data to offer all children a carefully tailored canteen of opportunities. The range of options should be based on the child's individual strengths, needs, and interests.

The empirical evidence for rGE has typically come in two forms. First, there are now more than 100 empirical studies showing that environmental measures (e.g., home and school experiences) are almost as heritable as behavioral measures (e.g., intelligence, anxiety or conduct problems) (Avinun & Knafo, 2013; Kendler & Baker, 2007; Plomin, 2014; Plomin & Bergeman, 1991). That is, differences between individuals in the environments they experience show evidence of genetic influence. These empirical studies include the first ever GWAS of an environmental measure, namely family chaos, that is, parent- and child-reported perceptions of order and routine in the home (Butcher & Plomin, 2008). This does not of course mean that environments, objectively speaking, have their own DNA. It merely reflects the fact that people select and modify their environments based on their genetic propensities. Secondly, it has consistently been found that genes mediate correlations between environments and behavior. So, for instance, if a study finds that children who watch a lot of TV score more poorly in standardized tests of achievement than children who never watch TV, then that relationship may well be mediated genetically rather than environmentally. That is, the relationship between TV viewing and achievement may rely on a genetic rather than an environmental mechanism. For instance, genes have been found to significantly mediate associations between the family environment and the development of children's psychopathology (Knafo & Jaffee, 2013); household chaos and disruptive behavior (Jaffee et al., 2012); perceptions of family chaos and academic achievement (Hanscombe et al., 2010, 2011), and perceptions of school and academic achievement (Asbury et al., 2008; Krapohl et al., 2014).

It is important to note that where rGE exist they are hidden within the heritability estimate calculated on the basis of a twin or adoption study. So, just as "environmental" variables are strongly influenced by genes, it is likely that genetic influence is often dependent on an environmental mechanism. Whether a relationship between an environment and an outcome, say teacher quality and reading ability, is mediated genetically, environmentally, or both is likely to be important. Robert Plomin (2014) reports an exchange between two eminent Professors, R. C. Roberts (1967) and John Loehlin (Plomin, DeFries & Loehlin, 1977) in which Roberts said: "it matters not one whit whether the effects of the genes are mediated through the external environment or directly through, say, the ribosomes." Loehlin, like us, disagreed: "Although formally it may not matter one whit in which way the effects of the genes are mediated, in practice it often matters quite a few whits, especially if one should happen to be interested in intervening in the process. Changing behavior by changing parental attitudes is a decidedly different proposition from tinkering with ribosomes, even though a similar behavioral change might conceivably be brought about by either means" (Plomin, 2014, p. 630). When we consider behavior in the context of rGE we are considering an active model of gene–environment interplay with a strong emphasis on choice, which is at odds with notions of both genetic and environmental determinism. Recent research has shown that both the choice to take A levels (advanced qualifications for over-sixteens in the United Kingdom, and the subjects they choose, are substantially influenced by genes; Rimfeld et al., 2016). We know that children actively select and modify their environments on the basis of their genetic propensities and we are optimistic that, given time and well-designed research, rGE may eventually offer practical pointers for creating truly personalized educational strategies.

Although there is a good and growing body of empirical support for the important role of rGE this body of research is not without its weaknesses. One is that, for such a complex and potentially important process, there simply has not been enough research. Commentators such as Plomin (2014) and Rutter (2006) have championed rGE as a major research priority but it remains, in fact, a much lower priority for most researchers than gene hunting. Both lines of inquiry need attention and, for education, rGE is likely to be key. Research needs to focus on how and why individuals differ in the environments they experience. We need to understand more about how our genetically influenced behavior affects our environment. Another problem has been that the twin design has not allowed us to address some very significant environmental influences head on. The twin design is a within-family design that does not allow us to look at genetic mediation of family-wide environmental influences such as neighborhood, school, family income, or household resources. GCTA, mentioned earlier, may be able to help here as it is a between-family design that does not rely on kinship coefficients. It estimates the net influence of genes using the DNA of unrelated individuals (Harlaar et al., 2014; Power et al., 2013; Zaitlen & Kraft, 2012) and can, therefore, be used to explore group level variables such as school quality and SES. Understanding when and how associations between these environmental measures and academic achievement are mediated by genes may have educational and policy implications. And indeed we have already shown that genetics influences individual differences in SES and also mediate the relationship between SES and academic achievement (Krapohl & Plomin, 2016; Trzaskowski et al., 2014).

There is a lot more work to be done on identifying *r*GE and understanding their implications for education. However, the clear existence of *r*GE – genetic influence on exposure to particular environments and experiences and on the links between a child's environment and their behavior – supports a push for genuine and widespread personalization of teaching and learning that is advocated by both educators and geneticists (Asbury & Plomin, 2013).

GENE × ENVIRONMENT INTERACTIONS (G×E)

Medical research has led the way with GxE and replicated findings have emerged in fields such as cardiology and psychiatry. GxE describes a process wherein a genetic effect is altered by the presence of a particular environment, or an environmental effect is altered by the presence of a particular genetic variant. The study of GxE is characterized by a strong emphasis on measured genes and measured environments. For example, heart disease researchers found that risk of coronary heart disease among smokers was related to their lipoprotein lipase genotype (Talmud et al., 2000) and/or their alipoprotein E4 (*ApoE4*) genotype (Humphries et al., 2001). This raises the question of whether health education, in this case focused on smoking, could perhaps be targeted more precisely at groups of young people along genetic lines and, of course, whether it should be. If we could identify risk alleles for difficulties with reading, say, or numeracy, similar implications and challenges may follow (Docherty et al., 2011).

Striking, albeit controversial and still hotly debated, results have also been found in studies of psychopathology (Duncan & Keller, 2011). For example, Caspi et al. (2003) found that the influence of stressful life events on depression was moderated by a functional polymorphism in the promoter region of the serotonin transporter 5-HTT. The researchers found that participants who carried one or two copies of the 5-HTT short allele experienced more depression and suicidality after experiencing stressful life events than individuals who were homozygous for the long allele. Other environmental risks considered in this way, with effects that were moderated by the presence or absence of a particular genetic variant include child maltreatment and its influence in the cycle of violence (Caspi et al., 2002; Kim-Cohen et al., 2006) and adolescent cannabis use (Caspi et al., 2005).

An educationally relevant GxE might involve, in a slightly far-fetched example, finding that a subset of children with a particular allele are likely to experience reading difficulties when exposed to, say, phonics instruction. The same may not be true if they were taught to read using whole words. If we could use genetic information to identify such a group, what a lot of heartache for children, parents, and teachers we might save. Until recently all replicated GxE findings have begun with an established environmental risk factor, such as smoking, child abuse, stress, or drug abuse, and a candidate gene, implying that GxE research is likely to be especially helpful in the context of known environmental risk mechanisms (Rutter, 2006). This suggests that we need to identify overtly "risky" educational environments. However, recent research has explored the differential susceptibility hypothesis – the idea that some people are more sensitive to both risky and positive environments – using measured genes (Belsky & Pluess, 2009). Researchers at the University of Leiden found that young children with the 7 repeat version of *Drd4* initially performed less well academically than children without this variant but, after a computerized book reading intervention, they significantly outperformed the other children (Plak et al., 2016). The children with the *Drd4–7* repeat variant benefited more from this particular educational intervention, appearing to be more sensitive to it. The steps involved in undertaking high quality GxE research have been clearly laid out by Moffitt et al. (2005), although there are remaining methodological concerns to be addressed (e.g., Duncan & Keller, 2011; Risch et al., 2009). Current research suggests that further testing of the differential susceptibility hypothesis, alongside the traditional diathesis-stress model, may be particularly beneficial to education.

Not all of what is described as GxE research involves measured genes, however. There is another body of research which looks at statistical interactions between heritability estimates (rather than single, measured genes) and environmental influences. In the case of the hypothetical phonics example outlined above, this kind of process might show that the heritability of reading was higher (or lower) in pupils who were exposed solely to formal phonics instruction than those who were not. This finding would have less practical utility than a formal GxE with measured genes in that it would not have the potential to identify particular children who should be singled out for a separate approach. However, as long as researchers checked that increased heritability was not just a consequence of reduced variance in one of the groups, it could potentially have value in helping to explain the mechanisms of genotype–environment interplay, and identifying sources of inequity in education that take their effect at a molecular as well as a social level.

Much of this type of heritability x environment research has found a trend for higher heritability estimates in more advantaged families, usually using factors such as parental education level or SES as indicators of advantage and disadvantage. This suggests that better-off families may be more effective at drawing out their children's genetic potential. One study, for instance, found that the heritability of verbal IQ was higher among children growing up in highly educated families (74 percent) than in those growing up with less well educated parents (26 percent) (Rowe et al., 1999). A similar result was found in a study that looked at the IQ of seven-year-old children (Turkheimer et al., 2003). This same team also found that genetic influence increases from infancy to two years for children in advantaged homes but not for less fortunate children (Tucker-Drob et al., 2011). However, a recent meta-analysis has shown that this effect does not seem to exist outside of the United States (Tucker-Drob & Bates, 2016). In all of these cases shared environmental influences – experiences that make children growing up in a family similar – were found to be more important in the groups where heritability was low, that is, the less advantaged groups.

One US study with direct relevance to education found that reading ability was more heritable among students with higher quality teachers, further evidence of heritability increasing in line with advantage (Taylor et al., 2010). However, as stated the effect has only been observed in the United States. In our own research in the United Kingdom, for instance, we found that children at higher levels of risk (in terms of household chaos) also showed higher heritability for cognitive ability than children growing up in less chaotic homes (Asbury et al., 2005; Price & Jaffee, 2008). More recently however, we found no evidence of a heritability × environment interaction for SES (Hanscombe et al., 2012), which may simply indicate insufficient power to detect an effect. It is likely that we need even larger sample sizes for interaction studies than is the case for traditional twin studies. The majority of G×E studies not using measured genes are underpowered and do not replicate. Contradictory findings could perhaps also reflect disparate levels of poverty, cultural differences, or differences in total variance in the two samples. It is possible, for instance, that environmental variation in the United Kingdom is reduced by a standardized national curriculum. Research has shown that heritability estimates are lower and shared environmental influences higher in decentralized educational systems, such as the United States, compared with centralized educational systems such as that in the United Kingdom (Samuelsson et al., 2005). This body of research, and even the inconsistencies that exist within it, remind us very clearly that heritability estimates are population specific. It may also be used, eventually, to indicate the best and most useful way of investing our resources. For instance, if high heritability can be seen as a full drawing out of a child's potential, and if heritability is indeed higher with better teachers, then that might have implications for how we select teachers and the schools to which they are recruited. Such a leap would, however, require studies that ensure the randomness of teaching quality in the population and a non-correlational research design to establish causality.

Interaction research with either specific genes or heritability estimates and experiences could potentially play a role in enabling personalized modification of the environment to improve achievement, or other educationally relevant traits such as attitude or well-being, for all. Research testing the differential susceptibility hypothesis appears particularly promising here. It could serve an important function in helping researchers to identify causal processes and use them as the basis for interventions where appropriate. As with gene hunting, we do not yet have enough evidence but we are optimistic that, in time, we will. If and when that time comes, we predict that an understanding of the interplay between genes and experiences is likely to be more useful to education than gene hunting in isolation.

CONCLUSION

What would it be like to live in a world in which everyone found their niche and earned their money doing something that fulfilled them, something they were good at and that they loved or at least liked? The fact is that not everyone wants to get a degree and a place on a graduate training scheme. Nor is everybody suited to doing so. There are people who may excel at fishing, spying, landscape gardening, game design, Formula 1 racing, furniture restoration, pottery, sugarcraft, or any number of things but never have the chance to discover their niche. One size fits all formulas in education systems fixated on means and league tables mitigate against them ever doing so, particularly if they do not have parents with the money, time, and social or cultural capital to create tailored opportunities outside of school. If an understanding of genetic influence can be used to help guide young people toward a niche that fits them perfectly, then surely that could only be a good thing. If, in time, we can identify genes that meaningfully (but not perfectly because genes are not deterministic; Haworth & Davis, 2014) predict learning and preference profiles we will be in a better position to make an informed "best guess" at the opportunities to which we should expose each child. If we can design techniques to pin down how children and young people make choices, and thereby understand the process of *r*GE better, then we can work toward a model of education and employment in which more natures are effectively nurtured. A big push on *r*GE and G×E research, and on developing new and more individual-sensitive methods of measuring experience and choice making in children and young people, will help with this goal. And if, after all, we never find the genes or fully understand how genetic and environmental risk mechanisms affect each other, we will still have done more good than harm. We do not have answers yet but we think we are looking in the right places. It is the responsibility of scientists to provide good evidence and of society to use that evidence well. We cannot personalize education along genetic lines right now but, if the time comes when we can, there are good reasons to think that we should. Most importantly though, we need to talk about the possibility so that science does not take society by surprise and catch us unprepared.

REFERENCES

Asbury, K., Almeida, D., Hibel, J., Harlaar, N., & Plomin, R. (2008). Clones in the classroom: A daily diary study of the non-shared environmental relationship between monozygotic twin differences in school experiences and achievement. *Twin Research and Human Genetics*, 11 (6), 586–595.

Asbury, K., & Plomin, R. (2013). *G is for genes. The impact of genetics on education and achievement*. Chichester, UK: Wiley-Blackwell.

Asbury, K., Wachs, T. D., & Plomin, R. (2005). Environmental moderators of genetic influence on verbal and nonverbal abilities in early childhood. *Intelligence*, 33(6), 643–661.

Avinun, R., & Knafo, A. (2013). Parenting as a reaction evoked by children's genotype: A meta-analysis of children-of-twins studies. *Personality and Social Psychology Review*, 18(1), 87–102.

Baron, R. M., & Kenny, D. A. (1986). The moderator–mediator variable distinction in social psychological research: Conceptual, strategic, and statistical considerations. *Journal of Personality and Social Psychology, 51*(6), 1173.

Belsky, J., & Pluess, M. (2009). Beyond diathesis stress: Differential susceptibility to environmental influences. *Psychological Bulletin, 135*(6), 885.

Barrans, J. D., Stamatiou, D., & Liew, C. C. (2001). Construction of a human cardiovascular cDNA microarray: Portrait of the failing heart. *Biochemical and Biophysical Research Communications, 280*(4), 964–969.

Butcher, L. M., & Plomin, R. (2008). The nature of nurture: A genomewide association scan for family chaos. *Behavioral Genetics, 38*(4), 361–371.

Byrne, B., Coventry, W. L., Olson, R. K., Hulslander, J., Wadsworth, S., DeFries, J., et al. (2008). A behaviour-genetic analysis of orthographic learning, spelling and decoding. *Journal of Research in Reading, 31*, 8–21.

Caspi, A., McClay, J., Moffitt, T. E., Mill, J., Martin, J., Craig, I. W., et al. (2002). Role of genotype in the cycle of violence in maltreated children. *Science, 297*(5582), 851–854.

Caspi, A., Moffitt, T. E., Cannon, M., McClay, J., Murray, R., Harrington, H., et al. (2005). Moderation of the effect of adolescent-onset cannabis use on adult psychosis by a functional polymorphism in the catechol-O-methyltransferase gene: Longitudinal evidence of a gene x environment interaction. *Biological Psychiatry, 57*(10), 1117–1127.

Caspi, A., Sugden, K., Moffitt, T. E., Taylor, A., Craig, I. W., Harrington, H., et al. (2003). Influence of life stress on depression: Moderation by a polymorphism in the 5-HTT gene. *Science, 301*(5631), 386–389.

Collins, F. (2010). *The language of life: DNA and the revolution in personalised medicine.* New York: Harper Collins.

Cortes, A., & Brown, M. A. (2011). Promise and pitfalls of the immunochip. *Arthritis Res Ther, 13*(1), 101.

Deary, I. J., Strand, S., Smith, P., & Fernandes, C. (2007). Intelligence and educational achievement. *Intelligence, 35*, 13–21.

Docherty, S. J., Kovas, Y., & Plomin, R. (2011). Gene-environment interaction in the etiology of mathematical ability using SNP sets. *Behavior Genetics, 41*(1), 141–154.

Duncan, L. E., & Keller, M. C. (2011). A critical review of the first 10 years of candidate gene-by-environment interaction research in psychiatry. *Perspectives, 168*(10).

Fidler, D. J., Hodapp, R. M., & Dykens, E. M. (2002). Behavioral phenotypes and special education parent report of educational issues for children with Down syndrome, Prader-Willi syndrome, and Williams syndrome. *The Journal of Special Education, 36*(2), 80–88.

Gould, S. J. (1996). *The mismeasure of man.* New York: W.W. Norton.

Hanscombe, K. B., Haworth, C. M. A., Davis, O. S. P., Jaffee, S. R., & Plomin, R. (2010). The nature (and nurture) of children's perceptions of family chaos. *Learning and Individual Differences, 20* (5), 549–553.

Hanscombe, K. B., Haworth, C., Davis, O. S., Jaffee, S. R., & Plomin, R. (2011). Chaotic homes and school achievement: A twin study. *Journal of Child Psychology and Psychiatry, 52*(11), 1212–1220.

Hanscombe, K. B., Trzaskowski, M., Haworth, C. M. A., Davis, O. S. P., Dale, P. S., & Plomin, R. (2012). Socioeconomic status (SES) and children's intelligence (IQ): In a UK-representative sample SES moderates the environmental, not genetic, effect on IQ. *PLoS One, 7*(2), e30320.

Harlaar, N., Hayiou-Thomas, M. E., Dale, P. S., & Plomin, R. (2008). Why do preschool language abilities correlate with later reading? A twin study. *Journal of Speech, Language, and Hearing Research, 51*, 688–705.

Harlaar, N., Trzaskowski, M., Dale, P. S., & Plomin, R. (2014). Word reading fluency: Role of genome-wide single-nucleotide polymorphisms in developmental stability and correlations with print exposure. *Child Development, 85*(3), 1190–1205.

Haworth, C. M., Dale, P., & Plomin, R. (2008). A twin study into the genetic and environmental influences on academic performance in science in nine-year-old boys and girls. *International Journal of Science Education, 30*(8), 1003–1025.

Haworth, C. M. A, & Davis, O. S. (2014). From observational to dynamic genetics. *Frontiers in Genetics.*

Hayiou-Thomas, M. E., Harlaar, N., Dale, P. S., & Plomin, R. (2010). Preschool speech, language skills, and reading at 7, 9 and 10 years: Etiology of the relationship. *Journal of Speech, Language and Hearing Research, 47*, 751–765.

Hernstein, R. J., & Murray, C. (2010). *The bell curve: Intelligence and class structure in American life.* New York: Free Press.

Humphries, S. E., Talmud, P. J., Hawe, E., Bolla, M., Day, I. N., & Miller, G. J. (2001). Apolipoprotein E4 and coronary heart disease in middle-aged men who smoke: A prospective study. *The Lancet, 358*(9276), 115–119.

Jaffee, S. R., Hanscombe, K. B., Haworth, C. M. A., Davis, O. S. P., & Plomin, R. (2012). Chaotic homes and children's disruptive behaviour: A longitudinal cross-lagged twin study. *Psychological Science, 23* (6), 643–650.

Johnson, W., Deary, I. J., & Iacono, W. G. (2009). Genetic and environmental transactions underlying educational attainment. *Intelligence, 37*, 466–478.

Kendler, K. S. (2005). "A gene for ...": The nature of gene action in psychiatric disorders. *American Journal of Psychiatry, 162*, 1243–1252.

Kendler, K. S., & Baker, J. H. (2007). Genetic influences on measures of the environment: A systematic review. *Psychological Medicine, 37*(5), 615–626.

Kendler, K. S., & Eaves, L. J. (1986). Models for the joint effects of genotype and environment on liability to psychiatric illness. *American Journal of Psychiatry.*

Kim-Cohen, J., Caspi, A., Taylor, A., Williams, B., Newcombe, R., Craig, I. W., et al. (2006). MAOA, maltreatment, and gene–environment interaction predicting children's mental health: New evidence and a meta-analysis. *Molecular Psychiatry, 11*(10), 903–913.

Knafo, A., & Jaffee, S. R. (2013). Gene-environment correlation in developmental psychopathology. *Development and Psychopathology, 25*(1), 1–6.

Kohl, H. (2009). The educational panopticon. *The Teachers College Record.*

Kovas, Y., Docherty, S., Davis, O., Meaburn, E., Dale, P. S., Petrill, S., et al. (2009). Generalist genes and mathematics: The latest quantitative and molecular genetic results from the TEDS study. *Behavior Genetics, 39*(6), 663–664.

Kovas, Y., Haworth, C. M. A., Dale, P. S., & Plomin, R. (2007). The genetic and environmental origins of learning abilities and disabilities in the early school years. *Monographs of the Society for Research in Child Development*, *72*, vii–160.

Kovas, Y., & Plomin, R. (2006). Generalist genes: Implications for the cognitive sciences. *Trends in Cognitive Sciences*, *10*(5), 198–203.

Krapohl, E., & Plomin, R. (2016). Genetic link between family socioeconomic status and children's educational achievement estimated from genome-wide SNPs. *Molecular Psychiatry*, *21*(3), 437–443.

Krapohl, E., Rimfeld, K., Shakeshaft, N. G., Trzaskowski, M., McMillan, A., Pingault, J. B., et al. (2014). The high heritability of educational achievement reflects many genetically influenced traits, not just intelligence. *Proceedings of the National Academy of Sciences*, 201408777

Maher, B. (2008). The case of the missing heritability. *Nature*, *456*(7218), 18–21.

Moffitt, T. E., Caspi, A., & Rutter, M. (2005). Strategy for investigating interactions between measured genes and measured environments. *Archives of General Psychiatry*, *62*(5), 473–481.

OECD. (2014), *PISA 2012 results: What students know and can do – Student performance in mathematics, reading and science (Vol. I, Revised Edition)*. Paris: OECD Publishing. doi:10.1787/9789264201118-en

Oliver, B. R., Dale, P. S., & Plomin, R. (2007). Writing and reading skills as assessed by teachers in 7-year-olds: A behavioural genetic approach. *Cognitive Development*, *22*(1), 77–95.

Olson, R. K. (2007). Introduction to the special issue on genes, environment and reading. *Reading and Writing*, *20*, 1–11.

Panofsky, A. L. (2009). Behavior genetics and the prospect of "personalized social policy." *Policy and Society*, *28*, 327–340.

Parens, E. (2004), Genetic differences and human identities. *Hastings Center Report*, *34*, S4–S35.

Plak, R. D., Merkelbach, I., Kegel, C. A., van IJzendoorn, M. H., & Bus, A. G. (2016). Brief computer interventions enhance emergent academic skills in susceptible children: A gene-by-environment experiment. *Learning and Instruction*, *45*, 1–8.

Plomin, R. (1994). *Genetics and experience: The interplay between nature and nurture*. Thousand Oaks, CA: Sage.

(2014). Genotype-environment correlation in the era of DNA. *Behavior Genetics*, *44*(6), 629–638.

Plomin, R., & Bergeman, C. S. (1991). The nature of nurture: Genetic influence on "environmental" measures (with open peer commentary). *Behavioral and Brain Sciences*, *14*(3), 373–414.

Plomin, R., DeFries, J. C., & Loehlin, J. C. (1977). Genotype-environment interaction and correlation in the analysis of human behavior. *Psychological Bulletin*, *84*(2), 309–322.

Plomin, R., & Kovas, Y. (2005). Generalist genes and learning disabilities. *Psychological Bulletin*, *131*(4), 592–617.

Plomin, R., & Simpson, M. A. (2013). The future of genomics for developmentalists. *Development and Psychopathology*, *25*(4), 1263–1278.

Power, R. A., Wingenbach, T., Cohen-Woods, S., Uher, R., Ng, M. Y., Butler, A. W., et al. (2013). Estimating the heritability of reporting stressful life events captured by common genetic variants. *Psychological Medicine*, *43*(9), 1965–1971.

Price, T. S., & Jaffe, S. R. (2008). Effects of the family environment: Gene-environment interaction and passive gene-environment correlation. *Developmental Psychology*, *44*(2), 305.

Rasbash, J., Leckie, G., Pillinger, R., & Jenkins, J. (2010). Children's educational progress: Partitioning family, school and area effects. *Journal of the Royal Statistical Society: Series A (Statistics in Society)*, *173*(3), 657–682.

Rimfeld, K., Ayorech, Z., Dale, P. S., Kovas, Y., & Plomin, R. (2016). Genetics affects choice of academic subjects as well as achievement. *Scientific Reports*, *6*.

Risch, N., Herrell, R., Lehner, T., Liang, K. Y., Eaves, L., Hoh, J., et al. (2009). Interaction between the serotonin transporter gene (5-HTTLPR), stressful life events, and risk of depression: A meta-analysis. *JAMA*, *301*(23), 2462–2471.

Roberts, R. C. (1967). Some concepts and methods in quantitative genetics. In J. Hirsch (Ed.), *Behavior-genetic analysis* (pp. 214–257). New York: McGraw-Hill.

Rondal, J., & Buckley, S. (2003). *Speech and language intervention in Down syndrome*. London: Whurr Publishers.

Rosenthal, R., & Jacobson, L. (1968). Pygmalion in the classroom. *The Urban Review*, *3*(1), 16–20.

Rowe, D. C., Jacobson, K. C., & Van den Oord, E. J. (1999). Genetic and environmental influences on vocabulary IQ: Parental education level as moderator. *Child Development*, *70*(5), 1151–1162.

Rutter, M. (2006). *Genes and behavior. Nature-nurture interplay explained*. Oxford: Blackwell.

Samuelsson, S., Byrne, B., Quain, P., Wadsworth, S., Corley, R., DeFries, J. C., et al. (2005). Environmental and genetic influences on prereading skills in Australia, Scandinavia, and the United States. *Journal of Educational Psychology*, *97*(4), 705.

Scarr, S., & McCartney, K. (1983). How people make their own environments: A theory of genotype greater than environment effects. *Child Development*, 424–435.

Selzam, S., Krapohl, E., von Stumm, S., O'Reilly, P. F., Rimfeld, K., Kovas, Y., et al. (2016). Predicting educational achievement from DNA. *Molecular Psychiatry*.

Shakeshaft, N. G., Trzaskowski, M., McMillan, A., Rimfeld, K., Krapohl, E., Haworth, C. M. A., et al. (2013). Strong genetic influence on a UK nationwide test of educational achievement at the end of compulsory education at age 16. *PloS One*, *8*(12), e80341.

Talmud, P. J., Bujac, S. R., Hall, S., Miller, G. J., & Humphries, S. E. (2000). Substitution of asparagine for aspartic acid at residue 9 (D9N) of lipoprotein lipase markedly augments risk of ischaemic heart disease in male smokers. *Atherosclerosis*, *149*(1), 75–81.

Taylor, J., Roehrig, A. D., Hensler, B. S., Connor, C. M., & Schatschneider, C. (2010). Teacher quality moderates the genetic effects on early reading. *Science*, *328* (5977), 512–514.

Tosto, M. G., Petrill, S. A., Halberda, J., Trzaskowski, M., Tikhomirova, T. N., Bogdanova, O. Y., et al. (2014). Why do we differ in number sense? Evidence from a genetically sensitive design. *Intelligence*, *43*, 35–46.

Trzaskowski, M., Harlaar, N., Arden, R., Krapohl, E., Rimfeld, K., McMillan, A., et al. (2014). Genetic influence on family socioeconomic status and children's intelligence. *Intelligence*, *42*, 83–88.

Tucker-Drob, E. M., & Bates, T. C. (2016). Large cross-national differences in gene× socioeconomic status interaction on intelligence. *Psychological Science, 27*(2), 138–149.

Tucker-Drob, E. M., Rhemtulla, M., Harden, K. P., Turkheimer, E., & Fask, D. (2011). Emergence of a gene× socioeconomic status interaction on infant mental ability between 10 months and 2 years. *Psychological Science, 22*(1), 125–133.

Turkheimer, E., Haley, A., Waldron, M., d'Onofrio, B., & Gottesman, I. I. (2003). Socioeconomic status modifies heritability of IQ in young children. *Psychological Science, 14*(6), 623–628.

Valentine, G. H. (1986). *The chromosomes and their disorders: An introduction for clinicians.* London: Heinemann.

Wadsworth, S. J., DeFries, J. C., Fulker, D. W., & Plomin, R. (1995). Cognitive ability and academic achievement in the Colorado Adoption Project: A multivariate genetic analysis of parent-offspring and sibling data. *Behav. Genet. 25*, 1–15.

Watson, J. D., & Crick, F. H. C. (1953). Genetical implications of the structure of deoxyribonucleic acid. *Nature, 171*, 964–967.

Yang, J., Benyamin, B., McEvoy, B. P., Gordon, S., Henders, A. K., Nyholt, D. R., et al. (2010). Common SNPs explain a large proportion of the heritability for human height. *Nature Genetics, 42*(7), 97–107.

Yang, J., Lee, S. H., Goddard, M. E., & Visscher, P. M. (2011). GCTA: A tool for genome-wide complex trait analysis. *American Journal of Human Genetics, 88*(1), 76–82.

(2013). Genome-wide complex trait analysis (GCTA): Methods, data analyses, and interpretations. *Methods in Molecular Biology, 1019*, 215–236.

Zaitlen, N., & Kraft, P. (2012). Heritability in the genome-wide association era. *Human Genetics, 131*(10), 1655–1664.

4

Early Adversity and Epigenetics:
Implications for Early Care and
Educational Policy

KATHERINE BECKMANN AND KIERAN O'DONNELL

Economic hardship has been linked to a myriad of adverse educational and developmental outcomes for children that limit future productivity (Barnett, 1998; Brooks-Gunn & Duncan, 1997). Today, with nearly 4.9 million children, under age five, living in families with incomes below the federal poverty level ($23,492 per year for a family of four in 2012), almost 25 percent of America's youngest children are at risk for untoward psychosocial, environmental, and economic conditions (Brooks-Gunn & Duncan, 1997; US Census Bureau, 2013). Further, families with low socioeconomic status (SES) report greater exposure to multiple stress factors and more severe stressful life events compared to those with higher SES (Liaw & Brooks-Gunn, 1994; Lupien, King, Meaney, & McEwen, 2000).

Although individuals differ in their physiological responsiveness and ability to adapt, research shows that chronic, elevated levels of stress are associated with biologic and behavioral consequences that negatively affect physical growth, onset and duration of puberty, metabolism, and susceptibility to illness as well as social, emotional, and cognitive functioning (Bradley & Corwyn, 2002; Center on the Developing Child at Harvard University, 2007; Harris & Seckl, 2011; McEwen & Seeman, 1999). There is growing evidence that early-life adversity predicts an individual's risk of later mental disorder. Moreover, there is evidence that history of early adversity, such as low SES, maternal/pre-natal anxiety and depression, childhood maltreatment, among other risk factors, predicts not only the risk of disorder but also the likelihood of treatment response (Nanni, Uher, & Danese, 2012). The stress diathesis model of psychopathology posits that dysregulation of an individual's stress physiology is one pathway to disorder.

The views expressed here are solely those of the authors and do not reflect the views of the Administration for Children and Families or the Department of Health and Human Services.

An important distinction is to be made here between the adaptive values of the acute stress response and potentially maladaptive effects of prolonged activation of an individual's stress physiology.

In this chapter, we will briefly review the main actors in the acute stress response system before discussing a framework proposed to describe the maladaptive effects of chronic stress. While outcomes associated with adverse early-life experiences are well documented, we are only beginning to understand how early adversity becomes biologically embedded. We will discuss how the emerging field of clinical epigenetics may explain, at least in part, how early-life experiences influence biology across the lifespan. This leads us to discuss ethical considerations for this new field of research and implications of recent findings for early care and education program and policy development. Identifying environmental factors that have persisting effects on stress physiology and the biological mechanisms which mediate these effects are attractive candidates for prevention and intervention strategies. In fact, interventions during the pre-natal period and early childhood may be the best means to mitigate the untoward effects of poverty, as well as other types of chronic stressors, on child development.

THE STRESS RESPONSE

Often referred to as the "fight or flight" response, the human body's acute stress response has been well characterized and is mediated by two distinct but interacting systems. Encompassing the sympathetic and parasympathetic branches, the autonomic nervous system responds most rapidly when faced with an acute stressor culminating with an increase in circulating catecholamines, such as epinephrine and norepinephrine. The second aspect of the acute stress response is activation of the hypothalamic-pituitary-adrenal (HPA) axis, which results in an increased synthesis and release of glucocorticoids, cortisol in humans, and corticosterone in most rodent species, which peak in concentration approximately twenty minutes post-stress. These steroid hormones have both acute and more persisting effects and are implicated in multiple processes including (but not limited to) gluconeogenesis and metabolism; immune function, bone growth; reproduction; metabolism; neuronal development, function and behavior; and cell growth and differentiation (Gluckman & Hanson, 2005; Meaney, 2010; Seckl & Holmes, 2007).

Unlike catecholamines, glucocorticoids, such as cortisol, have DNA-binding potential and can modulate expression of steroid-sensitive genes by binding to their related receptors, such as the glucocorticoid receptor or mineralocorticoid receptor. These receptor complexes recruit co-factors, translocate to the nucleus, and influence both gene activation (transcription) and gene repression (trans-repression). Importantly, the binding of these transcription factor complexes to DNA also have the potential to influence the epigenetic landscape of these target genes.

For these reasons, and possibly due to the ease of cortisol measurement in saliva, a large emphasis has been placed on the study of saliva cortisol in stress research; it should be noted that this is just one of several factors influenced by acute stress. Similarly, the terms "cortisol" and "stress" are often used interchangeably. However, like most biological processes, cortisol is modulated by multiple factors, including person-specific factors such as genotype and extrinsic factors such as the time of day. Indeed, one of the most consistent features of the cortisol diurnal profile is the steep rise in cortisol levels approximately thirty minutes post-wakening. Yet, the evidence linking this marked increase in cortisol to concurrent symptoms of subjective stress is weak at best (see Clow et al., 2009, for a review). As a result, caution is warranted when equating higher stress with high cortisol as this is an endpoint measurement of a complex system.

Somewhat more promising are recent studies focusing on hair cortisol as a measure of cumulative cortisol exposure. Cortisol sampled from hair provides a read-out of cortisol exposure over an approximate three month period. Similar to saliva cortisol measurements, hair cortisol is a single measure of a complex system but is less sensitive to potential confounders such as time of sample collection. For studies detailing more persistent effects on stress physiology, this cumulative measure may prove more informative. With these caveats in mind, we will now discuss the evidence that the adaptive nature of the acute stress is lost when faced with chronic unremitting psychosocial stress, with a special emphasis on the notion of "toxic stress" and its effects on the HPA axis and child development.

THE "TOXICITY" OF CHRONIC PSYCHOSOCIAL STRESS

There may be a cumulative effect of early social and environmental risk factors, similar to that of chemical and biological environmental insults, which lead to chronic, elevated levels of stress that may be an important pathway linking SES to health and developmental outcomes throughout the lifespan (Evans & Kim, 2010; Hackman, Farah, & Meaney, 2010). While protective in acutely stressful contexts, the aforementioned neurobiological and behavioral responses mobilized by the endocrine system throughout the lifespan can become maladaptive and pathogenic if consistently activated under chronic, overwhelming stress and adversity (McEwen & Seeman, 1999). For example, among the unfavorable effects of chronic, elevated levels of glucocorticoids on cognitive function are impaired hippocampal-dependent memory retention and retrieval as well as spatial memory in children and adults (McEwen, 2007).

The following framework categorizes stress exposure and coping strategies with an emphasis on the biological response to stress and its effects on an individual's physiology. These different types of experiences include (1) positive stress, which is moderate and short-lived, causing brief increases in heart rate or mild changes in stress hormone levels; (2) tolerable stress, which is severe enough to disrupt brain architecture if unchecked, but is buffered by supportive relationships that can facilitate adaptive coping and mitigate the damaging effects; and (3) toxic stress, which is severe and prolonged in the absence of the buffering protection of supportive relationships. Within this framework, the adaptive nature of the stress response is acknowledged; however, the persistent and prolonged activation of the stress system predicts the likelihood of adverse neurodevelopment as well as other organ and metabolic systems (Shonkoff et al., 2012).

For children from low income families, especially, these effects may begin before birth and continue throughout development. There is now good evidence that the maternal emotional state during pregnancy can influence child emotional/behavioral development (O'Donnell et al., 2014) with some evidence that such effects may be mediated at least in part by in utero cortisol exposure and moderated by the early care environment (Bergman et al., 2010; Buss et al., 2012).

Similarly, the effects of low SES on markers of inflammation in adulthood may be buffered by increased maternal warmth in the post-natal period (Chen et al., 2011). These studies and others emphasize the importance of the mother's emotional state during pregnancy and the early care environment as important predictors of both neurodevelopment and health risk. These findings mirror parallel themes in the field of cardiovascular disease (CVD) where links with adverse pre-natal and early-life environments are significant predictors of disease risk.

DEVELOPMENTAL ORIGINS OF HEALTH AND DISEASE

Although the human body's response to stress is essential for survival, changes to fetal endocrine, cardiovascular, metabolic, and behavior regulation resulting from chronic, elevated levels of maternal stress, while beneficial for immediate survival, may promote a developmental trajectory biased toward adverse health outcomes (Gluckman & Hanson, 2005; Meaney et al., 2007; Oitzl et al., 2010; Welberg & Seckl, 2001). This is exemplified by studies of fetal growth restriction adult onset CVD (Barker, 2007; Power & Tardif, 2005; Seckl, Drake, & Holmes, 2005). These and other studies have contributed to the developmental origins of health and disease (DOHaD) hypothesis, which posits that risk for adult onset pathology may be causally related to early-life factors. Indeed, while this hypothesis has been most commonly applied to the study of cardio-metabolic disease, there is growing evidence that both pre-natal and early-life factors influence susceptibility for adverse neurodevelopmental outcomes.

The potential for pre-natal and early-life factors to influence brain development is not surprising given the rapid growth experienced during this period. Fetal magnetic resonance imaging (MRI) studies illustrate that the fetal brain undergoes rapid proliferation across gestation with approximately seventeen-fold change in relative size from mid to late gestation with a further four-fold change observed from birth to early childhood (Huang, 2010). By the time a full-term infant is born, the basic wiring of the central nervous system has been completed. The pre-natal establishment of brain architecture provides the scaffolding on which one later builds the capacity to receive, interpret, and act upon information gathered from the surrounding world (Hammock & Levitt, 2006). Precursors of functional brain regions manifest during the first and second trimesters when neurons are produced and regional migration occurs. The fine-tuning of circuits within these regions takes place after birth through experience-dependent mechanisms such as synapse formation and pruning (Bourgeouis, Goldman-Rakic, & Rakic, 1999).

Pre-natal programming of the brain is particularly sensitive to the interplay of genetic and environmental factors. Evidence from animal models suggests that exposure to maternal stress in utero can result in low birth weight and later cognitive impairment such as learning deficits, increased anxiety, and reduced attention span (Hosseini-Sharifabad & Hadinedoushan, 2007; Yang et al., 2006). Pre- and peri-natal administration of glucocorticoids to pregnant rodents has resulted in reduced brain weight at birth as well as delayed neuronal maturation, myelination, gliogenesis, and synapse formation (Seckl, 2008). In non-human primates, prenatal exposure to elevated levels of glucocorticoids in rhesus monkeys has also been associated with reduced hippocampal volume later in life (Uno et al., 1989). In another study, female rhesus monkeys that were stressed during pregnancy gave birth to offspring that displayed decreased birth weight, impaired neuromotor development, attention deficits, and emotional dysregulation into adulthood (Schneider et al., 2002).

In humans, two MRI studies have examined the lasting effects of cortisol during pregnancy on child neuroanatomy (Buss et al., 2012; Sarkar et al., 2014). Buss and colleagues (2012) found maternal pre-natal cortisol levels (in saliva) were predictive of amygdala volume in seven-year-old girls but not boys. Interestingly, the authors found that these increases in amygdala volume partially mediated the association between maternal pre-natal cortisol and emotional problems in these children. Similarly, Sarkar et al. (2014) found that in utero cortisol, measured directly from the aminotic fluid of women undergoing amniocentesis, was predictive of white matter tract development in a similarly aged sample of children. Collectively, these studies provide some of the first evidence that maternal pre-natal cortisol may influence child neurological development and mediate associations between maternal pre-natal cortisol and child outcome. The question then becomes as to how these factors during the pre-natal and post-natal periods influence brain development and health outcomes across the lifespan.

CLINICAL EPIGENETICS

Genetic information is the blueprint for the structure and function of proteins that allow biological processes to occur (see Kornilov, this volume). The order of nucleotide sequences within DNA (e.g., ATCG) specifies which RNA products are transcribed (the process of a gene being "read" by cellular machinery). This RNA product is then translated into a particular protein, the building blocks of all cells and tissue. Variation at the level of the individual nucleotide is very common, termed single nucleotide polymorphism, and can yield markedly different RNA and protein products. Larger changes involving longer repeats of DNA can occur such as variable number tandem repeats or copy number variants (Ingles, 2004; Plomin, Defries, McClearn, & McGuffin, 2008). Despite these inter-individual differences in genotype, it is interesting to note that within any individual each of the 300 or more different cell types which are found in the human body share the same DNA and this DNA does not change during the process of development or aging. Instead, chemical marks or modifications that decorate the DNA impact how it is packaged in the nucleus of the cell and can strongly influence the expression of specific genes and phenotype of any given cell. Collectively, the study of these chemical marks is termed epigenetics stemming from the Greek word "*epi*" meaning on top of *genetics*, while epigenomics refers to the study of these modifications across the genome. Importantly, research suggests that early-life experiences – both positive and negative – may influence gene expression through epigenetic mechanisms (Meaney, 2010; Szyf, 2009a, 2009b).

DNA is often represented as a linear molecule to which transcription factors (the proteins that read the DNA sequence and initiate gene expression) gain unimpeded access. In reality, DNA is packaged or organized in a form that resembles beads lying along a string. The beads are termed "nucleosomes" and are comprised of approximately 146 base pairs wrapped around a protein core called a histone (Turner, 2001). The histones and DNA together are referred to as chromatin; the nucleosome is the organization of chromatin. In general, there is a tight physical relationship between the histone proteins and the accompanying DNA, resulting in a closed nucleosome configuration. This configuration is maintained, in part, by electrostatic bonds between the positively charged histones and the negatively charged DNA. Epigenetic modifications essentially favor a closed or open state of chromatin that either increases or decreases the ability of transcription factors to access and bind to the DNA, thus influencing the ability of the cell to read the underlying DNA.

Chromatin Modifications. Changes in chromatin structure are achieved in part through modifications to the histone proteins at the amino acids that form the histone protein tails. There are several examples of such modifications including, but not limited to acetylation, phosphorylation, methylation and ubiquitylation (see Maze, Noh, & Allis, 2013). These modifications are achieved through a series of enzymes that bind to the histone tails and modify the local chemical properties of specific amino acids (Grunstein, 1997; Jenuwein & Allis, 2001). For example, the enzyme histone acetyltransferase "transfers" an acetyl group onto specific lysine amino acids on the histone tails. The addition of the acetyl group diminishes the positive charge, loosening the connection between the histones and DNA, opening the chromatin, and improving the ability of transcription factors to access DNA sites. Thus, histone acetylation at specific lysine sites is commonly associated with active gene transcription.

DNA Methylation. Another level of regulation occurs directly on the DNA. Indeed, the classic and most well-characterized epigenetic alteration is DNA methylation, which involves the addition of a methyl group (CH_3) onto cytosines predominately bound to guanines (CpGs) in the DNA (Bird, 1986; Razin & Riggs, 1980). Recently, variations of DNA methylation such as hydroxymethylation and non-CG methylation, occurring on cytosines usually followed by a nucleotide other than a guanine, have been identified. Both non-CG methylation and hyroxymethylation are important for neuronal specification and brain development (Hahn et al., 2013; Lister et al., 2013). However, there is little data describing how these modifications may be influenced by early-life stress. As a result, this chapter will focus exclusively on CpG DNA methylation.

DNA methylation in the gene promoter, the region associated with the initiation of transcription, is typically associated with transcriptional repression or gene silencing. The repressive effect of DNA methylation on gene transcription appears to be mediated in one of two ways: (1) methylated DNA can directly impede transcription factor binding to DNA sites; and (2) methylated DNA can attract a group of proteins known as methylated DNA binding proteins (Deaton & Bird, 2011), which in turn attract an entire cluster of proteins that act as a repressor complex leading to gene silencing. Collectively, chromatin modifications alter the structure and chemical properties of the DNA, and thus gene expression. Modifications to the DNA and its chromatin environment can be considered phenotypes which provide an additional layer of information on genomic function. Furthermore, unlike the underlying DNA sequence, which remains static across development, epigenetic modifications are dynamically regulated and responsive to changes in the environment.

Evidence in Animals That Early Environment Influences Epigenetic Modifications. There is an abundant literature demonstrating persistent effects of the early care environment on epigenetic modifications such as DNA methylation (Bagot et al., 2012; Champagne, 2008; Mueller & Bale, 2008; Murgatroyd et al., 2009; Roth et al., 2009; Weaver et al., 2004, 2007; Zhang et al., 2010). For example, several studies by Meaney and colleagues (2007, 2010) demonstrate that naturally occurring variations in maternal care, indexed by the time a dam spends licking and grooming her offspring, are associated with marked differences in multiple phenotypes including learning and memory, obesity, sexual maturation and reproduction, fear/anxiety and depression-like behavior, immune function, and, indeed, the transmission of maternal care across generations.

Perhaps most relevant to our discussion of toxic stress are the findings of Weaver et al. (2004, 2007). In a series of studies, maternal care was shown to influence the methylation status of the glucocorticoid receptor gene (*Nr3c1*) in the hippocampus of her offspring. Evident by post-natal day 4 and persisting into adulthood, increased levels of maternal licking and grooming predicted decreased methylation of the *Nr3c1* promoter at a specific CpG. Importantly, these effects on DNA methylation co-varied with the corticosterone response to restraint stress; adult offspring of a low licking and grooming dams demonstrated a delayed recovery and showed elevated levels of corticosterone when compared to offspring from high licking and grooming dams. This group has now demonstrated that the early care environment influences the methylation of several genes related to neural plasticity suggesting wide-ranging effects of maternal care on the epigenetic regulation of several genes.

Consistent with the findings of the Meaney group, Murgatroyd and colleagues (2009) have shown that early-life stress can influence the methylation of the corticotrophin-releasing hormone gene, consistent with a lasting effect of early-life stress on HPA axis regulation. Finally, a number of non-human primate studies also demonstrate the pervasive effect of the early care environment on epigenetic regulation of gene expression. Provencal and colleagues found that a model of maternal deprivation (peer versus maternal rearing) was associated with widespread epigenetic changes in both peripheral tissue and the prefrontal cortex (Provençal et al., 2012).

Evidence in Humans That Early Environment Influences Epigenetic Modifications. Studies of the Dutch Hunger Winter, a brief period of famine which occurred in the Nazi-occupied region of the Netherlands during World War II, provide some of the first clinical evidence of the persisting effects of early adversity epigenetic mechanisms such as DNA methylation. Maternal periconceptual famine exposure predicted decreased methylation of *IGF2*, an important regulator of metabolism, in blood samples taken from adults pre-natally exposed to famine (Heijmans et al., 2008). Such results are consistent with the DOHaD model discussed earlier, whereby early-life experiences influence metabolic outcomes across development. Similarly, there is evidence that the early care environment influences epigenetic processes in humans.

Translating findings from rodent models, the Meaney group provided some of the first evidence that childhood trauma was associated with epigenetic regulation of gene expression (McGowan et al., 2009). In this study, the authors characterized the methylation status of the glucocorticoid receptor (GR) in human post-mortem hippocampal samples from suicide completers and matched controls who died of other causes. McGowan and colleagues found decreased expression of the GR in the hippocampus of those individuals with a history of child maltreatment; these findings were paralleled with elevated methylation of the GR promoter in these samples. In line with the work of Weaver et al. (2004), the authors found that the methylation sites of most relevance were situated in proximity to a binding site for the transcription factor NGFI-A, suggesting a role for this transcription factor in mediating these effects (Weaver et al., 2007). These findings bear considerable similarity to the maternal effect demonstrated in rats and are suggestive of early environment regulation of the neural epigenome in humans. A question remains as to whether or not the epigenetic changes that influence neural function may be evident in peripheral tissue, given the tissue specificity of epigenetic modifications (Davies et al., 2012). However, there is growing evidence that the effects of early adversity on the GR methylation may be evident in peripheral tissue (Radtke et al., 2011; Tyrka et al., 2012, and see below). Moreover, a number of studies have reported effects of early adversity on methylation across the genome.

Suggestive evidence that the early care environment has a persisting effect on the epigenome comes from a study of institutionalized children in Russian orphanages (Naumova et al., 2012). Naumova and colleagues assessed the methylation status of approximately 27,000 CpGs (~0.1 percent of CpGs in the human genome) in fourteen children raised in federal institutions and fourteen controls, ranging in age between seven and ten years of age. The authors observe a shift toward hypermethylation of a small subset of target genes associated with immune response and cellular signaling was observed in the institutionalized group. While the functional effects of these epigenetic changes on vulnerability for psychopathology remain to be clarified, this study provides proof-of-principle that the absence of parental care influences DNA methylation in humans. This study is in line with the work of Provencal et al., who show widespread changes in DNA methylation in blood and brain samples from non-human primates raised in a peer-rearing paradigm relative to maternally reared controls.

Likewise, at least three other studies have shown the persisting effect of childhood trauma on DNA methylation in peripheral tissue (Labonte et al., 2012; Mehta et al., 2013; Yang et al., 2013). It is not surprising that childhood trauma and institutionalization influence epigenetic modifications; however, there is also evidence that less severe stress can influence DNA methylation. Essex and colleagues (2011) find that maternal (and paternal) stress is associated with DNA methylation profiles assessed from samples taken from their adolescent children. These effects were most pronounced for maternal stress experienced in infancy and provide some indirect support for the notion of toxic stress having a lasting effect on child gene regulation (Essex et al., 2011). Indeed, a recent study shows evidence that maternal pre-natal depression in association with child genotype accounts for a significant portion of variance in genome-wide DNA methylation at birth, suggesting marked in utero effects on epigenetic processes (Teh et al.,

2014). It remains to be determined if such epigenetic modifications are predictive of later emotional/behavioral development; however, there is some evidence that this may be possible.

Mehta and colleagues (2013) show that childhood adversity associates with coordinated epigenetic and transcriptional changes in peripheral blood cells from adults with post-traumatic stress disorder (PTSD). Interestingly, despite a similar clinical presentation for both groups, early-life adversity was associated with an almost unique transcriptional profile, relative to PTSD patients without early-life exposure, suggesting differential regulation of gene transcription in this group. Collectively, these studies provide some evidence that early adversity may manifest as altered patterns of DNA methylation and/or gene expression. Moreover, there is at least one study to date, which reports an association between methylation of the candidate gene related to glucocorticoid signaling and symptom improvement in combat veterans undergoing psychotherapy for PTSD (Yehuda et al., 2013). Although this is a small study, the findings suggest an individual's epigenetic landscape could be informative for predicting likelihood of treatment response and could eventually impact therapy choice.

Less is known about how normal variations in the social environment influence epigenetic processes in typically developing children or how epigenetic biomarkers may be used to predict neurodevelopmental outcomes. However, genetic disorders such as Beckwith-Wiedemann, Silver-Russell, and Rett syndromes, characterized by dysregulation of epigenetic processes and varying degrees of mental retardation, support an association between epigenetic processes and child neurodevelopment. The field of clinical epigenetics is yet to determine a single epigenetic marker or profile which reliably associates with child cognitive or emotional/behavioral development and could be used to guide personalized medicine or intervention. This is in part due to our limited understanding of the stability (or dynamism) of DNA methylation across childhood/adolescence. However, the findings of Yehuda and colleagues provide proof-of-principle that epigenetic modifications, such as DNA methylation, could be used in the context of personalized medicine. A first step toward the use of epigenetic modifications for predicting child neurodevelopmental outcomes is to determine which sites of the epigenome show dynamic change across development.

Several studies have reported cross-sectional data on DNA methylation across ages and show a trend toward more variable DNA methylation profiles with advancing age (Heyn et al., 2012; Kaminsky et al., 2009). Two studies report longitudinal data samples collected during the first years of life and show moderate to high stability of DNA methylation profiles over the first years of life (Martino et al., 2011; Wang et al., 2012). However, it should be noted that both groups assessed methylation at <1 percent of all sites in the human genome. Indeed, many of the studies discussed above have focused solely on DNA methylation, which is just one of many epigenetic modifications. Before data generated from research studies can be applied to guide early intervention, we must understand how multiple epigenetic marks interact to regulate gene expression in health and disease states. Such issues will undoubtedly be addressed through the International Human Epigenome Consortium that will investigate multiple epigenetic markers in a wide variety of tissues and individuals, in a similar manner to the human genome project (see http://ihec-epigenomes.org/). These data will undoubtedly advance our understanding of how epigenetic modifications work together to regulate gene expression. This knowledge may help identify "typical" and aberrant epigenetic profiles, which could be used to identify at-risk individuals, perhaps those exposed to adverse early environments.

ETHICAL CONSIDERATIONS AND NEED FOR FUTURE RESEARCH

From a policy standpoint, the mechanisms by which environmental impacts on early development are biologically embedded and sustained into childhood will become increasingly important to understand as science is leveraged to design and implement effective prevention strategies and interventions in education and health. Integration of biological theory and research with social science gives a more complete picture of the ecological context in which families live and strive to flourish. The perinatal period is an especially important time to stimulate positive growth and alleviate the untoward effects of poverty, as well as other types of chronic stressors, on child development (Liaw & Brooks-Gunn, 1994). For young children who experience toxic stress, early interventions that provide specialized services targeted at mitigating environmental risk factors for the family, as a whole, may prevent disruption of brain architecture and promote better developmental outcomes (Center on the Developing Child at Harvard University, 2007).

Post-natal social environmental risk factors can have an independent effect on the well-being of the child and magnify pre-natal insults to stress reactivity and endocrine functioning (Meaney et al., 2007; Oitzl et al., 2010). Post-natal HPA function modification may result from neonatal handling, extended maternal separation (e.g., isolated or deprived rearing conditions), parental divorce, maternal mental illness, child abuse and neglect, and poverty (Gluckman & Hanson, 2005; Kapoor et al., 2006; Meaney et al., 2007). However, positive post-natal influences, such as social stimulation, support, and nurturance, can mitigate the untoward effects of early-life programming (Liu et al., 2000; Maccari et al., 1995; Weaver et al., 2004). In animal studies, environmental enrichment appears to support capacity for cognitive restoration or protection, reversing cognitive impairment resulting from early stress exposure (Hedges & Woon, 2011).

Although the protective effects of informal and formal social networks appear to be robust, little is known about the quality, timing, and duration of effective interventions targeted toward mitigating cumulative risk and toxic stress or how these positive interventions may influence epigenetic modifications (Cohen, Janicki-Deverts, & Miller, 2007; Collins, Dunkel-Schetter, Lobel, & Scrimshaw, 1993; Rozanski et al., 1999; Sameroff et al., 1987; Shonkoff, Boyce, & McEwen, 2009; Taylor & Repetti, 1997). Challenge remains in identifying specific influences of risk and developing strategies to diminish those influences while promoting protective factors. Future research should examine factors that promote resiliency by taking into account environmental and genetic measures of cumulative protective factors within the context of the individual and community in which he or she lives. In general, there is much more known about social and environmental risk and protective factors than genetic and epigenetic ones.

We are only beginning to understand the role epigenetics plays in who we are as humans. It is a hopeful notion that epigenetics could enhance our abilities to diagnose disease as well as identify other factors that impact child development, such as child maltreatment and poor nutrition; however, this requires continued research efforts. Advances in epigenetics could provide new opportunities for individualized environmental, behavioral, and/or biomedical prevention strategies that reverse maladaptive epigenetic modification. Catering to variability in phenotypic range and need, early interventions targeted at specific methylation patterns could alter the course of early adversity on later outcomes.

While questions remain regarding the biological pathways which link the early environment to epigenetic change and sustained changes in neurodevelopment, progress has been made in rodent studies (see Hellstrom et al., 2012; Liu et al., 2000). However, longitudinal data in clinical human studies is lacking. We are only beginning to understand the genomic basis of risk and resilience which may moderate the effects of early-life stress on both epigenetic change and clinical outcomes (see Klengel et al., 2013). We must better understand signaling pathways leading from early-life experience to DNA methylation changes (Szyf & Bick, 2013). In addition, early research has shown it may be possible that epigenetic programming can be transgenerational if chromosomal and epigenetic changes are stable (Harris & Seckl, 2011).

While none of the aforementioned adaptations and resulting outcomes are deterministic and absolute, it is possible that misinterpretation of this complex science could lead to misplaced justification to neglect much needed investment in education, housing, social services, and health care in areas of high poverty. If impacts of poverty result in epigenetic change and are biologically embedded over generations, why put greater investment in a population that is destined to fail? It would be a frightening notion that the bootstraps by which one is supposed to pull themselves up are actually faulty in some way. One could imagine this line of thinking as fodder for social stigma and discrimination. In truth, current science indicates a malleability that lends itself to the idea that policies can help to create environments that have substantial positive impact on human capitol. If we are to break the cycle of poverty, we must understand that these issues are complex and multifactorial and, therefore, need multidimensional supports.

IMPLICATIONS FOR EARLY CHILDHOOD POLICY AND PROGRAMS

What does the new science in toxic stress and epigenetics tell us to do differently to support healthy, happy children who are ready for school? What do we need to change in existing early childhood policy and programming to create the positive change in outcomes we need? Toxic stress and epigenetics presents an exciting new sphere in which to study the impacts of early adversity on domains of school readiness. However, recent advances in science do not make unique arguments for the relevance of early care and education and other early childhood interventions; they merely add a quantifiable dimension of inputs and outputs that reinforces current ideas presented by fields of education, health, and psychology.

We know that high-quality early care and education programs are strategic interventions to improve the life prospects of children with parents who have limited education and resources. Through opportunities in comprehensive early care and education settings, such as Early Head Start, Head Start, and home visiting programs, investment of resources in early intervention opportunities to promote family well-being is the best means to target the cumulative effects of a myriad of risk factors that may serve as pathways linking socioeconomic status to health and developmental outcomes (Shonkoff et al., 2009). Prevention is more efficient and produces higher returns on investment than later remediation with respect to human capacity (Carneiro & Heckman, 2003).

We also know that nurturing, supportive relationships make a difference. Economic self-sufficiency and positive self-worth make a difference.
Coping skills and protective factors such as parental resilience and social
connections make a difference. Recent science provides tangible measures
of biological impact at a molecular level that underscore the importance
of interventions focused on nurturing, supportive relationships, economic
self-sufficiency, supportive parenting, coping skills, and other protective
factors. These quantifiable measures often provide more convincing arguments to policy makers when determining resource allocation. After all, it
can be difficult to quantify and touch "nurturance."

Unfortunately, current early childhood and education policy does not
yet reflect recent advances in science – if it did, health, child development,
and education funding, policy, and programs would be better integrated
and mirror the comprehensive needs of the child and family as a unit. Far
greater investment would be made in supporting families before, during,
and after pregnancy. Family engagement as well as provision of health and
social services would be essential, funded aspects of a public school system
that begins at pre-kindergarten. In general, there would be greater investment in maternal and child health programs, mental health services, and
high-quality early care and education as well as comprehensive, paid family
leave policies (National Scientific Council on the Developing Child, 2010).
Paid family leave is one means to provide parental support while allowing
the opportunity for nurturing relationships to be built between new parents and infants that promote positive epigenetic change, cognitive development, social and emotional competence, and resiliency.

We cannot expect to eliminate the stressors of life nor would it be prudent if we could. The real issue emerges when life events and stress surpass
the coping abilities of families. Although the science behind it is complicated, competence and resiliency arise from basic adaptive attributes and
processes. Only through innovative, multidisciplinary approaches targeted
toward families as a unit can we face multidimensional problems and give
children the best foundations to reach their full potential.

REFERENCES

Bagot, R., Zhang, T., Wen, X., Nguyen, T., Nguyen, H., Diorio, J., et al. (2012).
Variations in postnatal maternal care and the epigenetic regulation of metabo-
tropic glutamate receptor 1 expression and hippocampal function in the rat.
Proceedings of the National Academy of Sciences, 109 (Suppl. 2), 17200–17207.
Barnett, W. (1998). Long-term cognitive and academic effects of early childhood
education on children in poverty. *Preventive Medicine, 27,* 204–207.

Barker, D. (2007). The origins of the developmental origins theory (Symposium). *Journal of Internal Medicine, 261*, 412–417.

Bergman, K., Sarkar, P., Glover, V., & O'Connor, T. (2010). Maternal prenatal cortisol and infant cognitive development: Moderation by infant-mother attachment. *Biological Psychiatry, 67*(11), 1026–32.

Bird, A. (1986). CpG-rich islands and the function of DNA methylation. *Nature, 321*(6067), 209–213.

Bourgeois, J., Goldman-Rakic, P., & Rakic, P. (1999). Formation, elimination, and stabilization of synapses in the primate cerebral cortex. In M. Gazzaniga (Ed.), *The New Cognitive Neurosciences* (pp. 45-53). Cambridge, MA: MIT Press.

Boyce, W., & Ellis, B. (2005). Biological sensitivity to context: I. An evolutionary-developmental theory of the origins and functions of stress reactivity. *Development and Psychopathology, 17*, 271–301.

Bradley, R., & Corwyn, R. (2002). Socioeconomic status and child development. *Annual Review of Psychology, 53*, 371–399.

Brooks-Gunn, J., & Duncan, G. (1997). The effects of poverty on children. *The Future of Children: Children and Poverty, 7*, 55–77.

Buss, C., Davis, E., Shahbaba, B., Pruessner, J., Head, K., & Sandman, C. (2012). Maternal cortisol over the course of pregnancy and subsequent child amygdala and hippocampus volumes and affective problems. *Proceedings of the National Academy of Sciences.*

Carneiro, P., & Heckman, J. (2003). *Human capital policy.* In J. Heckman, A. Krueger, & B. Friedman (Eds.). *Inequality in America: What role for human capital policies?* Cambridge, MA: MIT Press.

Center on the Developing Child at Harvard University. (2007). A science-based framework for early childhood policy: Using evidence to improve outcomes in learning, behavior, and health for vulnerable children. Retrieved from www.developingchild.harvard.edu.

Champagne, F. A. (2008). Epigenetic mechanisms and the transgenerational effects of maternal care. *Frontiers in Neuroendocrinology, 29*(3), 386–397.

(2010). Epigenetic influence of social experiences across the lifespan. *Developmental psychobiology, 52*(4), 299–311.

Clow, A., Hucklebridge, F., Stalder, T., Evans, P., & Thorn, L. (2009). The cortisol awakening response: More than a measure of HPA axis function. *Neuroscience Biobehavior Review, 35*(1), 97–103.

Cohen, S., Janicki-Deverts, D., & Miller, G. (2007). Psychological stress and disease. *Journal of the American Medical Association, 298*(14), 1685–1687.

Collins, N., Dunkel-Schetter, C., Lobel, M., & Scrimshaw, S. (1993). Social support in pregnancy: Psychosocial correlates of birth outcomes and postpartum depression. *Journal on Perspectives of Social Psychology, 65*, 1243–1258.

Chen, E., Miller, G., Kobor, M., & Cole, S. (2011). Maternal warmth buffers the effects of low early-life socioeconomic status on pro-inflammatory signaling in adulthood. *Molecular Psychiatry, 16*(7), 729–737.

Davies, M., Volta, M., Pidsley, R., Lunnon, K., Dixit, A., Lovestone, S., et al. (2012). Functional annotation of the human brain methylome identifies tissue-specific epigenetic variation across brain and blood. *Genome Biology, 13*(6), R43.

Deaton, A., & Bird, A. (2011). CpG islands and the regulation of transcription. *Genetic Development, 25*(10), 1010–22.

Essex, M. J., Boyce, W., Hertzman, C., Lam, L., Armstrong, J., Neumann, S., et al. (2011). Epigenetic vestiges of early developmental adversity: Childhood stress exposure and DNA methylation in adolescence. *Child Development, 84*(1), 58–75.

Evans, G., & Kim, P. (2010). Multiple risk exposure as a potential explanatory mechanism for the socioeconomic-health gradient. *Annuls of the New York Academy of Sciences, 1186,* 174–189.

Gluckman, P., & Hanson, M. (2005). *The fetal matrix: Evolution, development, and disease.* Cambridge: Cambridge University Press.

Grunstein, M. (1997). Histone acetylation in chromatin structure and transcription. *Nature, 389*(6649), 349–352.

Hackman, D., Farah, M., & Meaney, M. (2010). Socioeconomic status and the brain: Mechanistic insights from human and animal research. *Nature Reviews, 11,* 651–659.

Hahn, M., Qiu, R., Wu, X., Li, A., Zhang, H., Wang, J., et al. (2013). Dynamics of 5-hydroxymethylcytosine and chromatin marks in mammalian neurogenesis. *Cell Reports, 3*(2), 291–300.

Hammock, E., & Levitt, P. (2006). The discipline of neurobiological development: The emerging interface that builds processes and skills. *Human Development, 49,* 294–309.

Harris, A., & Seckl, J. (2011). Glucocorticoids, prenatal stress and the programming of disease. *Hormones and Behavior, 59,* 279–289.

Hedges, D., & Woon, F. (2011). Early-life stress and cognitive outcome. *Psychopharmacology, 214,* 121–130.

Heijmans, B., Tobi, E., Stein, A., Putter, H., Blauw, G., Susser, E., et al. (2008). Persistent epigenetic differences associated with prenatal exposure to famine in humans. *Proceedings of the National Academy of Sciences, 105*(44), 17046–17049.

Hellstrom, I., Dhir, S., Diorio, J., & Meaney, M.(2012). Maternal licking regulates hippocampal glucocorticoid receptor transcription through a thyroid hormone-serotonin-NGFI-A signalling cascade. *Philos Trans R Soc Lond B Biol Sci, 367*(1601), 2495–2510.

Heyn, H., Li, N., Ferreira, H., Moran, S., Pisano, D., Gomez, A., et al. (2012). Distinct DNA methylomes of newborns and centenarians. *Proceedings of the National Academy of Sciences, 109*(26), 10522–10527.

Hosseini-Sharifabad, M., & Hadinedoushan, H. (2007). Prenatal stress induces learning deficits and is associated with a decrease in granules and CA3 cell dendritic tree size in rat hippocampus. *Anatomical Science International, 82,* 211–217.

Huang, H. (2010). Structure of the fetal brain: What we are learning from diffusion tensor imaging. *The Neuroscientist, 16*(6), 634–49.

Ingles, S. (2004). *Statistical methods in genetic epidemiology.* New York: Oxford Press.

Jenuwein, T., & Allis, C. (2001). Translating the histone code. *Science 293*(5532), 1074–1080.

Kaminsky, Z., Tang, T., Wang, S., Ptak, C., Oh, G., Wong, A., et al. (2009). DNA methylation profiles in monozygotic and dizygotic twins. *Nature Genetics, 41*(2), 240–245.

Kapoor, A., Dunn, E., Kostaki, A., Andrews, M., & Matthews, S. (2006). Fetal programming of hypothalamo-pituitary-adrenal function: Prenatal stress and glucocortocoids. *Journal of Physiology, 572*, 31–44.

Klengel, T., Mehta, D., Anacker, C., Rex-Haffner, M., Pruessner, J., Pariante, C., et al. (2013). Allele-specific FKBP5 DNA demethylation mediates gene-childhood trauma interactions. *Nature Neuroscience, 16*(1), 33–41.

Labonte, B., Suderman, M., Maussion, G., Navaro, L., Yerko, V., Mahar, I., et al. (2012). Genome-wide epigenetic regulation by early-life trauma. *Archives of General Psychiatry, 69*(7), 722–731.

Liaw, F., & Brooks-Gunn, J. (1994). Cumulative familial risks and low-birth weight children's cognitive and behavioral development. *Journal of Clinical Child Psychology, 23*, 360–372.

Lister, R., Mukamel, E., Nery, J., Urich, M., Puddifoot, C., Johnson, N., et al. (2013). Global epigenomic reconfiguration during mammalian brain development. *Science, 341*(6146).

Liu, D., Diorio, J., Day, J., Francis, D., & Meaney, M. (2000). Maternal care, hippocampal synaptogenesis and cognitive development in rats. *Nature Neuroscience, 3*(8), 799–806.

Lupien, S. King, S., Meaney, M., & McEwen, B. (2000). Child's stress hormone levels correlate with mother's socioeconomic status and depressive state. *Biological Psychiatry, 48*, 976–980.

Maccari, S., Piazza, P., Kabbaj, M., Barbazanges, A., Simon, H., & Le Moal, M. (1995). Adoption reverses the long-term impairment in glucocorticoid feedback induced by prenatal stress. *Journal of Neuroscience, 15*, 110–116.

Martino, D., Tulic, M., Gordon, L., Hodder, M., Richman, T., Metcalfe, J., et al. (2011). Evidence for age-related and individual-specific changes in DNA methylation profile of mononuclear cells during early immune development in humans. *Epigenetics, 6*(9), 1085–1094.

Maze, I., Noh, K., & Allis, C. (2013). Histone regulation in the CNS: Basic principles of epigenetic plasticity. *Neuropsychopharmacology, 38*(1), 3–22.

McEwen, B. (2007). Physiology and neurobiology of stress and adaptation: Central role of the brain. *Physiological Reviews, 87*, 873–904.

McEwen, B., & Seeman, T. (1999). Protective and damaging effects of mediators of stress. *Annals of the New York Academy of Sciences, 896*, 30–47.

McGowan, P., Sasaki, A., D'Alessio, A., Dymov, S., Labonte, B., Szyf, M., et al. (2009). Epigenetic regulation of the glucocorticoid receptor in human brain associates with childhood abuse. *Nature Neuroscience, 12*(3), 342–8.

Meaney, M., Szyf, M., & Seckl, J. (2007). Epigenetic mechanisms of perinatal programming of hypothalamic-pituitary-adrenal function and health. *Trends in Molecular Medicine, 13*, 269–277.

Meaney, M. (2010). Epigenetics and the biological definition of gene x environment interactions. *Child Development, 81*(1), 41–79.

Mehta, D., Klengel, T., Conneely, K., Smith, A., Altmann, A., Pace, T., et al. (2013). Childhood maltreatment is associated with distinct genomic and epigenetic profiles in posttraumatic stress disorder. *Proceedings of the National Academy of Sciences.*

Mueller, B. & Bale, T. (2008). Sex-specific programming of offspring emotionality after stress early in pregnancy. *Journal of Neuroscience, 28*(36), 9055–65.

Murgatroyd, C., Patchev, A., Wu, Y., Micale, V., Bockmuhl, Y., Fischer, D., et al. (2009). Dynamic DNA methylation programs persistent adverse effects of early-life stress. *Nature Neuroscience, 12*(12), 1559–1566.

Nanni, V., Uher, R., & Danese, A. (2012). Childhood maltreatment predicts unfavorable course of illness and treatment outcome in depression: A meta-analysis. *American Journal of Psychiatry, 169*(2), 141–151.

National Scientific Council on the Developing Child. (2010). Early experiences can alter gene expression and affect long-term development. Working Paper No. 10. Retrieved from www.developingchild.net.

Naumova, O., Lee, M., Koposov, R., Szyf, M., Dozier, M., & Grigorenko, E. (2012). Differential patterns of whole-genome DNA methylation in institutionalized children and children raised by their biological parents. *Developmental Psychopathology, 24*(1), 143–155.

O'Donnell, K., Glover, V., Barker, E., & O'Connor, T. (2014). The persisting effect of maternal mood in pregnancy on childhood psychopathology. *Developmental Psychopathology, 26*(2), 393–403.

Oitzl, M., Champagne, D., van der Veen, R., & de Kloet, E. (2010). Brain development under stress: Hypotheses of glucocorticoid actions revisited. *Neuroscience and Biobehavioral Reviews, 34*, 853–866.

Plomin, R., Defries, J., McClearn, G, & McGuffin, P. (2008). *Behavioral genetics*, 5th ed. New York: Worth Publishers.

Power, M., & Tardiff, S. (2005). Maternal nutrition and metabolic control of pregnancy. In M. Power & J. Schulkin (Eds.), *Birth, distress and disease* (pp. 88–113). Cambridge, UK: Cambridge University Press.

Provençal, N., Suderman, M., Guillemin, C., Massart, R., Ruggiero, A., Wang, D., et al. (2012). The signature of maternal rearing in the methylome in rhesus macaque prefrontal cortex and T cells. *Journal of Neuroscience, 32*(44), 15626–15642.

Radtke, K., Ruf, M., Gunter, H., Dohrmann, K., Schauer, M., Meyer, A., et al. (2011). Transgenerational impact of intimate partner violence on methylation in the promoter of the glucocorticoid receptor. *Translational Psychiatry, 1*, e21.

Razin, A., & Riggs, A. (1980). DNA methylation and gene function. *Science, 210*(4470), 604–610.

Roth, T., Lubin, F., Funk, A., & Sweatt, J. (2009). Lasting epigenetic influence of early-life adversity on the BDNF gene. *Biol Psychiatry, 65*(9), 760–769.

Rozanski, A., Blumenthal, J., & Kaplan, J. (1999). Impact of psychological factors on the pathogenesis of cardiovascular disease and implications for therapy. *Circulation, 99*(16), 2192–2217.

Sameroff, A., Seifer, R., Barocas, R., Zax, M., & Greenspan, S. (1987). Intelligence quotient scores of 4-year-old children: Social-environmental risk factors. *Pediatrics, 79*, 343–350.

Sarkar, S., Craig, M., Dell'Acqua, F., O'Connor, T., Deeley, Q., Murphy, D., et al. (2014). Prenatal stress and limbic-prefrontal white matter microstructure in children aged 6–9 years: A preliminary diffusion tensor imaging study. *World Journal of Biological Psychiatry, 15*(4), 346–352.

Schneider, M., Moore, C., Kraemer, G., Roberts, A., & DeJesus, O. (2002). The impact of prenatal stress, fetal alcohol exposure, or both on development: Perspectives from a primate model. *Psychoneuroendocrinology, 27,* 285–298.

Seckl, J. (2008). Glucocorticoids, developmental "programming," and the risk of affective dysfunction. *Progress in Brain Research, 167,* 17–34.

Seckl, J., Cleasby, M., & Nyirenda, M. (2000). Glucocorticoids, 11β-hydroxysteroid dehydrogenase, and fetal programming. *Kidney International, 57,* 1412–1417.

Seckl, J., Drake, A., & Holmes, M. (2005). Prenatal glucocorticoids and the programming of adult disease. In M. Power & J. Schulkin (Eds.), *Birth, distress and disease* (pp. 88–113). Cambridge, UK: Cambridge University Press.

Seckl, J., & Holmes, M. (2007). Mechanisms of disease: Glucocorticoids, their placental metabolism and fetal "programming" of adult pathophysiology. *Nature Clinical Practice Endocrinology & Metabolism, 3,* 479–488.

Shonkoff, J., Boyce, W., & McEwen, B. (2009). Neuroscience, molecular biology, and the childhood roots of health disparities: Building a new framework for health promotion and disease prevention. *Journal of the American Medical Association, 301,* 2252–2259.

Shonkoff, J., Garner, A., Siegel, B., Dobbins, M., Earls, M., McGuinn, L., et al. (2012). The lifelong effects of early childhood adversity and toxic stress. *Pediatrics, 129*(1), e232–246.

Szyf, M. (2009a). Early life, the epigenome and human health. *Acta Paediatrica, 98*(7), 1082–1084.

(2009b). The early life environment and the epigenome. *Biochimica Biophysica Acta, 1790*(9), 878–885.

Szyf, M., & Bick, J. (2013). DNA methylation: A mechanism for embedding early life experiences in the genome. *Child Development, 84*(1), 49–57.

Taylor, S., & Repetti, R. (1997). Health psychology: What is an unhealthy environment and how does it get under the skin? *Annual Reviews of Psychology, 48,* 411–447.

Teh, A., Pan, H., Chen, L., Ong, M., Dogra, S., Wong, J., et al. (2014). The effect of genotype and in utero environment on inter-individual variation in neonate DNA methylomes. *Genome Research.*

Turner, B. (2001). *Chromatin and gene regulation: Mechanisms in epigenetics.* Cambridge, MA: Blackwell.

Tyrka, A., Price, L., Marsit, C., Walters, O., & Carpenter, L. (2012). Childhood adversity and epigenetic modulation of the leukocyte glucocorticoid receptor: Preliminary findings in healthy adults. *PLoS One, 7*(1), e30148.

Uno, H., Tarara, R., Else, G., Suleman, M., & Sapolsky, R. (1989). Hippocampal damage associated with prenatal glucorticoid exposure. *Journal of Neuroscience, 9,* 1705–1711.

US Census Bureau. (2013). *Current population survey, 2012 and 2013. Annual social and economic supplements.* Table 3. People in poverty by selected characteristics: 2011 and 2012.

Wang, D., Liu, X., Zhou, Y., Xie, H., Hong, X., Tsai, H., et al. (2012). Individual variation and longitudinal pattern of genome-wide DNA methylation from birth to the first two years of life. *Epigenetics, 7*(6), 594–605.

Weaver, I., Cervoni, N., Champagne, F., D'Alessio, A., Sharma, S., Seckl, J., et al. (2004). Epigenetic programming by maternal behavior. *Nature Neuroscience, 7*(8), 847–854.

Weaver, I., D'Alessio, A., Brown, S. E., Hellstrom, I., Dymov, S., Sharma, S., et al. (2007). The transcription factor nerve growth factor-inducible protein A mediates epigenetic programming: Altering epigenetic marks by immediate-early genes. *Journal of Neuroscience, 27*(7), 1756–1768.

Welberg, L., & Seckl, J. (2001). Prenatal stress, glucocorticoids, and the programming of the brain. *Journal of Neuroendocrinology, 13*, 113–128.

Yang, B., Zhang, H., Ge, W., Weder, N., Douglas-Palumberi, H., Perepletchikova, F., et al. (2013). Child abuse and epigenetic mechanisms of disease risk. *American Journal of Preventive Medicine, 44*(2), 101–107.

Yang, J., Han, H., Cao, J., Li, L., & Xu, L. (2006). Prenatal stress modifies hippocampal synaptic plasticity and spatial learning in young rat offspring. *Hippocampus, 16*, 431–436.

Yehuda, R., Daskalakis, N. P., Desarnaud, F., Makotkine, I., Lehrner, A. L., Koch, E., et al. (2013). Epigenetic biomarkers as predictors and correlates of symptom improvement following psychotherapy in combat veterans with PTSD. *Front Psychiatry, 4*, 118.

Zhang, T., Hellstrom, I., Bagot, R., Wen, X., Diorio, J., & Meaney, M. (2010). Maternal care and DNA methylation of a glutamic acid decarboxylase 1 promoter in rat hippocampus. *Journal of Neuroscience, 30*(39), 13130–13137.

5

Intelligence: The Ongoing Quest
for Its Etiology

ELENA L. GRIGORENKO AND SAMUEL D. MANDELMAN

"In China, DNA Tests on Kids ID Genetic Gifts, Careers."[1] This CNN.com/ Asia entry could certainly catch readers' attention! And it did, for at least two reasons. First, it concerned competition and high achievement. For the Chinese authorities who supported this initiative, it was about identifying "DNA prodigies" as early as possible and coming up with a specialized developmental plan for them. But the initiative was somewhat disconcerting; the use of genetics for societal stratification purposes has a long and controversial history, and seeing its resurgence, in yet another shape and form, triggered all kinds of ethical concerns. Second, it raised some important questions concerning the scientific validity of such practices, specifically: How much scientific evidence supported this initiative? What kinds of data might be generated by this initiative, and with what levels of certainty could they be interpreted? This initiative did not last long, for better or for worse. It was shut down by the Chinese government, along with other types of DNA testing.[2] The main reason for such an action was the concern that, although this and other initiatives might reflect the frontiers of research, they are not ready for the utilization in individual-based decision making, especially in uncontrolled market-driven settings.

---

[1] http://edition.cnn.com/2009/WORLD/asiapcf/08/03/china.dna.children.ability/#cnnSTCText

[2] www.forbes.com/sites/shuchingjeanchen/2014/03/03/china-cracks-down-on-dna-testing-2/#2b0fb6257407.

Preparation of this chapter was supported by grant from the Spencer Foundation and grant № 14.Z50.31.0027 from the Government of the Russian Federation. Grantees undertaking such projects are encouraged to express their professional judgment freely. Therefore, this article does not necessarily reflect the position or policies of the National Institutes of Health, and no official endorsement should be inferred. The content of this chapter partially overlaps with the content in Mandelman & Grigorenko (2010); updated with permission. We are thankful to Ms. Mei Tan for her editorial assistance.

This chapter seeks to scientifically establish the connection between "genetics" and "intelligence," the terms so easily linked by CNN, while, in reality, the etiological bases of intellectual abilities and disabilities have formed a central and not uncontroversial query within the sciences of psychology, philosophy, and education since the inception of these fields. The answers to this query have been highly variable, changing over time and cultures, and appear to be bracketed by two extreme positions.

A major proponent of the first polar position, Sir Francis Galton, advocated the genetic underpinning of human abilities (Galton, 1869). A major proponent of the second position, Dr. John Watson, argued for the overarching powers of environmental influences (Watson, 1924). The positions gathered between these two extremes are all the colors and shades of Newton's sevenfold rainbow, with the most balanced points of view acknowledging that both forces matter. Contemplating the etiology of human abilities and disabilities, one might first question its importance and, second, wonder why its pursuit has been so protracted. In this chapter, we attempt to broadly outline the current understanding of the etiology of intelligence and intelligence-related processes. Thus, we summarize the state of the field's understanding of the causes of intellectual abilities and disabilities. Finally, we provide a point of view on the Chinese initiative as presented in the CNN electronic publication, the reference that opened this chapter, concurring with the argument that any individualization of such DNA testing is premature.

VOCABULARY PREP: TERMS AND CONCEPTS

In this section we will describe the major concepts that have been and are still used to explore the connection between the genes and intelligence. We provide this brief overview to ensure that the content discussion presented in the section that follows is as clear as possible. Heritability is a statistic that describes the proportion of a given trait's variation (i.e., phenotypic[3] variation) within a population that is attributable to variation in the genes (see Tan, this volume). Higher heritability indicates higher levels of covariation between genetic and phenotypic variation; lower heritability indicates higher levels of covariation between environmental and phenotypic variation. As discussed in the following section of the chapter, heritability studies have, so far, dominated the field of studies connecting genes and intelligence. Yet, the last decade has generated a number of large-scale

---

[3] Phenotype: An observable trait or characteristic.

molecular-genetic studies of intelligence and related phenotypes. Generally speaking, heritability estimates of the majority of intellectual abilities fall in the range of 40–60 percent. Heritability estimates for intellectual abilities and disabilities have been estimated through numerous twin, adoption and family studies (see Tan, this volume).

Heritability estimates, however, represent only one type of statistic that may be used to estimate the degree of genetic endowment associated with a complex trait. Researchers have developed an impressive variety of relevant methodologies, designs, and statistics. One such statistic, for example, is the relative risk statistic[4] (Risch, 1990). This indicator can be estimated for different pairs of relatives (e.g., sibling pairs or parent–offspring pairs) and has been particularly informative in studies of clinical phenotypes.

In addition, there are methods of investigating patterns of familial transmission of a particular trait from generation to generation. These types of investigations are referred to as segregation analyses. Once again, there are varieties of statistics and approaches associated with such analyses. In some approaches these types of statistics might include not only estimates of main (genetic and environmental) and interactive (e.g., gene–gene) effects, but may also gauge the magnitudes of the effect sizes or the magnitude of these various effects, as well as the number of genes involved and the percent variance each gene might contribute to the overall genetic variance of the trait (e.g., Naples, Chang, Katz, & Grigorenko, 2009). Various investigations into the familial transmission of characteristics of intellectual functioning suggest that multiple genes are involved in the substrate of this transmission, and that the patterns of this transmission are rather complex (i.e., far from following simple Mendelian laws).

Heritability estimates, genetic risk ratios, and parameters of segregation analyses are all methodologies that capitalize on the availability of behavioral data only (i.e., indicators of a trait of interest collected from different types of relatives and the correlations between these indicators). Lately, however, much more interest has been given to combining these behavior indicators with measured genotypic information, i.e., genotypes as they are captured by structural variation in the DNA (see Kornilov, this volume). If information on genotypes (or genotyping information) is available, then this information is, broadly speaking, correlated with behavioral information. Two major data designs and analytic strategies are used for these purposes: linkage analyses and association analyses. Both linkage and

---

[4] Relative risk statistic: A statistic that is used to calculate the amount of risk in one population in relation to the risk in a different population.

association genetic studies have been carried out in the field; these studies are relatively novel, however, slowly but surely they are decreasing the accent on heritability studies of intellectual functioning.

INTELLIGENCE AND THE GENOME

In this main portion of the chapter, we discuss the evidence pertaining to observations that the genome is a major source of the variations in individuals' intellectual abilities and disabilities.

There are almost 300 monogenetic disorders that include symptoms of intellectual disability (Flint, 1999; Inlow & Restifo, 2004). These disorders are rather diverse, but they have five common features: (1) they are caused by disruptions of single genes (thus, the reference to *mono*genic disorders); (2) their presentation is typically severe, with a limited range of phenotypic variability and mental functioning that constitutes moderate to profound difficulty; (3) when considered individually they are rare (most at .01 percent), but together they account for a considerable portion of developmental disabilities; (4) they are highly pleiotropic, meaning that the disrupted gene appears to impact many brain-related pathways, and these effected pathways in turn cause large deviations from normative development; and (5) they require a high degree of individualization within the inclusive educational environment (see Hodapp & Fisher, this volume, for a discussion of this topic).

The important question here with regard to the literature on the genetic bases of severe intellectual disability is whether there are any findings or insights that can be brought to bear on the etiological bases of individual differences in intelligence as they are distributed in the general population. The answer to this question is still pending, but some recent meta-analyses (Reichenberg et al., 2016) arguably support the hypothesis proposed by Lionel Penrose almost a century ago (Penrose, 1938). According to these analyses, which involved more than 1,000,000 sibling pairs and 9,000 twin pairs, the etiology of severe (i.e., lowest 0.5 percent of IQ distribution) intellectual disability might be different from that of mild disability (i.e., lowest 3 percent of IQ distribution). Yet, factors influencing the latter are deemed to be similar to those shaping IQ in the normal range. Thus, the general conclusion of the field currently suggests that genes, in which mutations causing intellectual disability have been identified, might not be directly related to individual differences in intelligence; however, they might be involved in pathways (i.e., gene networks) that involve genes related to variation in intelligence.

There is a substantial body of literature dedicated to studies of the genetic bases of intelligence in the general population, that is, literature that draws on samples of individuals that are representative of their cultures and societies. As there is no single definition of intelligence, there is no single assessment that is used for its measurement (e.g., Cianciolo & Sternberg, 2004; Sternberg, 1996). In fact, there are probably hundreds of different assessments of intelligence, its different types, and its facets, all sharing some common aspects and all characterized by some specific features.

The fact that diverse cognitive abilities correlate among each other at a variety of values, ranging from low to high depending on the particulars of those abilities, has led to the formulation of the concept of the g-factor, Spearman's g (Spearman, 1904). Whereas nobody argues the veracity of these correlations, although estimated at the moderate value of ~.30 (Carroll, 1993) or slightly higher (Jensen, 1998), a variety of theoretical approaches have been employed in attempts to explain them. These explanations range from statements that these correlations are, indeed, driven by the g-factor, which is genetic in its nature and manifestation (Rijsdijk, Vernon, & Boomsma, 2002), to the view that the interdependency between cognitive abilities can be explained by the developmental, temporal, and functional (but not etiological!) dependencies of these abilities on each other (van der Maas et al., 2006).

Also of interest is that, regardless of the particular instrument or instruments used for the purposes of assessing intelligence or the IQ and the language in which such assessment is carried out, the findings on heritability, or the statistical estimate of the contributions of genetic variability to individual variability in intelligence, are quite consistent. Specifically, when summarized in reviews or meta-analyzed, the data suggest that IQ's heritability is ~.50 (Deary, Spinath, & Bates, 2006; Devlin, Daniels, & Roeder, 1997; Plomin & Spinath, 2004).

In fact, there have been so many studies on the heritability of intelligence that the flow of "plain vanilla" studies on the heritability of IQ, similar to those included in the meta-analyses and reviews mentioned above, has noticeably decreased. What is at the center of genetic and genomic studies of intelligence now are (1) studies that differentiate heritability patterns by some other third variable (e.g., age or environment); (2) studies that investigate the heritability of various intelligence-related componential cognitive processes that are correlated with intelligence but cannot substitute it; and (3) studies that attempt to "translate" the heritability of intelligence into the identification of specific genes that contribute to or form the genetic foundation of intelligence as it is captured in the concept of heritability. These topics will be briefly considered next.

Differentiating Heritability Estimates

It has been convincingly demonstrated by many studies that levels of heritability are not static – they differ throughout the lifespan and in different environmental conditions. Whereas it would be logical to assume that heritability would decrease with age due to accumulated life experience, thus minimizing the importance of the role of genetics, something rather different has been observed. In fact, heritability in infancy is estimated to be as low as 20 percent, while in adulthood it can be as high as 80 percent (Panizzon et al., 2014), although there is some, although contradictory evidence, that it might decrease again in the later years of life (Lee, Henry, Trollor, & Sachdev, 2010; McGue & Christensen, 2013; Reynolds et al., 2005). Based on results from twin (Bishop et al., 2003; Bouchard & McGue, 2003; Cardon & Fulker, 1993; McGue, Bouchard, Iacono, & Lykken, 1993; Patrick, 2000; Price et al., 2000; Reznick, Corley, & Robinson, 1997) and adoption (Petrill et al., 2004) studies, it appears that from birth onward, genetic variance becomes increasingly important in explaining individual differences in verbal and non-verbal intellectual abilities. To illustrate, a meta-analytic study that involved 11,000 pairs of twins (Haworth et al., 2010) registered a steady and significant increase in the estimates of the heritability of intelligence from 41 percent in childhood (age 9) to 55 percent in adolescence (age 12) and to 66 percent in young adulthood (age 17). Moreover, genetic influences appear not only to increase in their magnitude, but also to form the genetic foundation for the stability of intelligence across different stages of the lifespan (Bartels, Rietveld, Van Baal, & Boomsma, 2002; Polderman et al., 2006; Rietveld, Dolan, van Baal, & Boomsma, 2003). It seems that genetic variance in intelligence is predominated by innovative genetic influences in early childhood, which diminish and are replaced by amplified influences in middle childhood (Briley & Tucker-Drob, 2013), stabilizes in post-adolescence, and remains relatively high and constant until later in life (Brant et al., 2009; van der Sluis, Willemsen, de Geus, Boomsma, & Posthuma, 2008). These dynamics of heritability estimates across the life span have been of substantial interest to the field; their etiology is unknown, but they are, indeed, intriguing.

Similarly, there are studies indicating that heritability estimates differ substantially when they are sampled from different environments, emphasizing the importance of considering gene–environment interactions. For example, researchers (van Leeuwen, van den Berg, & Boomsma, 2008) carried out a study of families of twins, considering not only the heritability of IQ, but also indicators of assortative mating[5] occurring between parents. The results still indicated that the main source of variance in IQ was genetic (estimated at 67 percent). Yet, gene–environment interaction appeared to account for 9 percent of additional variance. These results suggested that environmental effects are larger for children with a genetic predisposition for low IQ, thus indicating that environmental influences do not affect all siblings uniformly.

The presence of gene–environment effects was also indicated by studies of differential heritabilities in families of different SES levels (Harden, Turkheimer, & Loehlin, 2007). Shared environmental influences were reported to be more powerful for adolescents from families of low SES, while genetic influences were reported to be more powerful for adolescents in high SES environments. Similarly, environmental influences were reported to be greater on the reading skills of children whose parents had less education, compared with children whose parents had higher levels of education (Friend, DeFries, & Olson, 2008). Notably, these effects appear to be more relevant in the United States. A nation-stratified meta-analysis has indicated that SES-based interactive effects are zero or reversed in societies where access to high-quality education and health care is more uniform (Tucker-Drob & Bates, 2016).

Thus, the field has moved from obtaining heritability estimates for intelligence and related skills per se to looking for "other" factors that differentiate these estimates and provide opportunities for formulating relevant mechanistic hypotheses.

Dissecting Intelligence into Its Componential Processes

Another "movement" in the research on understanding the etiology of individual differences in intelligence and its related processes is associated with the direction from molar to molecular, that is, from intelligence as a holistic construct to its components. A central question here investigates the presence and magnitude of genetic factors that influence *all* intelligence-related processes as opposed to genetic factors that influence only *some* of such processes.

---

[5] Assortative mating: Non-random mating in which people choose mates who are similar to themselves (in this case of similar intelligence).

Electrophysiological Measures

Since early on in the history of the field of intelligence, researchers have looked for ways to register and measure the brain's activity while it is engaged in intellectual tasks. One of the major lines of inquiry in this domain is related to the utilization of electrophysiological indicators obtained by scale recording.

Electroencephalography (EEG) is the measurement of the electrical activity produced by the brain at rest, when the brain, arguably, is not engaged in responding to any particular stimulus. The EEG is typically described through components of its rhythmic activity, divided into bands by frequency. EEG patterns also differ in their preferential registration location and in the activities that are associated with these locations. In general, states of low arousal are associated with a relatively high amount of slow activity; states of high arousal are indicated by faster activity. For example, the α-wave's frequency range is 8–12 Hz; it is typically registered in a condition of relaxation, with eyes closed. The β-wave frequency range is 12–30 Hz, and it is associated with active engagement in cognitive processing. The γ-wave frequency range is 30–100 Hz, and it is registered when the brain is performing certain cognitive and motor operations.

There is a history of research relating various EEG waves to various cognitive components, with a great amount of discussion regarding whether these measures do or do not relate to g (Deary, 2000; Ertl, 1971). There is also a substantial body of research investigating heritability estimates for different EEG peaks. This research has repeatedly reported moderate to high heritability estimates for different EEG peak frequencies (e.g., Posthuma, Neale, Boomsma, & de Geus, 2001), as well as for EEG coherence (i.e., the squared cross-correlation between two EEG signals at different scalp locations which is regarded as an index of brain interconnectivity; van Beijsterveldt, Molenaar, de Geus, & Boomsma, 1998a). Yet, there is a substantial amount of variability between these estimates, depending on the age of the subject and the part of the brain being registered.

For example, in a longitudinal investigation of stability and change in the genetic and environmental influences on EEG coherence in children aged five to seven years, researchers (van Baal, Boomsma, & de Geus, 2001) reported moderate heritability estimates for EEG coherence across all ages (the average value was at .58), but registered an increase in heritability for occipito-cortical connections of the right hemisphere, and a decrease in heritability in the prefronto-cortical connections in the left hemisphere. Modeling the continuity of genetic variance, they reported the presence of both stable (i.e., age-general) and novel (age-specific) genetic influences.

The heritability of α-peaks was also reported to be moderate to high (e.g., .66; Posthuma et al., 2001). It is notable that when this genetic variance was co-modeled with the genetic variance in IQ (as represented through verbal comprehension, working memory, perceptual organization, and processing speed, derived from the WAIS-IIIR), there was no evidence of shared genetic variance between the α-peak frequency and any of the four WAIS dimensions (Posthuma et al., 2001).

Methodologies that are based on *event-related potentials* (ERPs) record stereotyped electrophysiological responses to external (e.g., a stimulus) or internal (e.g., thought) events. ERPs reflect fluctuations in the pattern and/or amplitude of an EEG. Needless to say, these fluctuations are very small and, correspondingly, can be extrapolated from the background activity only (or mostly) within a framework of repeated measures, that is, the recordings of many trials presenting the same stimulus or stimuli. When dissected into its components, ERPs are typically classified into two broad categories – exogenous (auditory, visual, somatosensory EPs, N100, P200) and endogenous (P300, N400, P600/SPS) structural units (Fabiani, Gratton, & Federmeier, 2007). Early exogenous components are typically used to study information processing by primary sensory cortices (e.g., selective attention, early object recognition), whereas later endogenous components are utilized to investigate higher-order cognitive processes (e.g., working memory, executive control) (for a review, see de Geus, Wright, Martin, & Boomsma, 2001; Winterer & Goldman, 2003).

There have been numerous studies using different ERP units, particularly P300, which have been carried out in studies employing genetically informative designs. For example, it has been observed that both the amplitude and the latency of P300 are moderately heritable (e.g., Katsanis, Iacono, McGue, & Carlson, 1997; van Baal, van Beijsterveldt, Molenaar, Boomsma, & de Geus, 2001), although there are fluctuations in these estimates that have been attributed to task conditions (Winterer & Goldman, 2003), gender (van Beijsterveldt, Molenaar, de Geus, & Boomsma, 1998b), and age (van Baal, van Beijsterveldt, et al., 2001). Yet, the heritability of the amplitude and latency of P200 was reported to be relatively low (van Beijsterveldt & Boomsma, 1994). There is also some evidence of shared genetic variance among slow wave ERP units and working memory, but the amount of this variance appears to fluctuate regionally (e.g., ~35–37 percent at the prefrontal site and ~51–52 percent at the parietal site), and, most curiously, the sites showed no evidence of common genetic variance (Hansell et al., 2001).

Speed of Information Processing
Studies of various indicators of information processing speed have been prominent in the field of intelligence due to the observation that these indicators reliably (although not necessarily substantially) correlate with various aspects of intelligence, especially, with the g-factor (Deary, 2000). Correspondingly, many researchers have attempted to estimate heritability coefficients for these indicators. Here we will briefly summarize this work, but, prior to this summary, it is important to make the following comments.

First, the magnitudes of correlations between various types of indicators of speed of information processing obtained from different mental chronometric tasks differ. For example, correlations between g and reaction time were reported to be ~.3, whereas correlations between g and perceptual discrimination speed were reported to be ~.5 (Winterer & Goldman, 2003). Second, it is thought that there might be age- and gender-related differentiation in correlations between mental chronometric tasks and g (Beaujean, 2005). Both of these bits of information/hypotheses are important for interpreting findings regarding the heritability estimated for various indictors of speed of information processing.

In a meta-analytic study (Beaujean, 2005), a variety of indicators of performance differences in mental chronometric tasks were obtained within the context of genetically informative designs (i.e., designs that allow estimates of heritability). The results demonstrated that heritability estimates vary broadly (from ~30 percent to ~ 50 percent) and that they are somewhat dependent on task difficulty (i.e., increased task complexity is associated with higher heritability estimates). That is, heritability estimates of chronometric tasks are differentiated by their levels of difficulty. They are also differentiated by the age at which they are estimated: as information processing becomes more efficient in children, heritability estimates go up.

Researchers have also estimated the genetic overlap, or shared genetic variance, between various chronometric tasks, and then among these tasks and other intelligence-related indicators. For example, looking at the genetic overlap between IQ and indicators of inspection time and reaction time, researchers (Luciano et al., 2004) completed a series of model-fitting exercises using twin data. Results were interpreted as revealing the insufficiency of a unitary factor model for capturing the relationship between cognitive speed measures and all IQ subtests. Although there was some sharing of genetic variance, independent genetic effects were needed in the model to explain the associations between chronometric tasks and the various subtests of the utilized intelligence assessment. Based on these results, it is not surprising that different speed indicators show different amounts of genetic overlap (i.e., genetic correlations of different magnitude) with different intelligence-related indicators. For example, in one study, the overlapping genetic variance between inspection time and (a) performance IQ was ~30 percent and (b) verbal IQ ~7 percent (Edmonds et al., 2008). In yet another study, the average amount of shared genetic variance between three different choice reaction time tasks and (a) IQ was ~33 percent and (b) a working memory indicator ~18 percent (Luciano et al., 2001). Regardless, it appears that genetic variance in chronometric tasks (which is not highly shared) explains a moderate, although respectable amount of variance in intelligence and intelligence-related processes (Luciano et al., 2005). Yet, substantial specific and separate genetic factors appear to operate differently within different chronometric and intelligence tasks (Singer, MacGregor, Cherkas, & Spector, 2006).

Other Cognitive Processes
There are two large groups of cognitive processes that are often studied in conjunction with indicators of intelligence. These processes are captured by indicators of executive functioning and academic achievement. Executive functioning is an umbrella term for several related cognitive functions like selective and sustained attention, working memory, and inhibition. These processes are also related to intelligence (Friedman et al., 2006), although when they were first introduced as a concept, they were thought to account for the variance in cognitive performance that could not be explained by intelligence. Executive functioning is not a unidimensional construct and the processes (functions) that contribute to it are not homogeneous. Correspondingly, the literature contains differential heritability estimates for different executive functions. There is also evidence that there are different amounts of genetic variance shared between indicators of intelligence, the *g*-factor, and various executive functions. Specifically, it has been reported that genetic variance appears to be substantial and dominant in explaining individual differences in executive functioning in early and middle childhood (Polderman et al., 2007). When multiple executive functions (i.e., inhibiting dominant responses, updating working memory representations, and shifting between task sets) were considered in a twin study simultaneously, it was shown that behavioral correlations between these functions were attributable to the presence of a highly heritable common factor. Yet, each of these functions also appeared to be associated with a unique, substantial, function-specific genetic factor (Friedman et al., 2008). The literature also contains evidence of shared genetic variance between

short-term memory and executive functions; yet, it appeared that each of the investigated functions was also associated with its own source of genetic variance (Ando, Ono, & Wright, 2001). There are different groupings of various lower-level processes into higher entities, such as, e.g., relational processing (which, in turn, underlies reasoning, categorization, planning, etc.). Relational processing was reported (Hansell et al., 2015) to be highly heritable (67 percent), has considerable genetic overlap with IQ (59 percent), and is a major component of the genetic covariation between reasoning and working memory (72 percent).

Indicators of academic achievement are also often considered alongside indicators of intelligence. Indicators of achievement and intelligence share common genetic variance (e.g., Luciano et al., 2003). Yet, once again, the reports on the specifics of this sharing vary widely (Hart, Petrill, Thompson, & Plomin, 2009). For example, when academic achievement in reading and math as well as the g-factor were evaluated through internet tools, heritabilities were estimated at 0.38 for reading, 0.49 for mathematics, and 0.44 for g. Multivariate genetic analysis showed substantial genetic correlations between learning abilities: 0.57 between reading and mathematics, 0.61 between reading and g, and 0.75 between mathematics and g (Davis et al., 2008). Yet the degree of these genetic correlations and the traits' heritability estimates vary depending on a number of factors. For example, depending on whether the same or different teachers assess both members of a twin pair, a decrease in the heritability estimates by ~33–42 percent is observed (Walker, Petrill, Spinath, & Plomin, 2004). Similarly, heritability estimates depend on how broadly or narrowly the trait of interest is conceived and measured; a wider sampling net typically results in more variation among heritability estimates and lower values of shared genetic variance (Kremen et al., 2007).

Of note also are repeated references to the presence of achievement-specific genetic factors. For example, when a set of reading achievement indicators was considered alongside indicators from the WAIS-R (a specific intelligence test, the Wechsler Adult Intelligence Scale – Revised) in adolescent and young-adult twins, the resulting model supported one genetic general factor and three genetic group factors (verbal, performance, and reading). The genetic general factor accounted for 13–20 percent of reading performance, whereas "other" non-general factors accounted for the majority of the genetic variance, with specific reading factors explaining as much or more variance (~21 percent) than any of the other factors (Wainwright et al., 2004). Consistently, it appears that the observed phenotypic covariation between indicators of achievement and intelligence is primarily due to common genetic influence, but that the variance in the measure of academic achievement itself cannot be fully (or even mostly) explained by that common genetic factor (Wainwright, Wright, Geffen, Luciano, & Martin, 2005).

In summary, the results of quantitative genetic (also referred to as biometrical or behavior-genetic) research on the etiology of intelligence and related processes rule out the possibility of a single gene being behind the corresponding individual differences. Unlike intellectual disability, there are no few genes of major effect that are responsible for individual differences in intelligence. However, the quest for the number of genes involved (if they are at all countable), whether they all contribute to intelligence and intelligence-related traits or whether there are some *general* and *specific* genes, and the magnitudes of effect these genes have, is still unfolding (e.g., Butcher, Kennedy, & Plomin, 2006; Naples et al., 2009).

Grounding the Heritability of IQ

For the last two decades or so, researchers have been engaged in a search for the specific genes that are involved in the etiology of intelligence and intellectual abilities and disabilities (for a review, see Deary, Johnson, & Houlihan, 2009). Such searches usually unfold in one of two ways: as exploratory whole-genome investigations/screens (often also referred to as "scans"), or as hypothesis-driven studies of candidate regions in the genome or candidate genes[6] (see Kornilov, this volume).

Up until now, numerous genome-wide scans for genes contributing to intelligence and cognition have been completed. The first generation of such studies included linkage studies, which did not generate any conclusive results, but highlighted interesting partial overlaps. Specifically, the findings coincide in regions on chromosomes 2q, 6p, and 14q. These overlapping regions have been putatively interpreted as indicative of the presence of genes that could explain some of the variance in IQ. The inability of the field to generate robust findings were interpreted as indications that the architecture of IQ is complex, it involves many genes, each of which exerts small effect sizes, and the samples sizes of those linkage studies were simply too small to generate the needed statistical power for the reliable detection of these effects.

---

[6] Candidate genes: A gene whose function may be associated with a trait.

The second generation of such scans included (and some of such studies are still ongoing) genome-wide association (GWA) studies. These studies confirmed the hypothesis generated by the linkage studies that the sought-after effects are very small indeed. Thus, the largest up-to-date meta-analytic study of 17,989 children resulted in the detection of no individual single nucleotide polymorphisms (SNPs) with genome-wide significance (Benyamin et al., 2014), with the largest effect of a SNP accounting for .24 percent of the variance in IQ. Similarly, small effect sizes have been reported in other GWA studies of IQ (Desrivières et al., 2015) and its associated phenotypes, i.e., phenotypes robustly correlating with IQ, such as educational attainment (Rietveld et al., 2013).

The results of these scans are quite variable; to comprehend them, a series of meta-analyses with follow-up analyses have been carried out, which point to a relatively small set of genetic variants with replicated associations with intelligence and associated phenotypes. To illustrate such meta-results, ten (three with nominal and seven with suggestive significance) SNPs were detected as contributors to increasing educational attainment (Rietveld et al., 2013). The three top SNPs (rs9320913, rs11584700, and rs4851266) have been associated with the *g*-factor in a subsample of the sample used in the Rietveld (2013) report, and with indicators of reading and mathematics in an independent sample of school children (Ward et al., 2014). Yet three additional SNPs were subsequently found in the sample partially overlapping with that used in the Rietveld (2013) report, when other associated phenotypes were considered (Rietveld et al., 2014). A meta-analysis of fluid ability reported thirteen genetic variants exhibiting genome-wide significance, three of which (rs10457441, rs17522122, and rs10119) were strong and not redundant, i.e., represented different genes in different chromosomal regions (Davies et al., 2015). Yet another GWA study with associated phenotypes (executive function and processing speed) highlighted the SNP rs17518584 (Ibrahim-Verbaas et al., 2016).

A number of observations can be derived from this work. We will consider them in turn.

The first observation pertains to the variety of measures used in these studies. In fact, few studies utilize indicators referred to as intelligence. The studies used a range of indicators of both achievement and abilities and generated a wide spectrum of findings, allegedly implicating the majority of chromosomes in the human genome, with some, reportedly, demonstrating signals on both arms, short (p) and long (q). Thus, between all of these phenotypes and all of these regions, genes, and SNPs, the resulting picture is rather difficult to interpret.

Second, the magnitudes of the presented statistics and p values are rather modest. Although they are not indicative of the associated effect sizes, it is notable, that, when such effect sizes are estimated, they are reported to be very low (Plomin & Deary, 2015).

Third, these studies are not independent of each other. These studies have been collectively presented by a small number of groups, and it appears that there is a substantial overlap in the samples. Given that the presentations are split based on the availability of a complete (semi-complete) IQ battery versus the availability of specific subtests from IQ tests and/or other cognitive tests, and different inclusion/exclusion criteria, the question arises as to whether any of the reported signals would survive if a conservative but traditional approach to correcting for multiple comparisons were applied.

Fourth, these studies used a variety of designs and methodologies, analyzing both pooled DNAs for groups of individuals and individual DNAs, recruiting family members and singletons, and covering the genome with genetic markers at highly variable densities. All of these "differences and similarities" need to be carefully taken into account when considering the patterns of consistencies and inconsistencies in these findings. Fifth, none of these studies was specifically built to investigate the genetic bases of intelligence, however defined. In fact, the same genetic data were used to investigate linkage/association with multiple other phenotypes in different subsamples of the same samples. At this point, the impact of such reutilization of data on the inferential statistics has not been carefully appraised, but there have been concerns in the literature regarding the impact of such reutilization on p values, the definition of replicability, and the generalizability of the results (e.g., McCarthy et al., 2008).

In summary, although these scans present interesting data, the reported findings need to be interpreted with caution. In general, we tend to be somewhat less optimistic about the promise, stability, and replicability of these results as compared to what is present in the literature (Piffer, 2015), but consider them as interesting enough to argue that further investigations on the genetic bases of intelligence (broadly defined!) are warranted.

The fact that no genome scan has resulted in identifying specific genes for intelligence does not mean that there are no candidate genes for intelligence. To the contrary: there have been numerous studies that have investigated associations between intelligence, its various facets, and specific genes that were selected to be tested for such association for one reason or another. Some of these investigations are directly related to the scans discussed above and capitalize on the findings from those scans (e.g., Comings et al., 2003; Dick et al., 2007; Gosso, van Belzen, et al., 2006;

Jones et al., 2004 for association with the cholinergic muscarinic 2 receptor gene, *Chrm2*, at 7q33), whereas the majority of these candidate gene studies are totally unrelated to the scans, although they may come from the same research groups (e.g., Gosso et al., 2006, 2008 for association with the synaptosomal-associated protein of 25 kDa gene, *SNAP25*, at 20p12).

Here we briefly summarize the pattern of findings resulting from such investigations in general and discuss studies of only a number of selected genes in particular. In general, there have been numerous studies of a variety of candidate genes. This list of genes is inclusive of but not limited to (a) neurotransmitters and genes related to their metabolism (e.g., catechol-O-methyl transferase, *COMT* located at 22q11; monoamine oxidase A gene, *MAOA* at Xp11; cholinergic muscarinic 2 receptor, *CHRM2* at 7q33; dopamine D2 receptor, *DRD2* at 11q23; serotonin receptor 2A, *HTR2A* at 13q13; the serotonin transporter gene, *SLC6A4*, at 17q11.2; metabotrophic glutamate receptor, *GRM3* at 7q21; the glutathione transferase zeta 1 gene, *GSTz1*, at 14q24.3; the tryptophan hydroxylase 1 gene, *TPH1*, at 11p15.1; the tryptophan hydroxylase 2 gene, *TPH2*, at 12q21.1; the synapsin III gene, *SYN3*, at 22q12.3l and the adrenergic alpha 2A receptor gene, *ADRA2A* at 10q25); (b) genes related to developmental processes, broadly defined (e.g., cathepsin D, *CTSD* at 11p15; succinic semialdehyde dehydrogenase, *ALDH5A1* at 6p22; type I membrane protein related to beta-glucosidases, *klotho* at 13q13; brain-derived neurotrophic factor, *BDNF*, at 11p14; muscle segment homeobox 1, *MSX1* at 4p16; synaptosomal-associated protein 25, *SNAP25*, at 20p12; androgen receptor, *AR*, also known as *NR3C4*, at Xq11-12); and (c) genes of variable functions (e.g., heat shock 70 kDa protein 8, *HSPA8* at 11q24; insulin-like growth factor 2 receptor, *IGF2R* at 6q25; prion protein, *PRNP* at 20p13; dystrobrevin binding protein 1 or dysbinding 1, *DTNBP1* at 6p22; apolipoprotein E, *APOE* at 19q13; cystathionine-beta-synthase, *CBS* at 21q22; MHC class II antigen or major histocompatibility complex, class II, DR beta 1 gene, *HLA-DRB1* at 6p21). It is important to note, however, that in many of these studies of genes and cognition, the behavioral variables of interest are defined beyond IQ. In fact, they encompass the whole gamut of characteristics of intelligence and even cognition (e.g., executive functioning, creativity, working memory, and IQ itself). And, although findings from some of these studies have never been attempted to be replicated or have failed to be replicated, there is a certain amount of consistency in the findings for selected genes. We view the establishment of these specific associations between genes and intelligence (or cognition, however broadly defined) as a fundamental breakthrough, a switch from the hypothetical decomposition of variance that was characteristic of earlier heritability studies to a firm "grounding" of these heritabilities in the genome. The hope is that by understanding the functions of these genes and their interactive protein networks, the field will gain some additional understanding of how the general biological (and the specific genetic) machinery of intelligence works.

To exemplify this line of work, here we present brief comments on the research concerning one particular gene, *APOE*, which is relevant to research on both brain structure/function and intelligence, and is considered to be the only candidate gene demonstrating an association with cognitive functions, accounting for ~2 percent of the related variance (Wisdom, Callahan, & Hawkins, 2011).

The apolipoprotein E gene (*APOE*) is located on chromosome 19q13 and is responsible for the production of an apoprotein that is essential for the normal catabolism of triglyceride-rich lipoprotein components. This gene has been long studied in the context of research on neuronal development and repair; this research, in turn, is directly related to work on Alzheimer disease (AD) (Blackman, Worley, & Strittmatter, 2005; Buttini et al., 1999; Rapoport et al., 2008; Teasdale, Murray, & Nicoll, 2005; Teter & Ashford, 2002). The gene is polymorphic,[7] and there are three variants of *APOE* that have been studied extensively: *ApoE2*, *ApoE3*, and *ApoE4*. These variants are responsible for the production of three different isoforms (Apo-ε2, Apo-ε3, and Apo-ε4[8]) of the protein that differ only by single amino acid substitutions, but these substitutions have been shown to be associated with dramatic physiological outcomes. Of these three isoforms, ApoE-ε3 is associated with a normal protein, whereas Apo-ε2 and Apo-ε4 are related to abnormal proteins.

In the context of this discussion, the *ApoE4* allele[9] is of particular interest because it has been associated with atherosclerosis, AD, reduced neurite outgrowth, and impaired cognitive function. To illustrate, a meta-analysis of dozens of studies combining the data from ~20,000 individuals established that possession of the *ApoE4* allele in older people is associated with poorer performance on tests of global cognitive function, episodic memory, and executive function (Small, Rosnick, Fratiglioni, & Backman, 2004). Moreover, it has been shown that young healthy adults who carry the *ApoE4*

---

[7] Polymorphic: A locus with two or more alternative forms.

[8] These 3 allelic variants differ at 2 single-base variations located in exon 4 at codon positions 112 and 158. The T and C alleles of *APOE* 112T>C (rs429358) and *APOE* 158C>T (rs7412) encode arginine and cysteine, respectively. The variants differ such that *ApoE2* has a T allele at both positions 112 and 158; *ApoE3* has T and C alleles at positions 112 and 158, respectively; and *ApoE4* has C at both positions.

[9] Allele: An alternative form of a gene at a particular locus.

allele demonstrate altered patterns of brain activity both at rest and during cognitive challenges (Scarmeas & Stern, 2006).

In a pediatric cohort, carrying the *ApoE4* allele was related to having a thinned cortex in the region of the brain, the so-called entorhinal region, where the earliest AD-associated changes are typically registered (Shaw et al., 2007). However, an attempt to find an association between these polymorphisms and the *g*-factor in a case control sample of 101 high *g* and 101 average *g* children did not yield positive results (Turic, Fisher, Plomin, & Owen, 2001). Similarly, there are some studies that report a differential pattern of associations for the *ApoE4* allele in young adults. In particular, it has been reported that *ApoE4*, compared to both *ApoE2* and *ApoE3*, is associated with better episodic memory and a smaller neural investment (i.e., "economical" brain activity) in learning and retrieval (Mondadori et al., 2007).

There is also some evidence that the *ApoE2* allele may be protective; however, the mechanisms of this differential action of the variants in the *APOE* gene are not understood (Deary et al., 2002; Smith, 2002; Sundstrom et al., 2007). Also, it appears that even in familial AD only a relatively small portion of variation in memory is attributable to *APOE* (Lee, Flaquer, Stern, Tycko, & Mayeux, 2004). Thus, there are many unanswered questions with regard to the connections between the variation in this gene and differences in performance on memory and other cognitive tasks. It has been proposed that, by itself, the *ApoE4* allele does not influence any cognitive domains. Yet, when this allele co-occurs with other risk alleles,[10] such as the risk allele (allele T in the functional exon 2 polymorphism) in the cathepsin D gene (*CTSD*), the carriers of the two alleles demonstrate scores on cognitive tasks that are substantially lower than when either of the polymorphisms is considered independently (Payton et al., 2006). Thus, understanding this variation and its connection to individual differences in cognition and, subsequently, to the acquisition of AD or not is of great interest to researchers in a variety of fields.

In summary, there is a lot to sort out here. Although the importance of genetic factors to the development of intelligence and intelligence-related cognitive processing is widely acknowledged, and the field appears to be accepting of the role of specific genes such as *APOE*, the specific neurocognitive processes underlying their involvement continue to be a matter of debate. There could be multiple reasons for such a state of affairs.

---

[10] Risk allele: An alternate form of a gene that is associated with risk.

First, confirmation of the specific genes that form these genetic factors has proven difficult. While positive evidence of association has been reported for several interesting genes, thus far there has not been widespread success in replicating reported associations. Even though there are publications that present findings at borderline levels of p values (e.g., $p = .048$), these evaporate when corrections for multiple comparisons are introduced (e.g., Younger et al., 2005). In general, it is assumed that the effect sizes of specific genes involved in complex human traits are small (Greenwood & Parasuraman, 2003). Correspondingly, special attention needs to be given to designing powerful studies with a large N that displays as much genetic homogeneity as possible. Second, there are sometimes contradictory results with regard to an association of a particular gene/gene variant and cognition, albeit with different intelligence-related processes, as reported by the same or related groups of investigators (e.g., Reuter, Ott, Vaitl, & Hennig, 2007; Reuter et al., 2005). This suggests that findings might be presented partially, and such partiality might, once again, affect the corresponding p values. Third, looking at such a diverse picture of findings, it has been rather difficult to systematically distinguish between false positive findings, pleiotropic effects of genes on multiple cognitive processes, and the role of the g-factor (Starr, Fox, Harris, Deary, & Whalley, 2008). As mentioned above, very few studies actually limit themselves as "true" indicators of the g-factor (i.e., some kind of summative indicator of multiple intelligence-related measures). Most studies employ and analyze a variety of intelligence-related indicators. Thus, similar to the findings obtained from genome scans, the field unequivocally supports the idea of the involvement of genetic factors in the development of intelligence and abilities, but it is far from able to generate a cohesive picture of the genetic machinery behind these factors.

IN PLACE OF A CONCLUSION

In view of the lack of cohesiveness in our understanding of the genetic machinery of intelligence and intelligence-related processes, what can be said regarding the Chinese initiative described by CNN? Our answer to this question is that such an initiative is premature. Not only is it premature because there is no diagnostic tool to identify the DNA profile predisposing for intellectual giftedness, it is also premature because even if there were such a profile, it is unclear what kinds of environments should be formed for the individuals possessing such a profile. Most importantly, however, it is premature because it exhibits an underdeveloped attention to the very reason that we continue to value and study individual differences in cognitive functions in humans – to celebrate and promote human diversity, not to control or constrain it.

REFERENCES

Ando, J., Ono, Y., & Wright, M. J. (2001). Genetic structure of spatial and verbal working memory. *Behavior Genetics, 31,* 615–624.

Bartels, M., Rietveld, M. J. H., Van Baal, G. C. M., & Boomsma, D. I. (2002). Genetic and environmental influences on the development of intelligence. *Behavior Genetics, 32,* 237–249.

Beaujean, A. A. (2005). Heritability of cognitive abilities as measured by mental chronometric tasks: A meta-analysis. *Intelligence, 33,* 187–201.

Benyamin, B., Pourcain, B., Davis, O. S., Davies, G., Hansell, N. K., Brion, M. J., et al. (2014). Childhood intelligence is heritable, highly polygenic and associated with FNBP1L. *Molecular Psychiatry, 19,* 253–258. doi:10.1038/mp.2012.184

Bishop, E. G., Cherny, S. S., Corley, R., Plomin, R., DeFries, J. C., & Hewitt, J. K. (2003). Development genetic analysis of general cognitive ability from 1 to 12 years in a sample of adoptees, biological siblings, and twins. *Intelligence, 31,* 31–49.

Blackman, J. A., Worley, G., & Strittmatter, W. J. (2005). Apolipoprotein E and brain injury: Implications for children. *Developmental Medicine & Child Neurology, 47,* 64–70.

Bouchard, T. J., Jr., & McGue, M. (2003). Genetic and environmental influences on human psychological differences. *Journal of Neurobiology, 54,* 4–45.

Brant, A., Haberstick, B., Corley, R., Wadsworth, S., DeFries, J. C., & Hewitt, J. K. (2009). The developmental etiology of high IQ. *Behavior Genetics, 39,* 393–405.

Briley, D. A., & Tucker-Drob, E. M. (2013). Explaining the increasing heritability of cognitive ability across development: A meta-analysis of longitudinal twin and adoption studies. *Psychological Science, 24,* 1704–1713. doi:10.1177/0956797613478618

Butcher, L. M., Kennedy, J. K., & Plomin, R. (2006). Generalist genes and cognitive neuroscience. *Current Opinion in Neurobiology, 16,* 145–151.

Buttini, M., Orth, M., Bellosta, S., Akeefe, H., Pitas, R. E., Wyss-Coray, T., et al. (1999). Expression of human apolipoprotein E3 or E4 in the brains of Apoe-/- mice: Isoform-specific effects on neurodegeneration. *Journal of Neuroscience, 19,* 4867–4880.

Cardon, L. R., & Fulker, D. W. (1993). Genetics of specific cognitive abilities. In R. Plomin & G. E. McClearn (Eds.), *Nature, nurture & psychology* (pp. 99–120). Washington, DC: American Psychological Association.

Carroll, J. B. (1993). *Human cognitive abilities.* New York, NY: Cambridge University Press.

Cianciolo, A. T., & Sternberg, R. J. (2004). *A brief history of intelligence.* Malden, MA: Blackwell.

Comings, D. E., Wu, S., Rostamkhani, M., McGue, M., Iacono, W. G., Cheng, L. S., et al. (2003). Role of the cholinergic muscarinic 2 receptor (CHRM2) gene in cognition. *Molecular Psychiatry, 8,* 10–13.

Davies, G., Armstrong, N., Bis, J. C., Bressler, J., Chouraki, V., Giddaluru, S., et al. (2015). Genetic contributions to variation in general cognitive function: Ameta-analysis of genome-wide association studies in the CHARGE consortium (N=53,949). *Molecular Psychiatry, 20,* 183–192. doi:10.1038/mp.2014.188

Davis, O. S. P., Kovas, Y., Harlaar, N., Busfield, P., McMillan, A., Frances, J., et al. (2008). Generalist genes and the Internet generation: Etiology of learning abilities by web testing at age 10. *Genes, Brain and Behavior, 7,* 455–462.

de Geus, E., Wright, M., Martin, N., & Boomsma, D. (2001). Editorial: Genetics of brain function and cognition. *Behavior Genetics, 31*(6), 489–495.

Deary, I. J. (2000). *Looking down on human intelligence: From psychometrics to the brain.* Oxford, UK: Oxford University Press.

Deary, I. J., Johnson, W., & Houlihan, L. (2009). Genetic foundations of human intelligence. *Human Genetics, 126,* 215–232.

Deary, I. J., Spinath, F. M., & Bates, T. C. (2006). Genetics of intelligence. *European Journal of Human Genetics, 14,* 690–700.

Deary, I. J., Whiteman, M. C., Pattie, A., Starr, J. M., Hayward, C., Wright, A. F., et al. (2002). Cognitive change and the APOE epsilon 4 allele. *Nature, 481,* 932.

Desrivières, S., Lourdusamy, A., Tao, C., Toro, R., Jia, T., Loth, E., et al.. (2015). Single nucleotide polymorphism in the neuroplastin locus associates with cortical thickness and intellectual ability in adolescents. *Molecular Psychiatry, 20,* 263–274. doi:10.1038/mp.2013.197

Devlin, B., Daniels, M., & Roeder, K. (1997). The heritability of IQ. *Nature, 388,* 468–471.

Dick, D. M., Aliev, F., Kramer, J., Wang, J. C., Hinrichs, A., Bertelsen, S., et al. (2007). Association of CHRM2 with IQ: Converging evidence for a gene influencing intelligence. *Behavior Genetics, 37,* 265–272.

Edmonds, C. J., Isaacs, E. B., Visscher, P. M., Rogers, M., Lanigan, J., Singhal, A., et al. (2008). Inspection time and cognitive abilities in twins aged 7 to 17 years: Age-related changes, heritability and genetic covariance. *Intelligence, 36,* 210–225.

Ertl, J. P. (1971). Fourier analysis of evoked potentials and human intelligence. *Nature, 230,* 525–526.

Fabiani, M., Gratton, G., & Federmeier, K. D. (2007). Event-related brain potentials: Methods, theory, and applications. In J. T. Cacioppo, L. G. Tassinary, & G. G. Berntson (Eds.), *Handbook of psychophysiology* (3rd ed., pp. 85–119). New York, NY: Cambridge University Press.

Flint, J. (1999). The genetic basis of cognition. *Brain, 122,* 2015–2031.

Friedman, N. P., Miyake, A., Corley, R. P., Young, S. E., DeFries, J. C., & Hewitt, J. K. (2006). Not all executive functions are related to intelligence. *Psychological Science, 17,* 172–179.

Friedman, N. P., Miyake, A., Young, S. E., DeFries, J. C., Corley, R. P., & Hewitt, J. K. (2008). Individual differences in executive functions are almost entirely genetic in origin. *Journal of Experimental Psychology, 137,* 201–225.

Friend, A., DeFries, J. C., & Olson, R. K. (2008). Parental education moderates genetic influences on reading disability. *Psychological Science, 19,* 1–7.

Galton, F. (1869). *Hereditary genius. An inquiry into its laws and consequences.* London, UK: Macmillan.

Gosso, M. F., de Geus, E. J., van Belzen, M. J., Polderman, T. J., Heutink, P., Boomsma, D. I., et al. (2006). The SNAP-25 gene is associated with cognitive ability: Evidence from a family-based study in two independent Dutch cohorts. *Molecular Psychiatry, 11*, 878–886.

Gosso, M. F., de Geus, E. J. C., Polderman, T. J. C., Boomsma, D. I., Heutink, P., & Posthuma, D. (2008). Common variants underlying cognitive ability: Further evidence for association between the SNAP-25 gene and cognition using a family-based study in two independent Dutch cohorts. *Genes, Brain, & Behavior, 7*, 355–364.

Gosso, M. F., van Belzen, M., de Geus, E. J., Polderman, J. C., Heutink, P., Boomsma, D. I., et al. (2006). Association between the CHRM2 gene and intelligence in a sample of 304 Dutch families. *Genes, Brain, and Behavior, 5*, 577–584.

Greenwood, P. M., & Parasuraman, R. (2003). Normal genetic variation, cognition, and aging. *Behavioral & Cognitive Neuroscience Reviews, 2*, 278–306. doi:10.1177/1534582303260641

Hansell, N. K., Halford, G. S., Andrews, G., Shum, D. H. K., Harris, S. E., Davies, G., et al. (2015). Genetic basis of a cognitive complexity metric. *PLoS One, 10*, e0123886. doi:10.1371/journal.pone.0123886

Hansell, N. K., Wright, M. J., Geffen, G. M., Geffen, L. B., Smith, G. A., & Martin, N. G. (2001). Genetic influence on ERP slow wave measures of working memory. *Behavior Genetics, 31*, 603–614.

Harden, K. P., Turkheimer, E., & Loehlin, J. C. (2007). Genotype by environment interaction in adolescent's cognitive aptitude. *Behavior Genetics, 37*, 273–283.

Hart, S. A., Petrill, S. A., Thompson, L. A., & Plomin, R. (2009). The ABCs of math: A genetic analysis of mathematics and its links with reading ability and general cognitive ability. *Journal of Educational Psychology, 101*, 388–402.

Haworth, C. M. A., Wright, M. J., Luciano, M., Martin, N. G., de Geus, E. J. C., van Beijsterveldt, C. E. M., et al. (2010). The heritability of general cognitive ability increases linearly from childhood to young adulthood. *Molecular Psychiatry, 15*, 1112–1120. doi:10.1038/mp.2009.55

Ibrahim-Verbaas, C. A., Bressler, J., Debette, S., Schuur, M., Smith, A. V., Bis, J. C., et al. (2016). GWAS for executive function and processing speed suggests involvement of the CADM2 gene. *Molecular Psychiatry, 21*, 189–197. doi:10.1038/mp.2015.37

Inlow, J. K., & Restifo, L. L. (2004). Molecular and comparative genetics of mental retardation. *Genetics, 166*, 835–881.

Jensen, A. R. (1998). *The g factor: The science of mental ability*. New York, NY: Praeger.

Jones, K. A., Porjesz, B., Almasy, L., Bierut, L., Goate, A., Wang, J. C., et al. (2004). Linkage and linkage disequilibrium of evoked EEG oscillations with CHRM2 receptor gene polymorphisms: Implications for human brain dynamics and cognition. *International Journal of Psychophysiology, 53*, 75–90.

Katsanis, J., Iacono, W. G., McGue, M. K., & Carlson, S. R. (1997). P300 event-related potential heritability in monozygotic and dizygotic twins. *Psychophysiology, 34*, 47–58.

Kremen, W. S., Jacobsen, K., Xian, H., Eisen, S. A., Eaves, L. J., Tsuang, M. T., et al. (2007). Genetics of verbal working memory processes: A twin study of middle-aged men. *Neuropsychology, 21*, 569–580

Lee, J. H., Flaquer, A., Stern, Y., Tycko, B., & Mayeux, R. (2004). Genetic influences on memory performance in familial Alzheimer disease. *Neurology, 62,* 414–421.

Lee, T., Henry, J. D., Trollor, J. N., & Sachdev, P. S. (2010). Genetic influences on cognitive functions in the elderly: A selective review of twin studies. *Brain Research Reviews, 64,* 1–13. doi:10.1016/j.brainresrev.2010.02.001

Luciano, M., Posthuma, D., Wright, M. J., de Geus, E. J. C., Smith, G. A., Geffen, G. M., et al. (2005). Perceptual speed does not cause intelligence, and intelligence does not cause perceptual speed. *Biological Psychology, 70,* 1–8.

Luciano, M., Wright, M. J., Geffen, G. M., Geffen, L. B., Smith, G. A., Evans, D. M., et al. (2003). A genetic two-factor model of the covariation among a subset of Multidimensional Aptitude Battery and Wechsler Adult Intelligence Scale – Revised subtests. *Intelligence, 31,* 589–605.

Luciano, M., Wright, M. J., Geffen, G. M., Geffen, L. B., Smith, G. A., & Martin, N. G. (2004). A genetic investigation of the covariation among inspection time, choice reaction time, and IQ subtest scores. *Behavior Genetics, 34,* 41–50.

Luciano, M., Wright, M. J., Smith, G. A., Geffen, G. M., Geffen, L. B., & Martin, N. G. (2001). Genetic covariance among measures of information processing speed, working memory, and IQ. *Behavior Genetics, 31,* 581–592.

Mandelman, S. D., & Grigorenko, E. L. (2010). Intelligence: genes, environments, and everything in between. In R. J. Sternberg & S. Kaufman (Eds.), *The Cambridge handbook of intelligence* (pp. 85–106). New York, NY: Cambridge University Press.

McCarthy, M. I., Abecasis, G. R., Cardon, L. R., Goldstein, D. B., Little, J., Ioannidis, J. P., et al. (2008). Genome-wide association studies for complex traits: Consensus, uncertainty and challenges. *Nature Reviews Genetics, 9,* 356–369.

McGue, M., Bouchard, T. J., Jr., Iacono, W. G., & Lykken, D. T. (1993). Behavioral genetics of cognitive ability: A life-span perspective. In R. Plomin & G. E. McClearn (Eds.), *Nature, nurture, & psychology* (pp. 59–76). Washington, DC: American Psychological Association.

McGue, M., & Christensen, K. (2013). Growing old but not growing apart: Twin similarity in the latter half of the lifespan. *Behavior Genetics, 43,* 1–12. doi:10.1007/s10519-012-9559-5

Mondadori, C. R. A., de Quervain, D. J.-F., Buchmann, A., Mustovic, H., Wollmer, M. A., Schmidt, C. F., et al. (2007). Better memory and neural efficiency in young Apolipoprotein E e4 carriers. *Cerebral Cortex, 17,* 1934–1947.

Naples, A. J., Chang, J. T., Katz, L., & Grigorenko, E. L. (2009). Same or different? Insights into the etiology of phonological awareness and rapid naming. *Biological Psychology, 80,* 226–239.

Panizzon, M. S., Vuoksimaa, E., Spoon, K. M., Jacobson, K. C., Lyons, M. J., Franz, C. E., et al. (2014). Genetic and environmental influences of general cognitive ability: Is g a valid latent construct? *Intelligence, 43,* 65–76. doi:10.1016/j.intell.2014.01.008

Patrick, C. L. (2000). Genetic and environmental influences on the development of cognitive abilities: Evidence from the field of developmental behavior genetics. *Journal of School Psychology, 38,* 79–108.

Payton, A., Van den Boogerd, E., Davidson, Y., Gibbons, L., Ollier, W., Rabbitt, P., etal. (2006). Influence and interactions of cathepsin D, HLA-DRB1 and APOE on cognitive abilities in an older non-demented population. *Genes, Brain & Behavior, 5*, 23–31.

Penrose, L. S. (1938). *A clinical and genetic study of 1,280 cases of mental defect.* London, UK: H. M. Stationery Office.

Petrill, S. A., Lipton, P. A., Hewitt, J. K., Plomin, R., Cherny, S. S., Corley, R., et al. (2004). Genetic and environmental contributions to general cognitive ability through the first 16 years of life. *Developmental Psychology, 40*, 805–812. doi:10.1037/0012-1649.40.5.805

Piffer, D. (2015). A review of intelligence GWAS hits: Their relationship to country IQ and the issue of spatial autocorrelation. *Intelligence, 53*, 43–50. doi:10.1016/j.intell.2015.08.008

Plomin, R., & Deary, I. J. (2015). Genetics and intelligence differences: Five special findings. *Molecular Psychiatry, 20*, 98–108. doi:10.1038/mp.2014.105

Plomin, R., & Spinath, F. M. (2004). Intelligence: Genetics, genes, and genomics. *Journal of Personality & Social Psychology, 86*, 112–129.

Polderman, T. J. C., Gosso, M. F., Posthuma, D., Van Beijsterveldt, T. C. E. M., Heutink, P., Verhulst, F. C., et al. (2006). A longitudinal twin study on IQ, executive functioning, and attention problems during childhood and early adolescence. *Acta Neurologica Belgica, 106*, 191–207.

Polderman, T. J. C., Posthuma, D., De Sonneville, L. M. J., Stins, J. F., Verhulst, F. C., & Boomsma, D. I. (2007). Genetic analyses of the stability of executive functioning during childhood. *Biological Psychology, 76*, 11–20.

Posthuma, D., Neale, M. C., Boomsma, D. I., & de Geus, E. J. C. (2001). Are smarter brains running faster? Heritability of alpha peak frequency, IQ, and their interrelation. *Behavior Genetics, 31*, 567–579.

Price, T. S., Eley, T. C., Dale, P. S., Stevenson, J., Saudino, K., & Plomin, R. (2000). Genetic and environmental covariation between verbal and nonverbal cognitive development in infancy. *Child Development, 71*, 948–959.

Rapoport, M., Wolf, U., Herrmann, N., Kiss, A., Shammi, P., Reis, M., et al. (2008). Traumatic brain injury, Apolipoprotein E-epsilon4, and cognition in older adults: A two-year longitudinal study. *Journal of Neuropsychiatry & Clinical Neurosciences, 20*, 68–73.

Reichenberg, A., Cederlöf, M., McMillan, A., Trzaskowski, M., Kapara, O., Fruchter, E., et al. (2016). Discontinuity in the genetic and environmental causes of the intellectual disability spectrum. *Proceedings of the National Academy of Science, 113*, 1098–1103. doi:10.1073/pnas.1508093112

Reuter, M., Ott, U., Vaitl, D., & Hennig, J. (2007). Impaired executive control is associated with a variation in the promoter region of the Tryptophan Hydroxylase 2 gene. *Journal of Cognitive Neuroscience, 19*, 401–408. doi:10.1162/jocn.2007.19.3.401

Reuter, M., Peters, K., Schroeter, K., Koebke, W., Lenardon, D., Bloch, B., et al. (2005). The influence of the dopaminergic system on cognitive functioning: A molecular genetic approach. *Behavioural Brain Research, 164*, 93–99.

Reynolds, C. A., Finkel, D., McArdle, J. J., Gatz, M., Berg, S., & Pedersen, N. L. (2005). Quantitative genetic analysis of latent growth curve models of cognitive abilities in adulthood. *Developmental Psychology, 41*, 3–16. doi:10.1037/0012-1649.41.1.3

Reznick, J. S., Corley, R., & Robinson, J. A. (1997). A longitudinal twin study of intelligence in the second year. *Monographs of the Society for Research in Child Development, 249,* 62(1).

Rietveld, C. A., Esko, T., Davies, G., Pers, T. H., Turley, P., Benyamin, B., et al. (2014). Common genetic variants associated with cognitive performance identified using the proxy-phenotype method. *Proceedings of the National Academy of Sciences, 111,* 13790–13794. doi:10.1073/pnas.1404623111

Rietveld, C. A., Medland, S. E., Derringer, J., Yang, J., Esko, T., Martin, N. W., et al. (2013). GWAS of 126,559 individuals identifies genetic variants associated with educational attainment. *Science, 340,* 1467–1471. doi:10.1126/science.1235488

Rietveld, M. J. H., Dolan, C. V., van Baal, G. C. M., & Boomsma, D. I. (2003). A twin study of differentiation of cognitive abilities in childhood. *Behavior Genetics, 33,* 367–381.

Rijsdijk, F. V., Vernon, P. A., & Boomsma, D. I. (2002). Application of hierarchical genetic models to Raven and WAIS subtests: A Dutch twin study. *Behavior Genetics, 32,* 199–210.

Risch, N. (1990). Linkage strategies for genetically complex traits. II. The power of affected relative pairs. *American Journal of Human Genetics, 46*(2), 229–241.

Scarmeas, N., & Stern, Y. (2006). Imaging studies and APOE genotype in persons at risk for Alzheimer's disease. *Current Psychiatry Reports, 8,* 11–17.

Shaw, P., Lerch, J. P., Pruessner, J. C., Taylor, K. N., Rose, A. B., Greenstein, D., et al. (2007). Cortical morphology in children and adolescents with different apolipoprotein E gene polymorphisms: An observational study. *Lancet Neurology, 6,* 494–500.

Singer, J. J., MacGregor, A. J., Cherkas, L. F., & Spector, T. D. (2006). Genetic influences on cognitive function using The Cambridge Neuropsychological Test Automated Battery. *Intelligence, 34,* 421–428.

Small, B. J., Rosnick, C. B., Fratiglioni, L., & Backman, L. (2004). Apolipoprotein E and cognitive performance: A meta-analysis. *Psychology & Aging, 14,* 592–600.

Smith, J. D. (2002). Apolipoprotiens and aging: Emerging mechanisms. *Ageing Research Reviews, 1,* 345–365.

Spearman, C. (1904). General intelligence, objectively determined and measured. *American Journal of Psychology, 15,* 201–292.

Starr, J. M., Fox, H., Harris, S. E., Deary, I. J., & Whalley, L. J. (2008). GSTz1 genotype and cognitive ability. *Psychiatric Genetics, 18,* 211–212.

Sternberg, R. J. (1996). *Successful intelligence.* New York, NY: Simon & Schuster.

Sundstrom, A., Nilsson, L. G., Cruts, M., Adolfsson, R., Van Broeckhoven, C., & Nyberg, L. (2007). Fatigue before and after mild traumatic brain injury: Pre-post-injury comparisons in relation to Apolipoprotein E. *Brain Injury, 21,* 1049–1054.

Teasdale, G. M., Murray, G. D., & Nicoll, J. A. (2005). The association between APOE epsilon4, age and outcome after head injury: A prospective cohort study. *Brain, 128,* 2556–2561.

Teter, B., & Ashford, J. W. (2002). Neuroplasticity in Alzheimer's disease. *Journal of Neuroscience Research, 70,* 402–437.

Tucker-Drob, E. M., & Bates, T. C. (2016). Large cross-national differences in gene x socioeconomic status interaction on intelligence. *Psychological Science, 27,* 138–149. doi:10.1177/0956797615612727

Turic, D., Fisher, P. J., Plomin, R., & Owen, M. J. (2001). No association between apolipoprotein E polymorphisms and general cognitive ability in children. *Neuroscience Letters, 299*, 97–100.

van Baal, G. C. M., Boomsma, D. I., & de Geus, E. J. C. (2001). Longitudinal genetic analysis of EEG coherence in young twins. *Behavior Genetics, 31*, 637–651.

van Baal, G. C. M., van Beijsterveldt, C. E. M., Molenaar, P. C. M., Boomsma, D. I., & de Geus, E. J. C. (2001). A genetic perspective on the developing brain: Electrophysiological indices of neural functioning in young and adolescent twins. *European Psychologist, 6*, 254–263.

van Beijsterveldt, C. E., & Boomsma, D. I. (1994). Genetics of the human electro-encephalogram (EEG) and event-related brain potentials (ERPs): A review. *Human Genetics, 94*, 319–330.

van Beijsterveldt, C. E., Molenaar, P. C., de Geus, E. J., & Boomsma, D. I. (1998a). Genetic and environmental influences on EEG coherence. *Behavior Genetics, 28*, 443–453.

(1998b). Individual differences in P300 amplitude: A genetic study in adolescent twins. *Biological Psychology, 47*, 97–120.

van der Maas, H. L. J., Dolan, C. V., Grasman, R. P. P. P., Wicherts, J. M., Huizenga, H. M., & Raijmakers, M. E. J. (2006). A dynamical model of general intelligence: The positive manifold of intelligence by mutualism. *Psychological Review, 113*, 842–861.

van der Sluis, S., Willemsen, G., de Geus, E. J. C., Boomsma, D. I., & Posthuma, D. (2008). Gene-environment interaction in adults' IQ scores: Measure of past and present environment. *Behavior Genetics, 38*, 348–360.

van Leeuwen, M., van den Berg, S. M., & Boomsma, D. I. (2008). A twin-family study of general IQ. *Learning and Individual Differences, 18*, 76–88.

Wainwright, M. A., Wright, M. J., Geffen, G., Luciano, M., & Martin, N. (2005). The genetic basis of academic achievement on the Queensland Core Skills Test and its shared genetic variance with IQ. *Behavior Genetics, 35*(2), 133–145.

Wainwright, M. A., Wright, M. J., Geffen, G. M., Geffen, L. B., Luciano, M., et al. (2004). Genetic and environmental sources of covariance between reading tests used in neuropsychological assessment and IQ subtests. *Behavior Genetics, 34*, 365–376.

Walker, S. O., Petrill, S. A., Spinath, F. M., & Plomin, R. (2004). Nature, nurture and academic achievement: A twin study of teacher assessments of 7-year-olds. *British Journal of Educational Psychology, 74*, 323–342.

Ward, M. E., McMahon, G., St Pourcain, B., Evans, D. M., Rietveld, C. A., Benjamin, D. J., et al. (2014). Genetic variation associated with differential educational attainment in adults has anticipated associations with school performance in children. *PLoS One, 9*, e100248. doi:10.1371/journal.pone.0100248

Watson, J. B. (1924). *Behaviorism*. Chicago: University of Chicago Press.

Winterer, G., & Goldman, D. (2003). Genetics of human prefrontal function. *Brain Research Reviews, 43*, 134–163.

Wisdom, N. M., Callahan, J. L., & Hawkins, K. A. (2011). The effects of apolipoprotein E on non-impaired cognitive functioning: A meta-analysis. *Neurobiology of Aging, 32,* 63–74. doi:10.1016/j.neurobiolaging.2009.02.003

Younger, W. Y. Y., Shih-Jen, T., Chen-Jee, H., Ming-Chao, C., Chih-Wei, Y., & Tai-Jui, C. (2005). Association study of a functional MAOA-uVNTR gene polymorphism and cognitive function in healthy females. *Neuropsychobiology, 52,* 77–82.

6

A Behavioral Genetic Perspective
on Non-Cognitive Factors and
Academic Achievement

ELLIOT M. TUCKER-DROB AND K. PAIGE HARDEN

Interest in "non-cognitive" factors (also called) "soft" skills has surged in recent years. As suggested by their name, non-cognitive factors are typically defined by what they are not – they are not measures of intelligence or cognitive ability ("hard skills," which are most typically measured using performance-based psychometric tests). Implicit in the use of the term non-cognitive *skills* is also the assumption that they are useful for, or at least statistically predictive of, success and accomplishment in educational settings, and in life more generally. Although theorists have differed regarding which constructs fall under the rubric of non-cognitive skills (Farrington et al., 2012), these skills can be generally defined as systematic patterns of thinking, feeling, and behaving that are relevant for academic success and accomplishment. Non-cognitive factors, therefore, include intellectual interest, achievement motivation, conscientiousness, grit, academic self-concept, and attitudes toward education. In this chapter, we describe a transactional framework for understanding how individual differences in non-cognitive skills relate to cognitive development and academic achievement. Within this framework, research from social and educational psychology on non-cognitive skills can be integrated with research from behavior genetics on cognitive development and academic achievement. Considering these rather disparate lines of inquiry together points to future directions for understanding children's development.

TRANSACTIONAL MODELS
OF GENE–ENVIRONMENT CORRELATION

Historically, genetic influences on cognitive development and academic achievement were conceptualized as competing with environmental

During the time that this chapter was prepared, E. M. Tucker-Drob and K. P. Harden were supported by NIH grants HD081437, AA020588, and AA023322.

influences. Large genetic effects on cognition and achievement were thought to leave little room for environmental influence. Contemporary developmental behavioral genetic thinking sharply contrasts with this perspective. Rather than competing with environmental influences, genetic influences on cognitive development and academic achievement are thought to depend on reciprocal transactions between an individual's genetically influenced traits and inputs from his or her environment.

A central component of this transactional process is what is known as *gene–environment correlation*, or *r*GE. *r*GE refers simply to the fact that people with certain genotypes are systematically (i.e., non-randomly) more or less likely to experience certain environmental experiences. *r*GE can arise via three general mechanisms, which were proposed by Plomin, DeFries, and Loehlin (1977) and by Scarr and McCartney (1983). (See additional discussion of *r*GE in Asbury, Rimfled, and Krapohl, this volume.) *Passive r*GE arise when children are raised by their biological parents, and thus inherit genes from the same people who are also providing their rearing environments. For example, adults who are in more cognitively skilled jobs (e.g., lawyer) tend to make more money than adults in less cognitively demanding jobs (e.g., waiter). Consequently, children who inherit genes for higher cognitive ability from their parents are also more likely to grow up in socioeconomically advantaged homes. Passive *r*GE can thus represent a "double whammy," in that children who are at genetic risk for educational difficulties are also more likely to be raised in family environments that confer additional risk for educational difficulties.

*Active r*GE arises when children actively seek out different environmental experiences on the basis of their interests, preferences, proclivities, and aptitudes – which are, in part, influenced by genes. For example, an adolescent who likes to read, compared to one who does not, is more likely to enroll in a demanding literature course at school. In this way, two children who are ostensibly provided with equivalent learning opportunities (e.g., attending the same school) will differ in how actively they take advantage of those opportunities (e.g., enrolling in additional, or more challenging, courses).

*Evocative r*GE arises when children evoke different environmental experiences from others in their surroundings on the basis of observable patterns of behaving – which, again, are at least partially genetically influenced. For example, bright, motivated children are more likely to be placed in advanced classes (e.g., gifted and talented programs) and to receive praise from teachers, whereas children with behavioral problems are more likely to be placed in remedial classes and to be removed from the classroom for punishments (e.g., suspensions).

Finally, a fourth form of *r*GE – only briefly considered by Plomin et al. (1977) – arises when children exposed to the exact same environmental differentially experience, attend to, or interpret that environment. For instance, children attending the same classroom lecture may differ in the extent to which they attend to and engage in the lecture versus daydream, pass notes, or doodle in their notebook.[1] We suggest the term *attentional r*GE to refer to this process.

Transactional models of cognitive development build on the concept of *r*GE: Not only are individuals with particular genotypes posited to select, evoke, and attend to particular environments, but these environments are posited to have causal, reciprocal influences on cognitive abilities and academic achievement (Tucker-Drob, Briley, & Harden, 2013). Transactional models can, therefore, be contrasted with perspectives that conceptualize *r*GE simply as a source of confounding, in which largely impotent environmental experiences become associated with genetic differences, such that environments are inappropriately credited as causal for cognitive development and academic achievement in non-genetic designs (cf. Johnson, Turkheimer, Gottesman, & Bouchard, 2009). Rather, genotype–environment correlation is predicted to lead to differences in potent environmental experiences that promote (or possibly interfere with) cognitive development and academic achievement. According to transactional models, because the experiences relevant to cognition and achievement are non-randomly experienced as a function of genotype, the effect of environmental influence is to amplify heritable variation in achievement (Briley & Tucker-Drob, 2013; Tucker-Drob et al., 2013). In summary, although genetically informative research has conventionally sought to distinguish selection from causation, transactional models postulate that selection and causation work together in a dynamical system to affect psychological development.

NON-COGNITIVE FACTORS AS DRIVING FORCES IN ACADEMICALLY RELEVANT GENE– ENVIRONMENT TRANSACTIONS

What are the specific genetically influenced factors that lead children to differentially select, evoke, and attend to cognitively stimulating learning experiences? Dickens and Flynn (2001) proposed that earlier levels of

---

[1] In briefly considering this form of *r*GE, Plomin et al. (1977) gave the example of "the optimist selecting an environment suited to his genotype when he looks at his world through his rose-colored glasses."

cognitive ability are the driving forces of *r*GE for intelligence. Specifically, they reasoned that early, relatively weak, genetic influences on cognitive ability become amplified through a cascading process in which "higher IQ leads one into better environments causing still higher IQ, and so on." More recently, we have proposed that "noncognitive factors – including levels of scholastic motivation, drive for achievement, intellectual self-concept, and intellectual interest – are also critical for the process of selecting environmental niches" (Tucker-Drob & Harden, 2012a), and that "children select, evoke, and attend to learning experiences that are consistent with genetically influenced individual differences in their motivation to learn" (Tucker-Drob & Harden, 2012b).

Our proposal, illustrated in Figure 6.1, builds on several pre-existing theories. Cattell originally proposed *investment theory* in the context of his broader theory of fluid and crystallized intelligence. Cattell hypothesized that "this year's crystallized ability level is a function of last year's fluid ability level – and last year's interest in school work and abstract problems generally" (Cattell, 1987, p. 139). To elaborate, Cattell proposed that knowledge acquisition results from effortfully reasoning through cognitively challenging problems and ideas. Upon success, the procedural and the declarative knowledge gleaned "crystallizes." According to the investment hypothesis, learning is thought to depend both on non-cognitive factors that lead children to engage effortfully with challenging problems, and on cognitive factors that enable children to determine efficient procedures and mental schemas for solving and understanding those problems. Investment theory, which resembles Piaget's theory of *accommodation*, has been elaborated on by several authors, including Ackerman (1996) and Chamorro-Premuzic and Furnham (2004).

Although investment theory highlights the dual roles of both cognitive ability and non-cognitive factors, it does not specifically link non-cognitive factors to genetic differences. In contrast, Hayes' (1962) Experience-Producing Drive theory appears to have been one of the first to suggest that genetic influences on intelligence might be mediated by non-cognitive factors. Hayes wrote, "Intelligence is acquired by learning, and inherited motivational makeup influences the kind and amount of learning which occurs. The hereditary basis of intelligence consists of drives, rather than abilities as such" (p. 302).

Experience-Producing Drive theory has been elaborated on by Bouchard (1997) and, more recently, by Johnson (2010). Johnson (2010) wrote:

> genes exert their influences on the development of … traits through their control of motivations, preferences, and emotional responses. Over time, these motivations, preferences, and emotional responses drive the

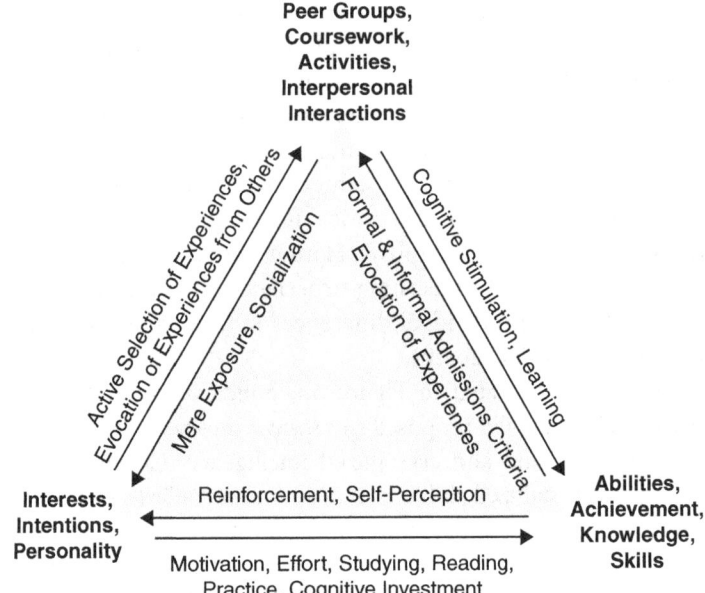

FIGURE 6.1 A conceptual model for the mutual relations between interests, proximal environments, and achievement. Reproduced from Tucker-Drob and Harden (2012a).

acquisition of experiences that result in the development, practice, and pursuance of skills, habits, patterns of response, and environmental circumstances, which in turn reinforce the underlying drivers.

In arguing for the importance of non-cognitive factors, Hayes (1962) wrote that "use of the plural, 'experience-producing drives' (EPD), raises the question of how many such mechanisms are involved." Indeed, many different academically relevant non-cognitive factors have been mentioned in the literature, and many different instruments have been developed to measure them. For example, Heckman and Ruinstein (2001) have written, "motivation, tenacity, trustworthiness, and perseverance are important traits for success in life." Similarly, Duckworth and Yeager (2015) have written, "There has been perennial interest in personal qualities other than cognitive ability that determine success, including self-control, grit, growth mindset, and many others." In the following section, we describe some of the key non-cognitive factors that have been investigated in the social, developmental, and educational psychology literatures. We evaluate whether there is evidence to support the hypothesis that each non-cognitive factor is involved in gene–environment transactions in the development of academic achievement.

CRITERIA FOR THE ROLE OF NON-COGNITIVE
FACTORS IN GENE–ENVIRONMENT
TRANSACTIONS

Before surveying the specific non-cognitive factors that have been routinely mentioned in the psychological literature, it is useful to consider what sorts of empirical evidence would be necessary to support the hypothesis that a non-cognitive factor is involved in academically relevant gene–environment transactions. We propose six general criteria:

Criterion 1. First, the non-cognitive factor should be correlated with academic achievement. In the absence of suppression effects (caused, for example, by a compensatory process), a correlation is a necessary (though not sufficient) condition of causality within the naturally occurring system under scrutiny. This sort of evidence is straightforward to obtain from observational data from representative samples and does need not come from genetically informative data.

Criterion 2. Second, it is important to consider whether the non-cognitive factor predicts achievement *incremental* to intelligence and the major dimensions of individual differences in personality known as the Big Five. The term "non-cognitive" specifically implies that the factor not only represents something other than cognitive ability, but has effects on achievement that are unique of cognitive ability. Moreover, as described later, myriad specific non-cognitive factors have been mentioned in the literature, and it would be useful to ascertain whether they represent something other than the Big Five personality traits or are conversely "a rose by any other name."

Criterion 3. Third, because we predict that genotypes become matched with educationally relevant environments via the effects of genes on non-cognitive factors, the non-cognitive factor should be, to some degree, heritable. This sort of evidence requires data from a genetically informed study, such as a twin or adoption study.

Criterion 4. Fourth, there should be a non-zero genetic correlation between the non-cognitive factor and academic achievement. In other words, the genetic variants that influence the non-cognitive factor should also be associated with variation in achievement. As formalized by Dickens and Flynn (2001), transactional models require that environmental experiences occur persistently and/or recurrently over time in order to have appreciable effects on achievement outcomes. Environments that are evoked, selected, and attended to on the basis of genetically influenced factors, which are themselves highly stable over time (Briley & Tucker-Drob, 2014; Tucker-Drob & Briley, 2014), are, therefore, most likely to occur with sufficient stability and persistence to have appreciable effects on achievement outcomes. This sort of evidence also requires data from a genetically informed study.

Criterion 5. Fifth, the direction of causation should be, at least partially, from the non-cognitive factor to achievement. Direction of causation can be evaluated, for example, via a cross-lagged panel model applied to longitudinal data on both the non-cognitive factor and achievement. Data that are both longitudinal and genetically informative are particularly valuable here because such data allow one to estimate the extent to which a longitudinal effect is genetically mediated (e.g., Tucker-Drob & Harden, 2012c). Additional, co-occurring causation from achievement to the non-cognitive factor would also be consistent with the transactional model, and such bidirectional causation could serve to amplify early genetic variation. Experimental evidence that interventions targeting a specific non-cognitive factor also influence achievement would also be relevant to Criterion 5, in demonstrating that the changes in the non-cognitive factor *can* cause changes in achievement. Such evidence, however, would not necessarily be indicative that naturally occurring genetic variation in the non-cognitive factor causes individual differences in achievement.

Criterion 6. Sixth, environmental experiences relevant for achievement should mediate the causal effect of non-cognitive factors on achievement *via genetic pathways*. In principle, this prediction can be tested in a multivariate behavioral genetic design, in which the non-cognitive factor, achievement, and the environmental experiences relevant for achievement are all measured. Unfortunately, however, identifying and measuring the many experiences relevant for achievement may be a herculean task, particularly, if the individual effects of specific environmental experiences are themselves small (cf. Turkheimer & Waldron, 2000). As Tucker-Drob and Briley (2014) have previously proposed, "if the environments relevant for cognition are those that result from stable and enduring gene–environment correlation processes (Dickens & Flynn, 2001), it may be the case that efforts to measure genetically influenced psychological tendencies to engage with a host of stimulating experiences [i.e. noncognitive factors] will prove to be more fruitful – albeit less direct – for indexing cumulative environmental effects than efforts to measure the experiences themselves."

With these criteria in mind, we now review current evidence regarding gene–environment transactions involving an array of non-cognitive factors, including: (1) Big Five personality traits, (2) intellectual interest, (3) academic interest, (4) self-perceived ability, (5) grit, (6) self-control/ impulse control; (7) achievement goal orientations (mastery/performance), (8) intelligence mindsets/implicit theories of intelligence, and (9) expectancies and values.

BIG FIVE PERSONALITY TRAITS

The "Big Five" personality trait taxonomy has become the dominant dimensional account of personality for the past quarter century (John, Naumann, & Soto, 2008). The factors (Openness, Conscientiousness, Extraversion, Agreeableness, and Neuroticism) represent individual differences in patterns of thinking, feeling, and behaving that are relatively stable across time and context. These factors, which have been well replicated across many different samples and testing formats, were not specifically derived from research on academic achievement, or research in any specific domain of life for that matter, but were instead derived from research seeking to describe the entire range of human personality across domains of life. The Big Five trait dimensions are, therefore, necessarily broad, and do not purport to capture all of the more nuanced, or context-specific, aspects of behavior. They serve as a sensible starting point in the search for non-cognitive correlates of academic achievement.

The two Big Five traits that have been consistently linked with achievement are Conscientiousness and Openness. Conscientiousness refers to a general tendency to be organized, planful, and effortful in one's goals and duties. Openness, also commonly termed Intellect, refers to a general tendency to be interested in ideas and creative pursuits. In a comprehensive meta-analysis of the relations between the Big Five and academic achievement, Poropat (2009) reported that Conscientiousness and Openness were correlated with achievement (typically course grades or overall grade point average – GPA) at .22 and .12, respectively. IQ was correlated with achievement at .25. Controlling for intelligence did not appreciably reduce personality–achievement associations. Thus, the evidence regarding Conscientiousness and Openness meets Criterion 1 (correlated with achievement) and Criterion 2 (correlated with achievement incremental to IQ).

Of all of the non-cognitive factors surveyed in the current article, it is likely that the Big Five are the factors that have been most extensively studied in behavioral genetic research designs. A number of surveys of the behavioral genetic literature on personality exist, and they largely agree in placing the overall magnitude of heritability for each of the Big Five traits at between approximately .40 and .50 (Bouchard & Mc Gue, 2003). Moreover, a recent meta-analysis of longitudinal behavioral genetic studies of personality (Briley & Tucker-Drob, 2014) indicated that genetic factors moderately contribute to longitudinal rank–order stability in Big Five personality traits across nearly the entire lifespan, from infancy to old age. Criterion 3 (the non-cognitive factor is heritable) is clearly met for all Big Five traits.

A handful of studies have presented evidence relevant to Criterion 4. In a twin and extended family design consisting of 650 families, Luciano, Wainwright, Wright, and Martin (2006) reported that variance shared between two facets of Conscientiousness (Competence and Dutifulness) and scores on measures of verbal IQ, performance IQ, and academic achievement was primarily explained by a common genetic factor. In what appears to have been a largely overlapping sample, Wainwright et al. (2008) reported positive genetic correlations between multiple facets of Openness and verbal IQ, performance IQ, academic achievement, and processing speed. Thus there is support for Criterion 4 (non-zero genetic correlation between the non-cognitive factor and achievement) for both Conscientiousness and Openness.

Evidence regarding Criteria 5 and 6 is more tentative. With respect to Criterion 5 (causation from the non-cognitive factor to achievement), there have been relatively few longitudinal studies of the relation between personality and later academic achievement, and it appears that those that do exist have not controlled for previous levels of academic achievement, as could be achieved in a cross-lagged approach (Chamorro-Premuzic & Frunham, 2003; Heaven, Ciarrochi, & Vialle, 2007). A meta-analysis by Richardson, Abraham, and Bond (2012) reported that Conscientiousness remained predictive of college GPA after controlling for both SAT/ACT and high school GPA, but it is not clear whether the studies contributing the meta-analysis measured Conscientiousness prior to, or concurrent with, college attendance. The former would be better evidence of a (prospective) directional association from Conscientiousness to achievement. With respect to Criterion 6 (environmental experiences relative to achievement should mediate the causal effect of non-cognitive factors on achievement via genetic pathways), we are not aware of any behavioral genetic studies of the Big Five personality traits and academically relevant environmental experiences. However, there is support for genetically mediated longitudinal associations between personality and life events more generally. For example, in a study of 338 adult twin pairs, Kandler, Bleidorn, Reimann, Angleitner, and Spinath (2012) found that associations between personality and life events were "primarily directional from personality to life events, and basically genetically mediated." Thus, there is circumstantial evidence in support of Criterion 6 for personality.

INTELLECTUAL INTEREST/INTELLECTUAL CURIOSITY

Although measures of the Big Five personality traits are the most popular personality measures in differential psychology, social and educational psychologists who are specifically interested in understanding cognitive development and academic achievement have developed and implemented an array of more narrow non-cognitive measures to specifically tap behavioral tendencies more directly tied to achievement.

Von Stumm, Hell, and Chamorro-Premuzic (2011) proposed that *intellectual curiosity* is the "third pillar of academic performance," after intelligence and conscientiousness. Intellectual curiosity refers to the desire to think, reason, learn, and understand. It also includes a preference for, enjoyment of, and history of engaging in thinking, reasoning, and learning behaviors. Intellectual curiosity is commonly measured with Ackerman's Typical Intellectual Engagement (TIE; Goff & Ackerman, 1992) scale and Cacioppo's Need for Cognition scale (Cacioppo & Petter, 1982). These scales lack discriminant validity from one another (Mussel, 2010; Woo, Harms, & Kuncel, 2007), but are distinguishable from (albeit correlated with) Openness. Regarding Criterion 1, Von Stumm et al. (2011) report a meta-analytic correlation of .33 between TIE and academic performance (GPA or an achievement test composite). Supporting Criterion 2, TIE remained uniquely associated with academic performance after controlling for openness, conscientiousness, and intelligence.

Compared to the Big Five, much less behavioral genetic research has been done on intellectual interest. One study (Tucker-Drob & Harden, 2012) reported that intellectual interest was 27 percent heritable (Criterion 3), and that genetic factors mediated approximately one third of the correlation ($r = .44$) between intellectual interest and academic achievement (Criterion 4). Regarding Criterion 5, there does not appear to be research that has employed longitudinal models to test for direction of causation between intellectual interest and achievement, although there has been evidence of this type for academic interest (Marsh et al., 2005, described below). With respect to Criterion 6, we are not aware of behavioral genetic studies that have directly examined associations between intellectual interest and academically relevant learning experiences.

ACADEMIC INTEREST

It is unclear whether academic interest is distinguishable from intellectual interest, and the two terms are often used interchangeably. It is possible that the term academic interest allows for more domain specificity, i.e., for different self concepts in math, science, and reading. Supporting Criterion 1, a meta-analysis by Schiefele et al. (1992) placed the magnitude of the domain-specific interest–achievement association at an average of .31, ranging from .16 to .35, with the magnitude differing somewhat by domain (e.g., mathematics vs. literature). In a large sample of approximately 375,000 American high-schoolers, Tucker-Drob and Briley (2012) similarly estimated the average magnitude of the domain-specific interest–achievement association to be .31, ranging from .22 to .44 depending on the domain (e.g., art, biological sciences, literature), with the exception of .07 for hunting/fishing, with associations persisting at nearly full strength after controlling for intelligence (Criterion 2).

In a study combining data from nearly 13,000 twins aged nine to sixteen years from six different countries, Kovas et al. (2015) reported that the heritability of subject-specific enjoyment (e.g., reading, math, and English) was consistently estimated at approximately 40 percent across subsamples (Criterion 3). There is also evidence from longitudinal research supporting Criterion 5. For instance, Marsh et al. (2005) used cross-lagged longitudinal models to test the direction of causation between academic interest (math interest) and academic achievement (math grades and test scores). Results indicated a stronger standardized effect from interest to later achievement (controlling for past achievement) than from achievement to later interest. There does not appear to have been research on academic interest directly relevant to Criteria 4 and 6.

SELF-PERCEIVED ABILITY

Self-perceived ability, self-assessed intelligence, and academic self concept are all closely related constructs that may be measured using domain-general or domain-specific (e.g., math self-concept, reading self-concept) levels. As well summarized by Chamorro-Premuzic, Harlaar, Greven, and Plomin (2010), these non-cognitive factors may relate to achievement through "both 'insight' (children's accounts of their previous performance) and self-efficacy (the self-fulfilling or motivational effects of self-beliefs)."

This general set of constructs is among the non-cognitive factors most studied using longitudinal methods. For instance, Marsh and Craven's (2006) review of results from a number of studies that have applied latent variable cross-lagged panel models presented evidence for a "reciprocal effects model," in which academic self-concept predicts later achievement above and beyond previous levels of achievement, *and* achievement predicts later academic self-concept above and beyond previous levels of academic self-concept (Criterion 5). Chamorro-Premuzic et al. (2010) have also reported evidence for reciprocal effects between self-perceived ability and achievement in a latent variable cross-lagged panel analysis, even after controlling for general cognitive ability at each wave (Criteria 2 and 5). Using data from nearly 13,000 children aged nine to sixteen years from six different countries, Kovas et al. (2015) reported that the heritability of subject-specific self-perceived ability was consistently estimated at approximately 40 percent across subsamples (Criterion 3). Moreover, consistent with the predictions of a transactional model, Greven, Harlaar, Kovas, Chamorro-Premuzic, and Plomin (2009) found that the associations between self-perceived ability and both concurrent achievement and later achievement, independent of IQ, were genetically mediated (Criteria 1, 2, 4, and 5). We are not aware of research on self-perceived ability relevant to Criterion 6.

GRIT

Grit refers to the tendency to take actions toward a prioritized goal (such as high academic achievement), in the face of challenges or temptations to do otherwise. Grit has been found to be highly related to, albeit distinguishable from, Conscientiousness (Duckworth et al., 2007), and has been linked with achievement even when IQ or SAT are controlled (Duckworth et al., 2007; Criteria 1 and 2), although this link may not persist after controlling for conscientiousness (Ivcevic and Brackett, 2014; possible failure of Criterion 2). We are aware of no genetically informed work on grit, so there is currently no evidence regarding Criteria 3, 4, or 6. We are also not aware of longitudinal research on the relation between grit and achievement that has controlled for previous levels of achievement, so there does not appear to be evidence relevant to Criterion 5.

IMPULSE CONTROL/SELF-CONTROL

Self-control (also described as impulse control, inhibitory control, inhibition, or – when conceptualized in the opposite valence – impulsivity) has been distinguished from grit primarily in terms of timescale. As Duckworth and Gross (2014) write, "self-control entails aligning actions with any valued goal despite momentarily more-alluring alternatives; grit, in contrast, entails having and working assiduously toward a single challenging superordinate goal through thick and thin, on a timescale of years or even decades." In addition, whereas grit is typically measured with a self-report instrument, self-control has been measured in myriad ways, ranging from self-report to performance-based measures. Importantly, measurement of impulse control has suffered from both the "jingle" and "jangle" fallacies (Whiteside & Lynam, 2001). Impulse control is, in fact, a highly complex construct that has been variably organized into four (Whiteside & Lynam, 2001), five (Cyders et al., 2007), or even eight different dimensions (Nigg, 2000). Moreover, self-report measures of impulse control correspond weakly – if at all – with performance-based measures (Cyders & Coskunpinar, 2011; Dick et al., 2010), and different performance measures are variably related to different underlying neural systems (Jentsch et al., 2014). Moffitt and colleagues (2011) combined multiple, longitudinal measures of self-control (including observational ratings of behavior in a testing situation in early childhood, and parent and teacher ratings on impulsivity scales in middle childhood) and found that self-control prospectively predicted academic achievement, even after controlling for IQ (Criteria 1, 2, and possibly 5). Duckworth and Seligman (2005) found that a "self-discipline" composite formed from self-report, parent-report, and teacher-report measures of impulsivity and self-control longitudinally predicted GPA above and beyond IQ and earlier GPA (Criteria 1, 2, and 5). In addition, various measures of self-control have been found to be heritable (Criterion 3), including both performance-based measures (Friedman et al., 2008) and self-report measures (Ellingson, Verges, Littlefield, Martin, & Slutske, 2013). We are not aware of research on self-control that is directly relevant to Criteria 4 or 6.

ACHIEVEMENT GOAL ORIENTATIONS

A number of related theoretical frameworks (Kaplan & Maehr, 2007; Meece, Anderman, & Anderman, 2006) generally conceptualize achievement goals as falling along two distinguishable dimensions. The *mastery goal orientation* dimension distinguishes individuals by how motivated they are to engage with education for the purposes of learning the material and increasing their competence, whereas the *performance goal orientation* dimension distinguishes individuals by how motivated they are to engage with education for the purposes of appearing competent to others and visibly outperforming their classmates. Mastery goals are generally found to be positively associated with academic achievement, whereas results have been inconsistent for performance goals, with many articles reporting negative associations and many others reporting positive associations. A meta-analysis by Hulleman, Schrager, Bodmann, and Harackiewicz (2010) indicated that inconsistences in previous research might be attributable to differences in the content of the specific measures used. Nevertheless, the absolute magnitudes of associations between goal orientations and academic performance were generally estimated to be low, i.e., in the vicinity of an absolute correlation of .10 (Criterion 1). A meta-analysis of correlates of university student's academic performance by Richardson et al. (2012) indicated that goal orientations were cross-sectionally associated with achievement at $r = .15$ and prospectively associated with achievement at $r = .09$, although the extent to which the studies contributing to the prospective meta-analytic estimate controlled for baseline achievement (as would be required for Criterion 5) is not clear.

Very little behavioral genetic work appears to have been conducted on achievement goal orientations. The only such study that we are aware of is that of Murayama, Elliot, and Yamagata (2011), who used data from Japanese twins to estimate the heritability of performance goals at approximately 40 percent (Criterion 3), with the remaining variation attributable to the non-shared environment. There appears to be no behavioral genetic evidence relevant to Criteria 4 and 6.

INTELLIGENCE MINDSETS/IMPLICIT THEORIES OF INTELLIGENCE

Dweck and colleagues (Dweck, 1999; Dweck & Leggett, 1988) have proposed a motivational model of achievement that is based on children's lay "theories" or "mindsets" about whether intelligence is fixed (an "entity theory") or malleable (an "incremental theory"). Intelligence mindsets are typically measured with a unidimensional scale, with entity and incremental mindsets occupying polar extremes. Because they view intelligence as something that can be changed with hard work, children with incremental mindsets are thought to put more effort into learning. Conversely, because they are more likely to view effort as futile for improving intelligence, children with entity mindsets are thought to put less effort into learning. Dweck and colleagues (e.g., Blackwell, Trzesniewski, & Dweck, 2007) have reported that naturally occurring variation in mindsets is predictive of later changes in academic achievement (Criteria 1 and 5), and that interventions to teach incremental theories of intelligence improve achievement outcomes. We are aware of very little behavioral genetic work on intelligence mindsets. For instance, a study by Spinath, Spinath, Riemann, and Angleitner (2003) examined intelligence mindsets in twins, but did not report results of behavioral genetic analyses. There, therefore, does not appear to be evidence relevant to Criteria 3, 4, and 6.

EXPECTANCIES AND VALUES

Eccles' theory of academic effort and motivation focuses on the intersection of expectancies and values (Nagengast, Marsh, Scalas, Xu, Hau, & Trautwein, 2011; Wigfield & Eccles, 2000). *Expectancies* refer to the student's expectations that they are capable of succeeding academically, and *values* refer to the student's perception that academic success is a desirable goal. Expectancies and values have been measured in a number of different ways, and it is likely that expectancies and values are each themselves multidimensional. In fact, a number of the constructs reviewed above have been classified as forms of expectancies and values (see, e.g., Hulleman, Barron, Kosovich, & Lazowski, 2014). For example, self-perceived ability can be construed as an expectancy regarding success. In addition, the Expectancy-Value theory highlights other sorts of constructs not previously discussed. For instance, students' perceptions about the value of education for success in life, their educational attainment goals, and their expectations regarding their ultimate levels of educational attainment are key constructs in the Expectancy-Value model that have been linked with academic achievement.

The Expectancy-Value model is supported by evidence linking expectancies and values (along with their interactions) to motivation and academic achievement (Nagengast et al., 2011; Wigfield & Eccles, 2000; Criterion 1), including cross-lagged longitudinal evidence of reciprocal associations between expectancies and values and achievement (Zhang et al., 2011; Criterion 5). There is also evidence that at least some expectancy and value measures predict achievement above and beyond intelligence (Spinath et al., 2006; Criterion 2). Perhaps because the model is largely set up as a socialization model, (Wigfield, Eccles, Fredricks, Simpkins, Roeser, & Schiefele, 2015), very little behavioral genetic work has been conducted under the Expectancy-Value framework. One exception comes from Briley, Harden, and Tucker-Drob (2014), who used both behavioral genetic and cross-lagged longitudinal models to demonstrate that *parental* educational expectations for their children's ultimate educational attainment (which, in Expectancy-Value theory is theorized to be key mechanism through which parents instill educational expectancies in their children; Jacobs & Eccles, 2000) are both predictive of their children's academic achievement and educationally relevant behaviors *and* sensitive to genetically influenced individual differences in child achievement and educationally relevant behaviors. However, we are aware of no behavioral genetic work on *child* non-cognitive factors that has been conducted explicitly in the context of the Expectancy-Value model. Tests of Criteria 3 and 4 are, therefore, lacking. One relative strength of research conducted in the context of the Expectancy-Value model is that researchers often used expectancies and values to predict academically relevant experiences, such as course enrollment and extra-curricular activity involvement (e.g., Meece, Wigfield, & Ecccles, 1990; Nagengast et al., 2011), partially fulfilling Criterion 6.

SUMMARIZING THE EMPIRICAL EVIDENCE ON NON-COGNITIVE FACTORS

The research supporting the role of non-cognitive factors in gene–environment transactions is summarized in Table 6.1. As described earlier, different constructs and measures vary in the extent to which they are correlated with academic achievement, the extent to which this correlation has been tested in longitudinal designs capable of disentangling causation from reverse causation, and the extent to which they have been examined using genetically informative methods. Currently, self-perceived ability and intellectual interest are the non-cognitive factors with the most evidence supportive of their roles in academically relevant gene–environment transactions. Perhaps not coincidentally, these are also the two non-cognitive factors that have been most systematically studied using both longitudinal and genetically informative designs.

SUGGESTIONS FOR FUTURE RESEARCH

Genetic differences between people matter for how well they perform in school settings. At the same time, children are clearly more likely to succeed academically if they experience high-quality home and school

TABLE 6.1 *Status of empirical support for specific non-cognitive factors as mechanisms of gene–environment transactions for academic achievement*

Non-Cognitive Factors	Criterion 1 Correlated with academic achievement	Criterion 2 Correlated with achievement incremental to IQ	Criterion 3 Heritable	Criterion 4 Genetic association with achievement	Criterion 5 Longitudinally predicts achievement (controlling for past achievement)	Criterion 6 Genes → non-cognitive factor → environment → achievement
Conscientiousness	+	+	+	+	~	~
Openness	+	+	+	+	~	
Intellectual Interest	+	+	+	+		
Academic Interest	+	+	+		+	
Self-Perceived Ability	+	+	+	+	+	
Grit	+	+				
Self-Control/Impulse Control	+	+	+		+	
Mastery Achievement Goal Orientations	+				~	
Performance Achievement Goal Orientations	~				~	
Intelligence Mindsets/ Implicit Theories of Intelligence	+		+		+	
Expectancies and Values	+	+			+	~

+ indicates positive empirical support for criterion identified in the current literature review. ~ indicates mixed or incomplete empirical support for criterion identified in the current literature review. Empty cells indicate no direct tests of the criterion identified in the current literature review.

environments. A rich and varied theoretical literature has proposed that genes, non-cognitive skills, cognitive ability, environmental experiences, and academic achievement are linked in a reciprocal and dynamic system. According to this proposal, children who are conscientious and open to new experiences, who like thinking about abstract ideas and who do not give up when frustrated, experience different types of environments than children without these skills. They are given more positive attention from teachers, are placed into more challenging classes, read more books, spend more time and effort on difficult problems, and interpret their successes and failures differently. Because individual differences in these non-cognitive skills are expected to be at least partially due to genetic differences between people, when non-cognitive skills shape a child's environmental experiences, this process results in *r*GE – children with certain genotypes are more likely to experience certain environments. If genotypes are matched to environments, and environments have causal effects on both achievement and non-cognitive skills, the net result of this process will be high heritability estimates for achievement – not in spite of the environment, but through the environment (Tucker-Drob et al., 2013).

Importantly, despite the considerable appeal of this theoretical model, and its recurrence in the literature over several decades, no study to date has tested a comprehensive model of the links between genes, non-cognitive skills, environmental inputs, and achievement in a longitudinal, genetically informative study. This hole in the empirical literature is probably due, at least in part, to disciplinary divides: Researchers in developmental, educational, and social psychology who focus on understanding children's motivations for learning and on measuring the quality of their academic experiences often view behavioral genetics research with some suspicion (if not outright hostility). At the same time, behavioral genetic researchers have commonly conceptualized *r*GE as a source of confounding that creates illusory correlations between environmental experiences and important life outcomes, rather than as a key mechanism for the emergence of heritable variation in academic achievement. Interdisciplinary research that pays attention to both behavioral genetic theory and research design, *and* to recent developments in the measurement of non-cognitive skills and academically relevant environments, will be necessary to understand the dynamic nuances of gene–environment transactions.

This chapter has focused on *r*GE, in which individuals with certain genotypes are more likely to experience certain environments. In contrast, gene–environment *interaction*, or GxE, implies that individuals with different genotypes differ in their response to environment inputs (e.g., are more vulnerable to adverse environments or sensitive to enriching ones), *and* that genotypes are more potent predictors of phenotypes in certain environments. Although the existence of *r*GE does guarantee the existence of gene–environment interaction (and vice versa), *r*GE may be, in some cases, a key mechanism driving G×E interactions on academic achievement (Tucker-Drob et al., 2013; Turkheimer & Horn, 2014). As we have described, the transactional model predicts that children will differentially select, evoke, and attend to environment inputs on the basis of their genetically influenced traits, including both initial cognitive abilities and non-cognitive skills. Importantly, however, this process of genotype–environment matching depends on there being an array of potential environmental inputs. For a motivated child to read, she needs access to a library. In macroenvironmental contexts (e.g., schools, neighborhoods, social classes) in which there is a more limited "cafeteria" of proximal learning experiences from which to choose, individual differences in preferences and interests are expected to have less relevance for the sorts of proximal environments that are experienced, and hence less relevance for cognitive development and academic achievement. This decoupling between genetically influenced non-cognitive factors and environmental inputs will result in diminished heritable variation in cognition and achievement in certain macroenvironmental contexts. Consistent with this proposal, we have found that both domain-general and domain-specific interest is more strongly predictive of knowledge and achievement outcomes in socioeconomically advantaged family, school, and national contexts (Tucker-Drob & Briley, 2012; Tucker-Drob, Cheung, & Briley, 2014). Moreover, in genetically informative samples, we have found that this interaction with socioeconomic status occurs on the genetic link between interest and achievement: genetic variation in interest is more strongly predictive of achievement under conditions of (family-level) socioeconomic advantage (Tucker-Drob & Harden 2012a, 2012b).

Importantly, it is not necessarily the case that all non-cognitive factors interact with socioeconomic status, or other markers of environmental opportunity, in the same way. We have focused our previous research primarily on interest, but other factors, such as grit or implicit theories of intelligence, could interact with macroenvironmental contexts according to altogether different patterns, or not at all. For instance, while interest is *more* predictive of achievement in more advantaged contexts, self-regulatory behaviors (e.g., grit, impulse control) may be *less* predictive of achievement in more advantaged contexts, which contain external behavioral scaffolds for positive approaches toward learning. Thus, we caution researchers interested in testing developmental theory against treating non-cognitive factors as interchangeable measures or treating any single measure as a "model phenotype" (Briley & Tucker-Drob, 2015).

CONCLUSIONS

Across the past half century, a variety of theorists – including Hayes (1962), Cattell (1987), Scarr and McCartney (1983), Bronfenbrenner and Ceci (1994), Dickens and Flynn (2001), Johnson (2010), Turkheimer and Horn (2014), and ourselves (Tucker-Drob et al., 2013; Tucker-Drob & Harden, 2012a, 2012b, 2012c) – have posited that heritable individual differences in cognitive ability and academic achievement emerge and widen via dynamic, reciprocal transactions between children's genetically influenced abilities and their specific environmental experiences. The current chapter contributes to this literature by integrating behavioral genetic theories of *r*GE with insights from social, developmental, and educational psychology regarding how to conceptualize and measure a diverse array of non-cognitive skills. These non-cognitive factors may operate as "experience-producing drives" (Hayes, 1962) and thus act as critical intermediaries in the process of gene–environment matching; however, as we have reviewed here, there is a paucity of behavioral genetic research on non-cognitive factors that directly tests this hypothesis. Our interdisciplinary synthesis points to the importance of longitudinal, genetically informed research that incorporates careful measurement not just of ability and achievement, but also non-cognitive factors and environmental experience. Such research is critically necessary to provide a more complete understanding of the developmental mechanisms that give rise to individual differences in cognitive development and academic achievement.

REFERENCES

Ackerman, P. L. (1996). A theory of adult intellectual development: Process, personality, interests, and knowledge. *Intelligence, 22,* 227–257.

Blackwell, L. S., Trzesniewski, K. H., & Dweck, C. S. (2007). Implicit theories of intelligence predict achievement across an adolescent transition: A longitudinal study and an intervention. *Child Development, 78,* 246–263.

Bouchard, T. J. (1997). Experience producing drive theory: How genes drive experience and shape personality. *Acta Paediatrica, 86,* 60–64.

Bouchard, T. J., & Mc Gue, M. (2003). Genetic and environmental influences on human psychological differences. *Journal of Neurobiology, 54*(1), 4–45.

Bronfenbrenner, U., & Ceci, S. J. (1994). Nature-nuture reconceptualized in developmental perspective: A bioecological model. *Psychological Review*, *101*, 568.

Briley, D. A., Harden, K. P., & Tucker-Drob, E. M. (2014). Child characteristics and parental educational expectations: Evidence for transmission with transaction. *Developmental Psychology, 50*, 2614–2632.

Briley, D. A., & Tucker-Drob, E. M. (2013). Explaining the increasing heritability of cognition over development: A meta-analysis of longitudinal twin and adoption studies. *Psychological Science, 24*, 1704–1713.

(2014). Genetic and environmental continuity in personality development: A meta-analysis. *Psychological Bulletin, 140*, 1303–1331.

(2015). Comparing the developmental genetics of cognition and personality over the lifespan (Invited submission). *Journal of Personality* (Special Issue: The New Look of Behavior Genetics in Social Inequality: Gene-Environment Interplay in Life Chances).

Cacioppo, J. T., & Petty, R. E. (1982). The need for cognition. *Journal of Personality and Social Psychology, 42*, 116.

Cattell, R. B. (1987). *Intelligence: Its structure, growth and action.* New York: Elsevier.

Chamorro-Premuzic, T., & Furnham, A. (2003). Personality predicts academic performance: Evidence from two longitudinal university samples. *Journal of Research in Personality, 37*, 319–338.

Chamorro-Premuzic, T., & Furnham, A. (2004). A possible model for understanding the personality-intelligence interface. *British Journal of Psychology, 95*, 249–264.

Chamorro-Premuzic, T., Harlaar, N., Greven, C. U., & Plomin, R. (2010). More than just IQ: A longitudinal examination of self-perceived abilities as predictors of academic performance in a large sample of UK twins. *Intelligence, 38*, 385–392.

Cyders, M. A., Smith, G. T., Spillane, N. S., Fischer, S., Annus, A. M., & Peterson, C. (2007). Integration of impulsivity and positive mood to predict risky behavior: Development and validation of a measure of positive urgency. *Psychological Assessment, 19*, 107.

Cyders, M. A., & Coskunpinar, A. (2011). Measurement of constructs using self-report and behavioral lab tasks: Is there overlap in nomothetic span and construct representation for impulsivity? *Clinical Psychology Review, 31*, 965–982.

Dick, D. M., Smith, G., Olausson, P., Mitchell, S. H., Leeman, R. F., O'Malley, S. S., et al. (2010). Review: Understanding the construct of impulsivity and its relationship to alcohol use disorders. *Addiction Biology, 15*, 217–226.

Dickens, W. T., & Flynn, J. R. (2001). Heritability estimates versus large environmental effects: The IQ paradox resolved. *Psychological Review, 108*(2), 346.

Duckworth, A., & Gross, J. J. (2014). Self-control and grit related but separable determinants of success. *Current Directions in Psychological Science, 23*, 319–325.

Duckworth, A. L., Peterson, C., Matthews, M. D., & Kelly, D. R. (2007). Grit: Perseverance and passion for long-term goals. *Journal of Personality and Social Psychology, 92*, 1087–1101.

Duckworth, A. L., & Seligman, M. E. (2005). Self-discipline outdoes IQ in predicting academic performance of adolescents. *Psychological Science, 16*, 939–944.

Duckworth, A. L., & Yeager, D. S. (2015). Measurement matters: Assessing personal qualities other than cognitive ability for educational purposes. *Educational Researcher, 44*, 237–251.

Dweck, C. S. (1999). *Self-theories: Their role in motivation, personality, and development.* Philadelphia, PA: Psychology Press.

Dweck, C. S., & Leggett, E. L. (1988). A social-cognitive approach to motivation and personality. *Psychological Review, 95*, 256.

Ellingson, J. M., Verges, A., Littlefield, A. K., Martin, N. G., & Slutske, W. S. (2013). Are bottom-up and top-down traits in dual-systems models of risky behavior genetically distinct? *Behavior Genetics, 43*, 480–490.

Farrington, C. A., Roderick, M., Allensworth, E., Nagaoka, J., Keyes, T. S., Johnson, D. W., et al. (2012). *Teaching adolescents to become learners: The role of non-cognitive factors in shaping school performance – A critical literature review.* Chicago, IL: Consortium on Chicago School Research.

Friedman, N. P., Miyake, A., Young, S. E., DeFries, J. C., Corley, R. P., & Hewitt, J. K. (2008). Individual differences in executive functions are almost entirely genetic in origin. *Journal of Experimental Psychology: General, 137*, 201.

Goff, M., & Ackerman, P. L. (1992). Personality-intelligence relations: Assessment of typical intellectual engagement. *Journal of Educational Psychology, 84*, 537.

Greven, C. U., Harlaar, N., Kovas, Y., Chamorro-Premuzic, T., & Plomin, R. (2009). More than just IQ school achievement is predicted by self-perceived abilities – But for genetic rather than environmental reasons. *Psychological Science, 20*(6), 753–762.

Heaven, P. C., Ciarrochi, J., & Vialle, W. (2007). Conscientiousness and Eysenckian psychoticism as predictors of school grades: A one-year longitudinal study. *Personality and Individual Differences, 42*, 535–546.

Hayes, K. J. (1962). Genes, drives, and intellect. *Psychological Reports, 10*, 299–342.

Heckman, J. J., & Rubinstein, Y. (2001). The importance of noncognitive skills: Lessons from the GED testing program. *American Economic Review, 91*(2), 145–149.

Hulleman, C. S., Barron, K. E., Kosovich, J. J., & Lazowski, R. A. (2014). Student motivation: Current theories, constructs, and interventions. In A. Lipnevich, F. Preckel, & R. Roberts (Eds.), *Psychosocial skills and school systems in the twenty-first century: Theory, research, and applications.* New York: Springer.

Hulleman, C. S., Schrager, S. M., Bodmann, S. M., & Harackiewicz, J. M. (2010). A meta-analytic review of achievement goal measures: Different labels for the same constructs or different constructs with similar labels? *Psychological Bulletin, 136*, 422.

Jacobs, J. E., & Eccles, J. S. (2000). Parents, task values, and real-life achievement-related choices. In C. Sansone & J. M. Harackiewicz (Eds.), *Intrinsic motivation* (pp. 405–439). San Diego, CA: Academic Press.

Jentsch, J. D., Ashenhurst, J. R., Cervantes, M. C., Groman, S. M., James, A. S., & Pennington, Z. T. (2014). Dissecting impulsivity and its relationships to drug addictions. *Annals of the New York Academy of Sciences, 1327*, 1–26.

John, O. P., Naumann, L. P., & Soto, C. J. (2008). Paradigm shift to the integrative big five trait taxonomy. In O. P. John, R. W. Robins, & L. A. Pervin (Eds.),

Handbook of personality: Theory and research (3rd ed., pp. 114–158). New York, NY: Guilford Press.

Johnson, W., Turkheimer, E., Gottesman, I. I., & Bouchard, T. J. (2009). Beyond heritability twin studies in behavioral research. *Current Directions in Psychological Science, 18*(4), 217–220.

Johnson, W. (2010). Extending and testing Tom Bouchard's experience producing drive theory. *Personality and Individual Differences, 49*, 296–301.

Kandler, C., Bleidorn, W., Riemann, R., Angleitner, A., & Spinath, F. M. (2012). Life events as environmental states and genetic traits and the role of personality: A longitudinal twin study. *Behavior Genetics, 42*, 57–72.

Kaplan, A., & Maehr, M. L. (2007). The contributions and prospects of goal orientation theory. *Educational Psychology Review, 19*, 141–184.

Kovas, Y., Garon-Carrier, G., Boivin, M., Petrill, S. A., Plomin, R., Malykh, S. B., et al. (2015). Why children differ in motivation to learn: Insights from over 13,000 twins from 6 countries. *Personality and Individual Differences, 80*, 51–63.

Luciano, M., Wainwright, M. A., Wright, M. J., & Martin, N. G. (2006). The heritability of conscientiousness facets and their relationship to IQ and academic achievement. *Personality and Individual Differences, 40*, 1189–1199.

Marsh, H. W., & Craven, R. G. (2006). Reciprocal effects of self-concept and performance from a multidimensional perspective: Beyond seductive pleasure and unidimensional perspectives. *Perspectives on Psychological Science, 1*(2), 133–163.

Marsh, H. W., Trautwein, U., Lüdtke, O., Köller, O., & Baumert, J. (2005). Academic self-concept, interest, grades, and standardized test scores: Reciprocal effects models of causal ordering. *Child Development, 76*, 397–416.

Meece, J. L., Anderman, E. M., & Anderman, L. H. (2006). Classroom goal structure, student motivation, and academic achievement. *Annual Review of Psychology, 57*, 487–503.

Meece, J. L., Wigfield, A., & Eccles, J. S. (1990). Predictors of math anxiety and its influence on young adolescents' course enrollment intentions and performance in mathematics. *Journal of Educational Psychology, 82*, 60.

Moffitt, T. E., Arseneault, L., Belsky, D., Dickson, N., Hancox, R. J., Harrington, H., et al. (2011). A gradient of childhood self-control predicts health, wealth, and public safety. *Proceedings of the National Academy of Sciences*, 201010076.

Murayama, K., Elliot, A. J., & Yamagata, S. (2011). Separation of performance-approach and performance-avoidance achievement goals: A broader analysis. *Journal of Educational Psychology, 103*, 238.

Mussel, P. (2010). Epistemic curiosity and related constructs: Lacking evidence of discriminant validity. *Personality and Individual Differences, 49*(5), 506–510.

Nagengast, B., Marsh, H. W., Scalas, L. F., Xu, M. K., Hau, K. T., & Trautwein, U. (2011). Who took the "×" out of expectancy-value theory? A psychological mystery, a substantive-methodological synergy, and a cross-national generalization. *Psychological Science, 22*(8), 1058–1066.

Plomin, R., DeFries, J. C., & Loehlin, J. C. (1977). Genotype-environment interaction and correlation in the analysis of human behavior. *Psychological Bulletin, 84*(2), 309.

Poropat, A. E. (2009). A meta-analysis of the five-factor model of personality and academic performance. *Psychological Bulletin, 135,* 322–338.

Scarr, S., & McCartney, K. (1983). How people make their own environments: A theory of genotype greater than environment effects. *Child Development,* 424–435.

Schiefele, U., Krapp, A., & Winteler, A. (1992). Interest as a predictor of academic achievement: A meta-analysis of research. In U. Schiefele, A. Krapp, & A. R. Winteler (Eds.), *The role of interest in learning and development* (pp. 183–212). Hillsdale, NJ: Lawrence Erlbaum Associates.

Spinath, B., Spinath, F. M., Harlaar, N., & Plomin, R. (2006). Predicting school achievement from general cognitive ability, self-perceived ability, and intrinsic value. *Intelligence, 34,* 363–374.

Spinath, B., Spinath, F. M., Riemann, R., & Angleitner, A. (2003). Implicit theories about personality and intelligence and their relationship to actual personality and intelligence. *Personality and Individual Differences, 35,* 939–951.

Tucker-Drob, E. M., & Briley, D. A. (2012). Socioeconomic status modifies interest-knowledge associations among adolescents. *Personality and Individual Differences, 53,* 9–15.

(2014). Continuity of genetic and environmental influences on cognition across the life span: A meta-analysis of longitudinal twin and adoption studies. *Psychological Bulletin, 140,* 949–979.

Tucker-Drob, E. M., Briley, D. A., & Harden, K. P. (2013). Genetic and environmental influences on cognition across development and context. *Current Directions in Psychological Science, 22,* 349–355.

Tucker-Drob, E. M., Cheung, A. K., & Briley, D. A. (2014). Gross domestic product, science interest, and science achievement: A person x nation interaction. *Psychological Science, 25,* 2047–2057.

Tucker-Drob, E. M., & Harden, K. P. (2012a). Intellectual interest mediates gene-by-socioeconomic status interaction on adolescent academic achievement. *Child Development, 83,* 743–757.

(2012b). Learning motivation mediates gene-by-socioeconomic status interaction on early mathematics achievement. *Learning and Individual Differences, 22,* 37–45.

(2012c). Early childhood cognitive development and parental cognitive stimulation: Evidence for reciprocal gene-environment transactions. *Developmental Science, 15,* 250–259.

Turkheimer, E., & Horn, E. E. (2014). Interactions between socioeconomic status and components of variation in cognitive ability. In D. Finkel & C. A. Reynolds (Eds.), *Behavior genetics of cognition across the lifespan* (pp. 41–68). New York: Springer.

Turkheimer, E., & Waldron, M. (2000). Nonshared environment: A theoretical, methodological, and quantitative review. *Psychological Bulletin, 126,* 78.

Von Stumm, S., Hell, B., & Chamorro-Premuzic, T. (2011). The hungry mind intellectual curiosity is the third pillar of academic performance. *Perspectives on Psychological Science, 6,* 574–588.

Wainwright, M. A., Wright, M. J., Luciano, M., Geffen, G. M., & Martin, N. G. (2008). Genetic covariation among facets of openness to experience and general cognitive ability. *Twin Research and Human Genetics, 11,* 275–286.

Whiteside, S. P., & Lynam, D. R. (2001). The five factor model and impulsivity: Using a structural model of personality to understand impulsivity. *Personality and Individual Differences, 30,* 669–689.

Wigfield, A., & Eccles, J. S. (2000). Expectancy–value theory of achievement motivation. *Contemporary educational psychology, 25,* 68–81.

Woo, S. E., Harms, P. D., & Kuncel, N. R. (2007). Integrating personality and intelligence: Typical intellectual engagement and need for cognition. *Personality and Individual Differences, 43,* 1635–1639.

Zhang, Y., Haddad, E., Torres, B., & Chen, C. (2011). The reciprocal relationships among parents' expectations, adolescents' expectations, and adolescents' achievement: A two-wave longitudinal analysis of the NELS data. *Journal of Youth and Adolescence, 40,* 479–489.

7

Precision Education Initiative: The Possibility of Personalized Education

CALLIE W. LITTLE, CONNIE BARROSO,
AND SARA A. HART

In partnership with the National Institutes of Health, in early 2015, President Obama unveiled the Precision Medicine Initiative, where he laid out plans to move precision medicine into the everyday clinical practice of physicians (Collins & Varmus, 2015; The White House, 2015). Precision medicine is an approach for disease treatment and prevention that takes into account individual variability in genes, environment, and lifestyle for each person (Collins & Varmus, 2015). The Precision Medicine Initiative recognizes the importance of using an evidence-based approach to guide more accurate and efficient diagnoses and decisions for the care and treatment of individuals (Sackett, 2000). We argue that the fundamental idea behind the Precision Medicine Initiative, such as individualizing treatment based on scientific evidence, clinical experience, and importantly patient-centered relevant factors, should not be limited solely to biomedical disease (Hart, 2016).

Specific learning disorder (SLD) is an overarching diagnosis based on deficits in skills related to academic achievement (American Psychiatric Association, 2013). It is considered a type of common neurodevelopmental disorder (American Psychiatric Association, 2013; 20 U.S.C. § 1401 (30)) with diagnosis issued through a clinical review of a variety of genetic and environmental risk factors and their interactions, such as family history and behavioral observations (Bishop, 2015). Treatment for SLD is complex, and different treatment outcomes are seen in response-to-treatment (Connor, Morrison, & Petrella, 2004). Comparably, common biomedical diseases (e.g., cardiovascular disorders or diabetes) are also diagnosed through a similar analysis of genetic and environmental factors that are considered individually and together; the complex treatment of biomedical disease also varies depending on the individual patient's needs and responses (Caspi, & Bell, 2004).

Although these features between SLD and biomedical disease are similar, some important differences exist. In particular, much more research has been conducted in the biomedical field to disentangle underlying mechanisms involved in the etiology and development of diseases. However, we argue that with more time and funding devoted to the corresponding research, these knowledge differences will close. We propose that the personalized medicine approach, applied through what we call "precision education," would be ideal toward improving the classification and treatment of SLD. Early identification of risk factors associated with SLD can serve to inform preventative intervention and reduce the negative impact that these disabilities can have on success and health throughout the lifespan (Berkman, Sheridan, Donahue, Halpern, & Crotty, 2011; Vogler, DeFries, & Decker, 1985). A precision education approach can provide more than increased understanding of SLD; it can also improve effectiveness of education, making resources and practices related to understanding SLD available to clinicians and educators. With this chapter, we will familiarize the reader with SLD and its associated risk factors, in order to show the reader how the underlying model of SLD is similar to that underlying biomedical disease and lends itself to using a precision medicine approach. From there, we will briefly outline how the evidence-based approach of precision education can improve how students are served within the educational system.

HISTORY OF DEFINING LEARNING DISABILITY

In 1975, the United States enacted the precursor for the Individuals with Disabilities Education Act (IDEA), a special education act that provides regulations for funding and services to individuals aged three to twenty-one with defined educationally handicapped conditions, including SLD (20 U.S.C. § 1401 (30)). Under the IDEA, 2.4 million US public school students have been identified as having a learning disability (LD), the broader term encompassing any of the thirteen individual SLDs defined by the Act (Cortiella & Horowitz, 2014). Cumulative research evidence across longitudinal studies indicates that most students who were identified with SLD during elementary school have problems into middle school, high school, and beyond (Lyon, 1998; Shaywitz et al., 1999). Studies have also suggested that students with SLD have higher rates of lower grades and course failure compared to those without SLD (Cortiella & Horowitz, 2014). In addition, one in every two students with SLD encounter school disciplinary actions such as suspension or expulsion, second only to students in the emotional disturbance category (Cortiella & Horowitz, 2014).

Of the identified SLDs, reading-based disorders are found to be the most prevalent (Lyon, 1996; Shaywitz & Shaywitz, 2003) and the concerns over lack of intervention are significant (Compton, Miller, Elleman, & Steacy, 2014; Fuller et al., 2004; Mortimore & Crozier, 2006; Torgesen, 2002). Deficits in reading are associated with lowered academic success, higher rates of drop-out and incarceration, as well as reduced career success and financial stability (Fletcher, Lyon, Fuchs, & Barnes, 2006; Hernandez, 2011; Reynolds, Temple, Robertson, & Mann, 2002). Due to the prevalence and significance of reading-based disorder, this chapter will focus on these sub-types of SLD, and the term SLD within this chapter will represent reading-based SLD unless explicitly labeled otherwise. The evident and persistent consequences of SLD has yielded a push in research toward understanding the nature of SLD: using investigative efforts to draft a more concrete definition of SLD, searching for indicators that may be critical in diagnosis, and generating quality and efficient reading interventions targeted toward treatment of SLD.

Historically, definitions of SLD have been thought to include difficulty with reading in the absence of disability in general cognitive ability, and unexplained by extraneous factors such as socioeconomic status (SES) or sensory and neurological deficits (Rutter & Yule, 1975; Vellutino, Fletcher, Snowling, & Scanlon, 2004). Despite a long history of attempts to define SLD, there is still no consensus (Fletcher et al., 2013). Disagreements among early SLD researchers included notions that SLD was caused by visual representation issues rooted in the brain, resulting in reversals of letters or words while reading and deficiencies in phonological skills (see Vellutino, Fletcher, Snowling, & Scanlon, 2004, for a review). Other early researchers described attentional issues as the cause, while some assigned causality to motor development problems. In spite of these early disagreements on how to properly define SLD, there were commonalities among many theorized etiologies: most researchers generally concluded that SLD was biological in nature, that it was outside of the realm of cognitive impairment, and initial descriptions were concentrated on definitions that were exclusionary in nature, such that they listed what it was not rather than what it was.

To improve upon previous definitions, inclusionary criterion for SLD classification was added to increase accuracy in diagnosis. The most recent *Diagnostic and Statistical Manual for Mental Disorders* (DSM-V) suggests the umbrella term of "specific learning disability" that includes specifiers to describe the particular academic field of the learning difficulty, reading, writing, and mathematics. In addition, four criteria must also be met to be clinically diagnosed under the DSM-V: criterion A indicates that at least one of six symptoms of the learning difficulty must persist for at least six months notwithstanding tutoring or extra instruction; criterion B states that these specific learning difficulties are quantifiably lower than age-expected scores and cause academic, vocational, or everyday impairment; criterion C considers the onset age of the learning difficulty; and criterion D refers to the disorders and unfavorable conditions that must be ruled out before a SLD diagnosis can be made (American Psychiatric Association, 2013).

Despite the gains made by implementing an inclusionary definition, there are still many uncertainties about the specifics of the components involved with SLD, such as fluency, comprehension, or even more global issues that present no specific component problems. The historical definitions, and even some still used today, are out-of-date because many do not account for the direct influence of non-biological factors, such as social or educational problems, on reading disability. Without a clear definition, efforts to ameliorate the persistence and consequence of SLD remain difficult and we argue that developing a clear definition should be the major near-term focus of precision education.

Two component process theories are currently supported in the literature: the Phonological Core Deficit Model and the Phonological Core Variable-Difference Model (Adams, 1994; Wagner & Torgeson, 1987; Stanovich, 1988; Stanovich & Siegel, 1994). Both theories discuss the relevance of component processes to an SLD diagnosis; however, differential strengths and limitations exist between them, creating uncertainty. The Phonological Core Deficit hypothesizes that word-level reading problems are due to phonological processing deficits (Adams, 1994; Wagner & Torgesen, 1987). The hypothesis suggests a neurological explanation of SLD, pointing to a possible functional or structural deficit in the left hemisphere of the brain, which is often linked with processing language sounds (Ramus, 2001). However, a major limitation of this hypothesis is that it may not be specific enough in its definition of SLD. For example, the same issues in phonological processing are also seen in "garden-variety poor readers" (Gough & Tunmer, 1986), who are individuals with low cognitive ability coupled with their reading deficits (Fletcher et al., 1994).

In contrast, the Phonological Core Variable-Difference Model assumes a continuous distribution of SLD and its correlates, such as general cognitive processing (Stanovich, 1988; Stanovich & Siegel, 1994). Based on the idea that "garden-variety poor readers" (i.e., children who simply struggle learning to read) actually have deficits beyond phonological problems, which truly reading disabled individuals do not have, it surmises that there are multiple specific deficits within phonological processing which represent the SLD. A limitation of this theory is that some empirical work has indicated that there are no discernable differences in reading difficulties between SLD and garden-variety poor readers (Catts, Hogan, & Fey, 2003). However, within the literature, there is also ample support toward a definition of SLD with phonological processing difficulties at its core (Morris et al., 1998; Ramus, 2003; Shaywitz, Morris, & Shaywitz, 2008).

Despite some inconsistency in the field with identifying the component processes that may be associated with SLD, a broad definition has garnered relative agreement, but is still limited in explanatory power, consistency and accuracy (Lyon, Shaywitz, & Shaywitz, 2003). This definition describes SLD as neurobiological in origin, characterized by difficulties with accurate and/or fluent word recognition and by poor spelling and decoding which are the results of a deficit in phonological processing, and is *unexplained* in relation to other cognitive abilities or to classroom instruction. The "unexplained" components in common between previous and current descriptions of the deficits in phonological processing have left several points to be clarified.

In response to these limitations, two groups of models have been proposed over the years to form alternative operational definitions of SLD. First, the "IQ discrepancy model" views these "unexplained" difficulties as those that are seen in the absence of any deficits in general cognitive ability. Some research has suggested that individuals who had been diagnosed with SLD but have higher cognitive ability actually scored lower on non-word reading than their counterparts with both reading and general cognitive disability (Rutter & Yule, 1975). Moreover, this discrepancy model has been supported through direct comparison of raw scores on ability/achievement tests (Stanovich, 1988). However, when extended to a regression-based procedure, where individuals were labeled as SLD when their predicted reading scores exceed their actual reading scores by a certain standard error, these effects were not seen (see Stanovich & Siegel, 1994). Widely criticized by practitioners in the education field, this model has several limitations, including: the similarity between reading-disabled and garden-variety poor readers, measurement issues such as only using one time point, and arbitrary cutoff points, as well as no suggestion for intervention and/or remediation.

Instead, the education community has championed the use of another alternative operational definition, one that includes as SLD all children who fail to respond to an effective literacy-based instruction (Fuchs et al., 2003). The model from which this definition stems, the "Response to Intervention" (RTI) model, uses a three-tier system with students moving up or down through the tiers based on their responses to the differentiated levels of instruction. Tier one represents core classroom instruction, tier two represents more intensive group instruction, and tier three represents the most intensive instruction, usually provided to smaller groups than tier two or at the individual level (Fuchs & Fuchs, 2006). Those children who do not respond to phonology-based interventions, otherwise known as treatment resisters, seem to have a discernibly different profile from children without reading difficulties and the non-treatment resisters who have reading difficulties (Torgeson, 2000; Vellutino, Scanlon, & Lyon, 2000; Vellutino et al., 1996). Some limitations to note with this model are that child groups which are largely non-overlapping and have different aspects of reading highlighted for intervention are dependent on the definition in use for "responding to intervention"; varying protocol and timing of the movement through the intervention tiers have been used in the literature to identify successful response-to-treatment (Al Otaiba et al., 2014). In addition to the inconsistency of group classification and forms of intervention, there is also the potential to overidentify students with poor performance (i.e., slow learners) as learning disabled (Spencer et al., 2014).

What has become clear from years of research on SLD is that it is likely best thought of as a spectrum of disability. In general, reading ability and disability may be viewed simply as being part of the same continuum, with reading disability located at the low end of the spectrum (Fletcher et al., 2013). In addition, the behavioral and molecular genetics literatures on reading disability have supported the idea that reading disability is a quantitative disorder (e.g., multifactorial inheritance), or that there are likely hundreds of inherited genetic variants that influence each individual's position on the spectrum of reading ability (Plomin, DeFries, Knopik, & Neiderhiser, 2013). Twin and adoption studies and molecular genetics methods have found that the genetic variants that influence SLD are commonly found within the entire population, suggesting that there is little to no genetic difference between reading ability and reading disability (Bishop, 2015). The heritability of SLD has typically been found to be approximately 50–60 percent (Bishop, 2015); however, the influence of any given genetic variant has been found to be small, indicating that the genetics underlying SLD are quantitative (Grigorenko, 2013). Complicating the matter is the evidence that genetic variants have been found to interact with each other, and also genetic and specific environmental influences interact (Grigorenko et al., 2007; Hart, Soden, Johnson, Schatschneider, & Taylor, 2013; Scarr & McCartney, 1983; Taylor, Roehrig, Hensler, Connor, & Schatschneider, 2010). The compendium of results across twin studies supports the conceptualization of reading ability and disability falling along the same continuum; they are, therefore, influenced by similar elements and characterized by parallel risk factors (e.g., Plomin & Kovas, 2005).

Overall, while progress has been made in the classification and identification of SLD, substantial limitations remain across the proposed definitions. We propose that these limitations can be addressed and improved within an individualized, evidence-based, precision education approach.

INDICATORS OF LD

There is a growing consensus that SLD as a diagnosis is representative of the low end of the disability–ability spectrum and can be characterized as a constellation of deficits that are influenced by both genetic and environmental factors (van Bergen, van der Leij, & de Jong, 2014). This consensus has been extended empirically to so-called hybrid models of SLD, which combine elements of traditional reading disability approaches and an RTI framework to immediately address existing reading difficulties early on (Wagner, 2008). Hybrid models consider multiple possible indicators of LD, such as phonological processing deficits and rapid word naming, that should improve accuracy of classification of SLD. The hybrid model indicators create a vast framework that can be used for both the classification and treatment of SLD (Fletcher et al., 2013; Spencer et al., 2014). These are, however, limited by the research evidence for each tested hybrid model (Spencer et al., 2014). Certain hybrid models may use diagnosis guidelines that do not align with the current understanding and knowledge of the mechanisms of SLD. Even more, hybrid models in the reading disability literature have been primarily tested using behavioral measures, instead of non-behavioral measures, such as family history and group differences (Spencer et al., 2014). Interestingly, an integral part of precision medicine is that it considers measurement of all modalities for diagnosis and treatment (i.e., not just clinical and genetic but also lifestyle, environmental, and others; Kohane, 2015). It is thought that this more comprehensive picture of the patient allows for better individualization of therapy by distinguishing subgroups of patients and treating them based on this group membership. For example, Schatschneider, Wagner, Hart, and Tighe (2016) found that among several different classifications of reading disability, including low expected reading scores, low unexpected reading scores, and an initial configuration of the hybrid model with several indicators, the hybrid model with students exhibiting at least two of the five "symptoms" was the most stable in classifying students with reading disability across time. The emerging evidence that SLD might be best defined by indicators across modalities suggests that the precision education approach is the best way to accelerate progress in the diagnosis and treatment of LD.

Due to a long-term investment into research on SLD, there is a list of potential indicators which have been empirically supported as predictors of SLD. So far, this list is remarkably similar to those thought to underlie biomedical disorders. Sex (e.g., Shaywitz & Shaywitz, 2008), cognitive factors (e.g., phonological processing; Fletcher et al., 2002), cognitive-behavioral factors (e.g., attention-deficit hyperactivity disorder [ADHD]; Rapport, Scanlan, & Denney, 1999), previous classroom performance (e.g., Fuchs, Mock, Morgan, & Young, 2003), environmental risk (e.g., Taylor, Hart, Mikolajewski, & Schatschneider, 2012), and brain-based risk (e.g., Hoeft et al., 2011; Pugh et al., 2000), family history risk (e.g., van Bergen, Bishop, van Zuigen, & de Jong, 2015), and specific genetic variant risk (e.g., Kegel, Bus, & van Ijzendoorn, 2011), are all possible indicators that may help classify an individual as having SLD (Spencer et al., 2014). It is likely that they are not all uniquely important in classifying SLD, but some differing combination will be important for each individual. It is also likely that there is not a unique combination for each individual, but instead that groups of similar combinations will emerge.

Sex. There are fairly clear sex differences in SLD diagnosis, in that boys, on average, have greater vulnerability to reading disability (not accounted for by referral bias; Quinn & Wagner, 2013). Since the mid-1970s, epidemiological studies have suggested a significant frequency of males to be more likely to have SLD compared to females (Hawke et al., 2007; Limbrick, Wheldall, & Madelaine, 2008). Berger, Yule, and Rutter (1975) investigated two culturally different locations through surveys and found a sex difference in SLD, even between IQ-referenced (adjusted) and non–IQ-referenced low achievement in reading. Yule et al. used students who had an IQ-referenced SLD, and found a 2:1 ratio of males to females who had SLD, with a higher frequency of males being categorized as having a severe SLD. Rutter et al. (2004) provided more confirmation of this phenomenon, with SLD being more common in males than females in their four large-scale studies. These results, taken together suggest that males show a higher average occurrence of SLD within the population (likely 2-to-1 risk; Cortiella & Horowitz, 2014), suggesting that sex should be taken into consideration when identifying potential risk factors for SLD.

Cognitive and Cognitive-Behavioral Factors. Alongside sex differences, child traits such as cognitive and cognitive-behavioral factors have been considered as indicators of SLD. For instance, Fletcher et al. (1996) identified eight different cognitive constructs that are intrinsic to children's achievement: phonological awareness, rapid naming, verbal short-term memory, non-verbal short-term memory, lexical/vocabulary, speech production, perceptual-motor functions, and visual attention. These cognitive factors are usually associated with SLD when children display manifest disability, or the inability to meet age-appropriate expectations for performance (Fletcher et al., 2002). Unexpected patterns of achievement may indicate that one or more types of SLD are present. In line with Fletcher's model for classifying SLD, cumulative evidence has indicated that SLD is associated with deficits in cognitive factors such as executive functioning or working memory (Brady, 1991; Jerman, Reynolds, & Swanson, 2012). In addition, cognitive-behavioral factors such as ADHD may influence SLD status. Characterized by higher than average hyperactivity and inattention, ADHD can lead to problems concentrating and externalized behaviors which can produce a discrepancy between a child's cognitive factors and observed performance (Rapport et al., 2008). In addition to SLD status, ADHD has been associated with other factors influencing educational attainment (e.g., homework behavior) and has been found to be moderately to highly heritable (Little, Hart, Schatschneider, & Taylor, 2014; Willcutt, Pennington, & DeFries, 2000), suggesting ADHD as a salient risk factor for SLD considered under a precision education approach.

Previous Classroom Performance. Previous classroom performance has also been evidenced to be an indicator of SLD. Reading interventions that cater to individual student academic needs or examine different treatment applications of teaching reading skills endure a continuous cyclical process (Fuchs, 2003). An intervention is implemented, progress is monitored, student responsiveness is assessed, and then it begins all over again. The progress of this cycle is driven by the student response to the treatment or the child's previous classroom performance. Students are more likely to be identified as needing special education services for reading when poor classroom performance persists across multiple school years (Al Otaiba & Fuchs, 2006). Moreover, poor classroom performance is thought to indicate that the reading problem is not simply just a symptom of limited instruction, but is likely an indicator of an SLD (Spencer et al., 2014). The relation between previous classroom performance and SLD status and between previous classroom performance and future academic success indicate previous classroom performance as a key indicator to consider at an individualized level, consistent with the goals of precision education services.

Environmental Risk. Beyond risk factors at the child and family level, environmental risk has also been identified an important indicator of SLD, as the environment provides many resources directly and indirectly to children. Sirin (2005) found that a parent's location in the environmental variable of SES structure has a strong impact on their child's academic achievement. Higher levels of SES within a family allow its members to have more resources in the home and incidentally provide the potential for social capital (e.g., access to social support systems in the community) and help in succeeding in school (Coleman, 1988; Dufur et al., 2013). SES defines the accessibility of school, classroom, and neighborhood environments (Reynolds & Wahlberg, 1992). Cases where these resources are lacking could place a child at a high risk for developing SLD.

Brain-Based Risk. Brain-based risk examines the neurophysiological processes and brain structure connectivity that is activated when children learn. Most of the previous studies have focused on identifying neural correlates of SLD. Some studies have focused on pinpointing the neural systems that could potentially mediate treatments for SLD (Hoeft et al., 2011; Pugh et al., 2000). For example, research using multiple imaging has shown that in children and adults with SLD, the brain is hypoactivated during reading tasks (Cao, Bitan, & Booth, 2008; Shaywitz et al., 1998). Specifically, the hypoactivated regions are those requiring the use of the left parietotemporal and occipitotemporal regions for phonological analyses of print. Findings like this provide indicators of SLD through brain-related functions (Pugh et al., 2000; Tanaka et al., 2011).

Family History Risk. One key indicator of SLD is family history. Family history is an important component of precision medicine, and is a commonly overlooked indicator of SLD. Knowledge of family history for SLD status has the potential to assist educational researchers in identifying children at high risk for SLD and implementing treatments to prevent children's learning disabilities.

Within the research on different types of SLD, several methods and measures have been used to ascertain family history, including self-report, questionnaires, family pedigrees, and psychometric tests. Also due to the varying definitions of SLD, multiple strategies for diagnosis have been used. In general, the estimated familial risk for SLD is usually higher than the population risk. The overall affected rate for first-degree relatives is 30–70 percent, which suggests a possibly substantial genetic effect (Hallgren, 1950; Kovas et al., 2013; Volger et al., 1985). Retrospective studies have suggested that there is a high probability that family members of a child with SLD will also be learning disabled themselves or have had issues learning to read (Owen, Adams, Forrest, Stolz, & Fisher, 1971; DeFries, Singer, Foch, & Lewitter, 1978). This finding has been supported by other studies; results have shown that parents of SLD children have a 25–60 percent incident rate of reading difficulty themselves (Pennington, 2009).

Prospective studies have investigated the potential for children to develop SLD when a family history of SLD is known (Gilger, Pennington, & DeFries, 1991; Vogler, DeFries, & Decker, 1985). Using the inverse principles of Bayesian probability to quantify the likelihood of SLD, Gilger et al. (1991) reported significant findings that an offspring's risk of developing SLD increases exponentially if at least one parent reported a history of SLD. Volger et al. (1985) found similar results, and furthermore found that there were no parental sex differences that contributed an increase in risk. Other more recent research has supported their results, expressing findings such as children who have at least one parent with reading difficulties are eigth times more likely to develop SLD (Vellutino et al., 2004).

In addition, longitudinal family history studies have shed light on the relationship between genetically based cognitive defects and the lifespan development of SLD. Scarborough's (1990) study was influential in that it was one of the first to examine the differences between two-year-olds at family risk of dyslexia and children from control families who were not at risk. During follow-up at eight years old, 65 percent of the participants with a family history of SLD were identified with SLD. In addition, high-risk SLD participants exhibited several indicators of reading deficits as they developed compared to their control group counterparts: at thirty months, they made more speech production errors and used a smaller range of syntactic devices; at thirty-six and forty-two months, their vocabulary was less developed and they still faced syntactic issues; and at five years of age their letter knowledge, phonological awareness, and expressive naming skills were all poor. Snowling, Gallagher, & Frith (2003) study followed the development of a group of three-year-olds with at least one dyslexic parent and compared them with a no-risk but similar age group. Between the groups, 66 percent of the high-risk children had reading skills that were one standard deviation below the mean than the control group. Furthermore, after conducting retrospective analyses, they found that high-risk children who developed SLD had poorer narrative skills and slower vocabulary development than those high-risk children who developed normal reading abilities.

Similarly, Carroll, Mundy, and Cunningham (2014) conducted one of the first studies to examine poor readers with and without family history of SLD. Their findings were similar to Snowling et al.'s (2003): poor readers with a family history of SLD typically had lower performance, but typical performers with a family history of SLD had weak early literacy skills although their language skills were not impaired. Moreover, van Bergen, van der Liej, and de Jong (2014) also found that intergenerational transmission of SLD was due to not only strong genetic influences, but also to environmental influences and complex interactions between genes and environment. Even more striking was that regardless of reading outcome, students with a family history of SLD had poorer speech production skills than typically developing students. Similar results have been found in a recent longitudinal study, where student family risk status is predictive of dyslexia at ages 3.5, 4.5, 5.5, and 6–7 (Thompson et al., 2015). These results provide strong evidence that family history should be considered an important risk factor for LD and efforts should be taken to accurately and reliably measure family history within a precision education framework.

Specific Genetic Variant Risk. In addition to the key risk factors we have mentioned, molecular genetics has further examined the role of specific genes in the manifestation of SLD, making genomics a possible source of indicators of SLD. Researchers have attempted to identify genes that are associated with SLD, through genetic linkage analysis (i.e., powerful analysis used to detect chromosomal location of disease genes; Pulst, 1999) and genome-wide association studies. This research has proved to be complex; however, with the advent of newer genomic methodologies and better strategies to characterize the phenotypes of SLD, there has been progress in understanding the genetic influences on SLD.

Thus far, in the search for genetic associations to SLD, at least nine candidate regions (the so-called *DYX1* to *DYX9* regions) have been identified and mapped, with varying empirical support (Carrion-Castillo, Franke, & Fisher, 2013). The following is a brief outline of these most studied regions and their overall functions. The *DYX1*, *DYX2*, *DYX3*, and *DYX5* loci are the most studied.

Chromosome 15 in the *DYX1* locus was one of the first genes to be found to be linked with a susceptibility to SLD. Originally the linkage was pinpointed to the centromere of chromosome 15, but replication studies did not support this location (Fisher et al., 2002; Grigorenko et al., 2000). However, a different locus on the same chromosome, *15q15.1* to *15q21.3*, proved to be supported by other studies (Chapman et al., 2004; Schumacher et al., 2008). The translocation breakpoint for chromosome 15 disrupts a gene *DYX1C1* between exons 8 and 9, which has been labeled SLD susceptibility 1 candidate 1. Subsequent work, to find support for the associations and alleles has produced mixed results. The cause of SLD thus remains uncertain and a there remains a need for further research.

Studies of the *DYX2* region have been the most replicated thus far. Cardon et al. (1994) provided the first evidence for SLD susceptibility as evidenced by a quantitative trait locus analysis (method used to link phenotypic and genotypic data to explain genetic basis of variation in complex traits; Lynch & Walsh, 1998) on chromosome 6 in the DYX2 region. This same linkage has been reported and supported in other studies on reading-related traits (Grigorenko et al., 1997); however, several others have not found support for the involvement of this region (Schulte-Körne et al., 1998). The *KIAA0319* gene in this locus has been associated with the categorical diagnosis of SLD in not only SLD populations, but also in general population studies (Couto et al., 2010). On the other hand, the *DCDC2* gene in this locus has only been implicated in individuals with SLD (Scerri et al., 2011).

The *DYX3* locus was identified by Fagerheim et al. (1999) while studying a Norwegian multigenerational family with a long history of SLD. Many investigations found several haplotypes in this locus to be associated with SLD in two different populations (Anthoni et al., 2007). Moreover, markers on *2p12* have been significantly associated with IQ in verbal and non-verbal performance domains (Scerri et al., 2012). However, another study has proposed that genes from this locus are likely to be more related to general cognition than specific reading or language skills because of its association with white matter volume in certain areas of the brain.

The *DYX5* region encompasses several candidate genes that might be related to the manifestation of SLD. In particular, one of the genes in this region is the *ROBO1* gene in *3p12* was found to be disrupted in an SLD case; a de novo chromosomal translocation had affected the locus (Hannula-Jouppi et al., 2005). Furthermore, it was discovered that reduced levels of *ROBO1* can affect auditory processing as well as brain development (Lamminmäki, Massinen, Nopola-Hemmi, Kere, & Hari, 2012). This region also contains genes that have exhibited linkage to speech-sound disorder (e.g., speech-sound production errors), which is related to deficits in phonological processing in developmental SLD (Stein et al., 2004).

MOVING BEYOND IDENTIFICATION
OF LEARNING DISABILITIES

At this moment, research concerning the proposed key indicators of SLD, along with additional possible indicators, is still growing, similar to the search for genetic influences on biomedical outcomes. For example, new neuroscience research has found that there are pre-existing connection differences in brain regions that are associated with reading skills before children learn to read (Saygin et al., 2016). New methods and technologies have to be developed before we are able to identify with certainty children who may be at risk based on a genetic factor. The same can be said for educational neuroscience, which is still in its infancy but is growing rapidly. Future researchers will need to continually update the hybrid model of the classification of SLD with new and more specific knowledge. This includes proposing other indicators of SLD not considered here, or dismissing some currently thought to be important.

Clearly, biomedical research and practice is closer to the precision medicine approach than educational research and practice are to achieving precision education. However, just as within medicine there are some diseases further along in using precision medicine than others, this does not discount the potential impact of precision medicine as a useful model (e.g., cancers; Pelham et al., 2011). Thus, we suggest that precision education should still be used as a framework for future research and practice in education.

Specifically, mirroring the Precision Medicine Initiative, we should pursue research that takes the knowledge gained from relevant work concerning the classification of SLD and begin developing, testing, and implementing optimal individual-centric interventions for SLD (Pelham et al., 2011). With these precision education interventions, the goal would be to identify an individual with SLD early, and provide a tailored intervention with the goal of treating SLD. There is already a history in the literature of tailoring treatment, specifically with attention paid to individualizing, or differentiating, the time spent on instruction and/or the type of instruction given based on cognitive factors and previous classroom performance (Connor et al., 2004; Fuchs et al., 2003). Efforts to determine which instructional practices delivered in which ways are most effective to specific groups of students have been present throughout the history of formal education (Juel & Minden-Cupp, 2000; Slavin, 1982); however, recently more rigorous research methods have begun to be applied to understanding these processes (e.g., Al Otaiba et al., 2011; Connor et al., 2011, 2013; Suppes, Holland, Hu, & Vu, 2013). Studies focusing on interactions between individual student characteristics and classroom instruction in reading have generally found positive results (Al Otaiba et al., 2011; Connor et al., 2009, 2011). Interventions designed to account for child-by-instruction interactions use previous student performance to produce recommendations for both amount and type of instruction needed in order to achieve target achievement levels. Results of these studies have indicated that the more closely practitioners follow the recommended instructional guidelines the more likely students reach or exceed target achievement levels (Al Otaiba et al., 2011; Connor et al., 2004, 2009, 2011), although these gains may be potentially moderated by the risk factors we have outlined along with other individual and environmental influences.

An important next step is to expand this work to individualize instruction based on all of the actual indicators determined to be influential predictors of SLD. Using molecular genetic techniques to identify and replicate identification of specific genes associated with SLD is an important step in developing a storehouse of indicators for consideration with SLD intervention and treatment. Inherited genetic variants contribute to increased risk for SLD, moreover, combining these specific genetic variants into risk indices can facilitate early indication of SLD and inform related interventions. Within the Precision Medicine Initiative, cancer treatment has been at the forefront, using increased knowledge of genetic variants to make improvements in early detection of risk, diagnostic categories, as well as targeted treatments and therapies, including more closely specified targets for drug therapies (e.g., Collins & Varmus, 2015; Jameson & Longo, 2015). Mirroring this, precision education can combine existing information on specific genetic variants of SLD with new genomic studies to advance knowledge of what approach works for which students, when to apply certain techniques or treatments and for how long, while minimizing risk and maximizing efficacy. Similarly, environmental indicators can be studied in isolation and in combination with genetic variants to improve classification and treatment of SLD at the individual level or by distinguishing among potential SLD profiles. Although it may be the case that each individual is truly different, it is likely the case that subgroups of individuals exist who are similar on both indicators and their RTI. For example, within the established literature on individualized student instruction, some groups of children benefit from more direct, teacher-managed instruction, whereas others benefit more from self-managed instructional time (Connor et al., 2013). These groups have been categorized by assessment scores in reading and subcomponent skills; however, including other genetic, environmental and psycho-social indicators could markedly improve classification and treatment. This will mean that individualization of education may not be at the individual level but at the small group level, which will make this personalized education approach more feasible in the classroom setting. However, more investigation is necessary to determine which indicators are most successful for accurate classification. In general, the research and subsequent application to getting to this point will not be easy or quick. The next generation of educational researchers will need to be trained in new ways to detect, measure, and analyze the factors important to understanding precision education, including the behavioral, practical, methodological, genetic, and environmental (e.g., Thomas et al., 2015). Multidisciplinary educational research will be crucial to the success of expanding educational practices into a precision-based framework.

The present educational system of uniform instruction, broad assessment, and the inconsistent classification of SLD is not serving the public well anymore. A large part of this is due to a disconnect between evidence-based practice and what actually occurs in schools. Work across fields studying SLD has grown to a watershed moment supporting the individualization of education, importantly including individualizing the classification and treatment of SLD. We believe that a precision education approach should be driving not only educational research but also educational practice.

REFERENCES

Adams, M. J. (1994). *Beginning to read: Thinking and learning about print.* Cambridge, MA: MIT Press.

Al Otaiba, S., Connor, C. M., Folsom, J. S., Greulich, L., Meadows, J., & Li, Z. (2011). Assessment data-informed guidance to individualize kindergarten reading instruction: Findings from a cluster-randomized control field trial. *The Elementary School Journal, 111*(4), 535–560.

Al Otaiba, S., Connor, C. M., Folsom, J. S., Wanzek, J., Greulich, L., Schatschneider, C., et al. (2014). To wait in tier 1 or intervene immediately a randomized experiment examining first-grade response to intervention in reading. *Exceptional Children, 81*(1), 11–27.

Al Otaiba, S., & Fuchs, D. (2006). Who are the young children for whom best practices in reading are ineffective? An experimental and longitudinal study. *Journal of Learning Disabilities, 39*(5), 414–431.

American Psychiatric Association. (2013). *Diagnostic and statistical manual of mental disorders (DSM-5).* Arlington, VA: American Psychiatric Association.

Anthoni, H., Zucchelli, M., Matsson, H., Müller-Myhsok, B., Fransson, I., Schumacher, J., et al. (2007). A locus on 2p12 containing the co-regulated MRPL19 and C2ORF3 genes is associated to dyslexia. *Human Molecular Genetics, 16*(6), 667–677.

Artelt, C., Baumert, J., Julius-McElvany, N., & Peschar, J. (2001). *Learners for life: Student approaches to learning results from PISA 2000.* Paris: OECD.

Berger, M., Yule, W., & Rutter, M. (1975). Attainment and adjustment in two geographical areas. *The British Journal of Psychiatry, 126*(6), 510–519.

Berkman, N. D., Sheridan, S. L., Donahue, K. E., Halpern, D. J., & Crotty, K. (2011). Low health literacy and health outcomes: An updated systematic review. *Annals of Internal Medicine, 155*(2), 97–107.

Bishop, D. V. M. (2015). The interface between genetics and psychology: Lessons from developmental dyslexia. *Proceedings of the Royal Society B*, 282, 1–8.

Brady, S. A. (1991). The role of working memory in reading disability. In S. A. Brady & D. P. Shankweiler (Eds.), *Phonological processes in literacy* (pp. 129–152). Hillsdale, NJ: Erlbaum.

Cao, F., Bitan, T., & Booth, J. R. (2008). Effective brain connectivity in children with reading difficulties during phonological processing. *Brain and Language*, 107(2), 91–101.

Cardon, L. R., Smith, S., Fulker, D. W., Kimberling, W. J., Pennington, B. F., & DeFries, J. C. (1994). Quantitative trait locus for reading disability on chromosome 6. *Science*, 266(5183), 276–280.

Carrion-Castillo, A., Franke, B., & Fisher, S. E. (2013). Molecular genetics of dyslexia: an overview. *Dyslexia*, 19(4), 214–240.

Carroll, J. M., Mundy, I. R., & Cunningham, A. J. (2014). The roles of family history of dyslexia, language, speech production and phonological processing in predicting literacy progress. *Developmental Science*, 17(5), 727–742.

Caspi, O., & Bell, I. R. (2004). One size does not fit all: Aptitude x treatment interaction (ATI) as a conceptual framework for complementary and alternative medicine outcome research. Part II – Research designs and their applications. *Journal of Alternative & Complementary Medicine*, 10, 698–705.

Catts, H. W., Hogan, T. P., & Fey, M. E. (2003). Subgrouping poor readers on the basis of individual differences in reading-related abilities. *Journal of Learning Disabilities*, 36(2), 151–164.

Chapman, N. H., Igo, R. P., Thomson, J. B., Matsushita, M., Brkanac, Z., Holzman, T., et al. (2004). Linkage analyses of four regions previously implicated in dyslexia: Confirmation of a locus on chromosome 15q. *American Journal of Medical Genetics Part B: Neuropsychiatric Genetics*, 131(1), 67–75.

Collins, F. S., & Varmus, H. (2015). A new initiative on precision medicine. *New England Journal of Medicine*, 372(9), 793–795.

Compton, D. L., Miller, A. C., Elleman, A. M., & Steacy, L. M. (2014). Have we forsaken reading theory in the name of "quick fix" interventions for children with reading disability? *Scientific Studies of Reading*, 18(1), 55–73.

Compton, D. L., Fuchs, D., Fuchs, L. S., & Bryant, J. D. (2006). Selecting at-risk readers in first grade for early intervention: A two-year longitudinal study of decision rules and procedures. *Journal of Educational Psychology*, 98(2), 394.

Connor, C. M., Morrison, F. J., Fishman, B., Crowe, E. C., Al Otaiba, S., & Schatschneider, C. (2013). A longitudinal cluster-randomized controlled study on the accumulating effects of individualized literacy instruction on students' reading from first through third grade. *Psychological Science*, 24(8), 1408–1419.

Connor, C. M., Morrison, F. J., Fishman, B., Giuliani, S., Luck, M., Underwood, P., et al. (2011). Testing the impact of child characteristics x intstruction on third graders' reading comprehension by differentiating literacy instruction. *Reading Research Quarterly*, 46(3), 189–221.

Connor, C. M., Morrison, F. J., Fishman, B. J., Ponitz, C. C., Glasney, S., Underwood, P. S., et al. (2009). The ISI classroom observation system: Examining the literacy instruction provided to individual students. *Educational Researcher*, 38(2), 85–99. doi:10.3102/0013189x09332373

Connor, C. M., Morrison, F. J., & Petrella, J. N. (2004). Effective reading comprehension instruction: Examining child x instruction interactions. *Journal of Educational Psychology, 96*(4), 682–698.

Connor, C. M., Morrison, F. J., Schatschneider, C., Toste, J., Lundblom, E., Crowe, E., et al. (2011). Effective classroom instruction: Implications of child characteristics by reading instruction interactions on first graders' word reading achievement. *Journal for Research on Educational Effectiveness, 4*, 1–35.

Connor, C. M., Piasta, S. B., Fishman, B., Glasney, S., Schatschneider, C., Crowe, E., et al. (2009). Individualizing student instruction precisely: Effects of child by instruction interactions on first graders' literacy development. *Child Development, 80*(1), 77–100.

Coleman, J. S. (1988). Social capital in the creation of human capital. *American Journal of Sociology*, S95–120.

Collins, F. S., & Varmus, H. (2015). A new initiative on precision medicine. *New England Journal of Medicine, 372*, 793–795.

Cortiella, C., & Horowitz, S. H. (2014). *The state of learning disabilities: Facts, trends and emerging issues*. New York: National Center for Learning Disabilities.

Couto, J. M., Livne-Bar, I., Huang, K., Xu, Z., Cate-Carter, T., Feng, Y., et al. (2010). Association of reading disabilities with regions marked by acetylated H3 histones in KIAA0319. *American Journal of Medical Genetics Part B: Neuropsychiatric Genetics, 153*(2), 447–462.

Curran, P. J., & Hussong, A. M. (2009). Integrative data analysis: The simultaneous analysis of multiple data sets. *Psychological Methods, 14*(2), 81–100. doi:10.1037/a0015914

DeFries, J. C., Singer, S. M., Foch, T. T., & Lewitter, F. I. (1978). Familial nature of reading disability. *The British Journal of Psychiatry, 132*(4), 361–367.

Doyle, M., & Furnham, A. (2012). The distracting effects of music on the cognitive test performance of creative and non-creative individuals. *Thinking Skills and Creativity, 7*(1), 1–7.

Dufur, M. J., Parcel, T. L., & Troutman, K. P. (2013). Does capital at home matter more than capital at school? Social capital effects on academic achievement. *Research in Social Stratification and Mobility, 31*, 1–21.

Fagerheim, T., Raeymaekers, P., Tønnessen, F. E., Pedersen, M., Tranebjærg, L., & Lubs, H. A. (1999). A new gene (DYX3) for dyslexia is located on chromosome 2. *Journal of Medical Genetics, 36*(9), 664–669.

Fisher, S. E., Francks, C., Marlow, A. J., MacPhie, I. L., Newbury, D. F., Cardon, L. R., et al. (2002). Independent genome-wide scans identify a chromosome 18 quantitative-trait locus influencing dyslexia. *Nature Genetics, 30*(1), 86–91.

Fletcher, J. M., Lyon, G. R., Fuchs, L. S., & Barnes, M. A. (2006). Learning disabilities: From identification to intervention. Guilford Press.

Fletcher, J. M., Foorman, B. R., Boudousquie, A., Barnes, M. A., Schatschneider, C., & Francis, D. J. (2002). Assessment of reading and learning disabilities a research-based intervention-oriented approach. *Journal of School Psychology, 40*(1), 27–63.

Fletcher, J. M., Stuebing, K. K., Barth, A. E., Miciak, J., Francis, D. J., & Denton, C. A. (2014). Agreement and coverage of indicators of response to intervention: A multi-method comparison and simulation. *Topics in Language Disorders, 34*(1), 74.

Fletcher, J. M., Stuebing, K. K., Morris, R. D., & Lyon, G. R. (2013). Classification and definition of learning disabilities: A hybrid model. In H. L. Swanson & K. Harris (Eds.), *Handbook of learning disabilities* (2nd ed., pp. 33–50). New York, NY: Guilford Press.

Fletcher, J. M., Francis, D. J., Stuebing, K. K., Shaywitz, B. A., Shaywitz, S. E., Shankweiler, D. P., et al. (1996). Conceptual and methodological issues in construct definition. In G. R. Lyon (Ed.), Validating the constructs of attention, memory, and executive functions (pp. 17–42). Baltimore: Brookes.

Fuchs, D., Mock, D., Morgan, P. L., & Young, C. L. (2003). Responsiveness-to-intervention: Definitions, evidence, and implications for the learning disabilities construct. *Learning Disabilities Research & Practice, 18*(3), 157–171.

Fuchs, D., & Fuchs, L. S. (2006). Introduction to response to intervention: What, why, and how valid is it? *Reading Research Quarterly, 41*(1), 93–99.

Fuller, M., Healey, M., Bradley, A., & Hall, T. (2004) Barriers to learning: A systematic study of the experience of disabled students in one university, *Studies in Higher Education, 29*, 303–318.

Gilger, J. W., Pennington, B. F., & DeFries, J. C. (1991). Risk for reading disability as a function of parental history in three family studies. *Reading and Writing: An Interdisciplinary Journal, 3*, 299–313

Gough, P. B., & Tunmer, W. E. (1986). Decoding, reading, and reading disability. *Remedial and Special Education, 7*(1), 6–10.

Grigorenko, E. L. (2013). What we know (or do not know) about the genetics of reading comprehension and other reading-related processes. In Miller, B., Cutting, L., McCardle, P. (Eds.), *Unraveling reading comprehension* (pp. 293–313). Baltimore, MD: Brookes.

Grigorenko, E. L., Wood, F. B., Meyer, M. S., Hart, L. A., Speed, W. C., Shuster, A., et al. (1997). Susceptibility loci for distinct components of developmental dyslexia on chromosomes 6 and 15. *American Journal of Human Genetics, 60*(1), 27.

Grigorenko, E. L., Deyoung, C. G., Getchell, M., Haeffel, G. J., Klinteberg, B. A., Koposov, R. A., et al. (2007). Exploring interactive effects of genes and environments in etiology of individual differences in reading comprehension. *Development and Psychopathology, 19*(4), 1089–1103.

Grigorenko, E. L., Wood, F. B., Meyer, M. S., & Pauls, D. L. (2000). Chromosome 6p influences on different dyslexia-related cognitive processes: Further confirmation. *American Journal of Human Genetics, 66*(2), 715–723.

Hallgren, B. (1950). Specific dyslexia (congenital word-blindness); a clinical and genetic study. *Acta Psychiatrica et Neurologica. Supplementum, 65*, 1.

Hannula-Jouppi, K., Kaminen-Ahola, N., Taipale, M., Eklund, R., Nopola-Hemmi, J., Kääriäinen, H., et al. (2005). The axon guidance receptor gene ROBO1 is a candidate gene for developmental dyslexia. *PLoS Genet, 1*(4), e50.

Hart, S. A. (2016). Precision education initiative: Moving toward personalized education. *Mind, Brain, and Education, 10*(4), 209–211.

Hart, S. A., Soden, B., Johnson, W., Schatschneider, C., & Taylor, J. (2013). Expanding the environment: Gene x school-level SES interaction on reading comprehension. *Journal of Child Psychology and Psychiatry, 54*(10), 1047–1055. doi:10.1111/jcpp.12083

Hawke, J. L., Wadsworth, S. J., Olson, R. K., & DeFries, J. C. (2007). Etiology of reading difficulties as a function of gender and severity. *Reading and Writing, 20*(1–2), 13–25.

Hernandez, D. J. (2011). Double jeopardy: How third-grade reading skills and poverty influence high school graduation. Annie E. Casey Foundation.

Hoeft, F., McCandliss, B. D., Black, J. M., Gantman, A., Zakerani, N., Hulme, C., et al. (2011). Neural systems predicting long-term outcome in dyslexia. *Proceedings of the National Academy of Sciences, 108*(1), 361–366.

Jameson, J. L., & Longo, D. L. (2015). Precision medicine – Personalized, problematic, and promising. *Obstetrical & Gynecological Survey, 70*(10), 612–614.

Jerman, O., Reynolds, C., & Swanson, H. L. (2012). Does growth in working memory span or executive processes predict growth in reading and math in children with reading disabilities? *Learning Disability Quarterly, 35*(3), 144–157.

Johnson, C., & Kritsonis, W. A. (2007). National school debate: Banning cell phones in public schools: analyzing a national school and community relations problem. *National Forum of Educational Administration and Supervision Journals, 25*(4), 1–6.

Juel, C., & Minden-Cupp, C. (2000). Learning to read words: Linguistic units and instructional strategies. *Reading Research Quarterly, 35*(4), 458–492.

Kegel, C. A. T., Bus, A. G., & van Ijzendoorn, M. H. (2011). Differential susceptibility in early literacy instruction through computer games: The role of the dopamine D4 receptor gene (DRD4). *Mind, Brain, and Education, 5*(2), 71–78. doi:10.1111/j.1751-228X.2011.01112.x

Klinger, D. A., Shulha, L. A., & Wade-Woolley, L. (2010). *Towards an understanding of gender differences in literacy achievement.* Toronto: Education Quality and Accountability Office.

Kohane, I. S. (2015). Ten things we have to do to achieve precision medicine. *Science, 349*(6243), 37–38.

Kovas, Y., Voronin, I., Kaydalov, A., Malykh, S. B., Dale, P. S., & Plomin, R. (2013). Literacy and numeracy are more heritable than intelligence in primary school. *Psychological Science, 24*(10), 2048–2056.

Lamminmäki, S., Massinen, S., Nopola-Hemmi, J., Kere, J., & Hari, R. (2012). Human ROBO1 regulates interaural interaction in auditory pathways. *Journal of Neuroscience, 32*(3), 966–971.

Limbrick, L., Wheldall, K., & Madelaine, A. (2008). Gender ratios for reading disability: Are there really more boys than girls who are low-progress readers? *Australian Journal of Learning Difficulties, 13*(2), 161–179.

Little, C. W., Hart, S. A., Schatschneider, C., & Taylor, J. (2014). Examining associations among ADHD, homework behavior, and reading comprehension: A twin study. *Journal of Learning Disabilities,* 0022219414555715.

Lynch, M., & Walsh, B. (1998). *Genetics and analysis of quantitative traits.* Sunderland, MA: Sinauer.

Lyon, G. R. (1996). Learning disabilities. *The Future of Children*, 6, 54–76.

Lyon, R. G. (1998). The NICHD research program in reading development, reading disorders and reading instruction: A summary of research findings. Keys to successful learning. A National Summit on Research in Learning Disabilities.

Lyon, G. R., Shaywitz, S. E., & Shaywitz, B. A. (2003). A definition of dyslexia. *Annals of Dyslexia*, 53, 1–14.

Morris, R. D., Stuebing, K. K., Fletcher, J. M., Shaywitz, S. E., Lyon, G. R., Shankweiler, D. P., et al. (1998). Subtypes of reading disability: Variability around a phonological core. *Journal of Educational Psychology*, 90(3), 347–373.

Mortimore, T., & Crozier, W. R. (2006). Dyslexia and difficulties with study skills in higher education. *Studies in Higher Education*, 31(2), 235–251.

Owen, F. W., Adams, P. A., Forrest, T., Stolz, L. M., & Fisher, S. (1971). Learning disorders in children: Sibling studies. *Monographs of the Society for Research in Child Development*, 1–77.

Pelham, W. E., Jr., Waschbusch, D. A., Hoza, B., Gnagy, E. M., Greiner, A. R., Sams, S. E., et al. (2011). Music and video as distractors for boys with ADHD in the classroom: Comparison with controls, individual differences, and medication effects. *Journal of Abnormal Child Psychology*, 39(8), 1085–1098.

Pennington, B. F. (2009). *Diagnosing learning disorders: A neuropsychological framework*, 2nd ed. New York: Guilford Press.

Plomin, R., DeFries, J. C., Knopik, V. S., & Neiderheiser, J. (2013). *Behavioral genetics*. Palgrave Macmillan.

Plomin, R., & Kovas, Y. (2005). Generalist genes and learning disabilities. *Psychological Bulletin*, 131(4), 592–617.

Ponitz, C. C., Rimm-Kaufman, S. E., Brock, L. L., & Nathanson, L. (2009). Early adjustment, gender differences, and classroom organizational climate in first grade. *The Elementary School Journal*, 110(2), 142–162.

Pugh, K. R., Mencl, W. E., Jenner, A. R., Katz, L., Frost, S. J., Lee, J. R., et al. (2000). Functional neuroimaging studies of reading and reading disability(developmental dyslexia). *Mental Retardation and Developmental Disabilities Research Reviews*, 6(3), 207–213.

Pulst, S. M. (1999). Genetic linkage analysis. *Archives of Neurology*, 56(6), 667–672.

Quinn, J. M., & Wagner, R. K. (2013). Gender differences in reading impairment and in the identification of impaired readers: Results from a large scale study of at-risk readers. *Journal of Learning Disabilities*. doi:10.1177/0022219413508323

Ramus, F. (2001). Dyslexia: Talk to two theories. *Nature*, 412, 393–395.

(2003). Developmental dyslexia: Specific phonological deficit or general sensorimotor dysfunction? *Current Opinion in Neurobiology*, 13(2), 212–218.

Rapport, M. D., Alderson, R. M., Kofler, M. J., Sarver, D. E., Bolden, J., & Sims, V. (2008). Working memory deficits in boys with attention-deficit/hyperactivity disorder (ADHD): The contribution of central executive and subsystem processes. *Journal of Abnormal Child Psychology*, 36(6), 825–837. doi:10.1007/s10802-008-9215-y

Rapport, M. D., Scanlan, S. W., & Denney, C. B. (1999). Attention-deficit/hyperactivity disorder and scholastic achievement: A model of dual developmental pathways. *Journal of Child Psychology and Psychiatry and Allied Disciplines*, 40(8), 1169–1183.

Reynolds, A. J., Temple, J. A., Robertson, D. L., & Mann, E. A. (2002). Age 21 cost-benefit analysis of the Title I Chicago child-parent centers. *Educational Evaluation and Policy Analysis, 24*(4), 267–303.

Reynolds, A. J., & Walberg, H. J. (1992). A process model of mathematics achievement and attitude. *Journal for Research in Mathematics Education, 23*(4), 306–328.

Rutter, M., Caspi, A., Fergusson, D., Horwood, L. J., Goodman, R., Maughan, B., ... & Carroll, J. (2004). Sex differences in developmental reading disability: new findings from 4 epidemiological studies. *Jama, 291*(16), 2007–2012.

Rutter, M., & Yule, W. (1975). The concept of specific reading retardation. *Journal of Child Psychology and Psychiatry, 16*(3), 181–197.

Sackett, D. L. (2000). *Evidence-based medicine.* London: Churchill-Livingstone, Wiley Online Library.

Saygin, Z. M., Osher, D. E., Norton, E. S., Youssoufian, D. A., Beach, S. D., Feather, J., et al. (2016). Connectivity precedes function in the development of the visual word form area. *Nature Neuroscience, 19*(9), 1250–1255.

Scarborough, H. S. (1990). Very early language deficits in dyslexic children. *Child development, 61*(6), 1728–1743.

Scarr, S., & McCartney, K. (1983). How people make their own environments: A theory of genotype greater than environment effects. *Child Development, 54*(2), 424–435.

Scerri, T. S., Darki, F., Newbury, D. F., Whitehouse, A. J., Peyrard-Janvid, M., Matsson, H., et al. (2012). The dyslexia candidate locus on 2p12 is associated with general cognitive ability and white matter structure. *PloS One, 7*(11), e50321.

Scerri, T. S., Morris, A. P., Buckingham, L. L., Newbury, D. F., Miller, L. L., Monaco, A. P., et al. (2011). DCDC2, KIAA0319 and CMIP are associated with reading-related traits. *Biological Psychiatry, 70*(3), 237–245.

Schatschneider, C., Wagner, R. K., Hart, S. A., & Tighe, E. L. (2016). Using simulations to investigate the longitudinal stability of alternative schemes for classifying and identifying children with reading disabilities. *Scientific Studies of Reading, 20*(1), 34–48.

Schulte-Körne, G., Grimm, T., Nöthen, M. M., Müller-Myhsok, B., Cichon, S., Vogt, I. R., ... & Remschmidt, H. (1998). Evidence for linkage of spelling disability to chromosome 15. *American Journal of Human Genetics, 63*(1), 279.

Shaywitz, S., Fletcher, J., Holahan, J., Shneider, A. E., Marchione, K. E., Stuebing, K. K., et al. (1999). Persistence of dyslexia: The Connecticut Longitudinal Study at adolescence. *Pediatrics, 104*, 1351–1359.

Shaywitz, S. E., Morris, R., & Shaywitz, B. A. (2008). The education of dyslexic children from childhood to young adulthood. *Annual Review ofPsychology., 59*, 451–475.

Shaywitz, S. E., & Shaywitz, B. A. (2003). The science of reading and dyslexia. *Journal of American Association for Pediatric Ophthalmology and Strabismus, 7*(3), 158–166.

(2008). Paying attention to reading: The neurobiology of reading and dyslexia. *Development and Psychopathology, 20*(4), 1329–1349.

Shaywitz, S. E., Shaywitz, B. A., Pugh, K. R., Fulbright, R. K., Constable, R. T., Mencl, W. E., et al. (1998). Functional disruption in the organization of the brain for reading in dyslexia. *Proceedings of the National Academy of Sciences*, 95(5), 2636–2641.

Sirin, S. R. (2005). Socioeconomic status and academic achievement: A meta-analytic review of research. *Review of educational research*, 75(3), 417–453.

Schumacher, J., König, I. R., Schröder, T., Duell, M., Plume, E., Propping, P., et al. (2008). Further evidence for a susceptibility locus contributing to reading disability on chromosome 15q15–q21. *Psychiatric Genetics*, 18(3), 137–142.

Slavin, R. E., Leavey, M. B., & Madde, N. A. (1982). Combining cooperative learning and individualized instruction: Effects on student mathematics achievement, attitudes and behaviors. *The Elementary School Journal*, 84(4) 409–422.

Snowling, M. J., Gallagher, A., & Frith, U. (2003). Family risk of dyslexia is continuous: Individual differences in the precursors of reading skill. *Child Development*, 74(2), 358–373.

Spencer, M., Wagner, R. K., Schatschneider, C., Quinn, J. M., Lopez, D., & Petscher, Y. (2014). Incorporating RTI in a hybrid model of reading disability. *Learning Disability Quarterly*, 37(3), 161–171. 0731948714530967.

Spira, E. G., Bracken, S. S., & Fischel, J. E. (2005). Predicting improvement after first-grade reading difficulties: The effects of oral language, emergent literacy, and behavior skills. *Developmental Psychology*, 41(1), 225.

Stanovich, K. E. (1988). Explaining the differences between the dyslexic and the garden-variety poor reader: The phonological-core variable-difference model. *Journal of learning disabilities*, 21(10), 590–604.

Stanovich, K. E., & Siegel, L. S. (1994). Phenotypic performance profile of children with reading disabilities: A regression-based test of the phonological-core variable-difference model. *Journal of Educational Psychology*, 86(1), 24.

Stein, C. M., Schick, J. H., Taylor, H. G., Shriberg, L. D., Millard, C., Kundtz-Kluge, A., et al. (2004). Pleiotropic effects of a chromosome 3 locus on speech-sound disorder and reading. *American Journal of Human Genetics*, 74(2), 283–297.

Suppes, P., Holland, P. W., Hu, Y., & Vu, M. T. (2013). Effectiveness of an individualized computer-driven online math K-5 course in eight California title I elementary schools. *Educational Assessment*, 18(3), 162–181.

Tanaka, H., Black, J. M., Hulme, C., Stanley, L. M., Kesler, S. R., Whitfield-Gabrieli, S., et al. (2011). The brain basis of the phonological deficit in dyslexia is independent of IQ. *Psychological Science*, 22(11), 1442–1451.

Taylor, J., Roehrig, A. D., Hensler, B. S., Connor, C. M., & Schatschneider, C. (2010). Teacher quality moderates the genetic effects on early reading. *Science*, 328(5977), 512.

Taylor, J. E., Hart, S. A., Mikolajewski, A. J., & Schatschneider, C. (2012). An update on the Florida State Twin Registry. *Twin Research and Human Genetics*, 1(1), 1–5.

The White House, Office of the Press Secretary. (2015). Fact Sheet: President Obama's Precision Medicine Initiative. Retrieved April 9, 2015, from www.whitehouse.gov/the-press-office/2015/01/30/fact-sheet-president-obama-s-precision-medicine-initiative

Thomas, M. S., Kovas, Y., Meaburn, E. L., & Tolmie, A. (2015). What can the study of genetics offer to educators? *Mind, Brain, and Education*, 9(2), 72–80.

Thompson, P. A., Hulme, C., Nash, H. M., Gooch, D., Hayiou-Thomas, E., & Snowling, M. J. (2015). Developmental dyslexia: Predicting individual risk. *Journal of Child Psychology and Psychiatry, 56*(9), 976–987.

Torgesen, J. K. (2002). The prevention of reading difficulties. *Journal of School Psychology, 40*(1), 7–26.

(2000). Individual differences in response to early interventions in reading: The lingering problem of treatment resisters. *Learning Disabilities Research & Practice, 15*(1), 55–64.

van Bergen, E., Bishop, D. V. M., van Zuigen, T., & de Jong, P. F. (2015). How does parental reading influence children's reading? A study of cognitive mediation. *Scientific Studies of Reading.*

van Bergen, E., van der Leij, A., & de Jong, P. F. (2014). The intergenerational multiple deficit model and the case of dyslexia. *Frontiers in Human Neuroscience, 8*, 346.

Vellutino, F. R., Fletcher, J. M., Snowling, M. J., & Scanlon, D. M. (2004). Specific reading disability (dyslexia): What have we learned in the past four decades? *Journal of Child Psychology and Psychiatry, 45*(1), 2–40.

Vellutino, F. R., Scanlon, D. M., & Reid Lyon, G. (2000). Differentiating between difficult-to-remediate and readily remediated poor readers: More evidence against the IQ-achievement discrepancy definition of reading disability. *Journal of Learning Disabilities, 33*(3), 223–238.

Vellutino, F. R., Scanlon, D. M., Sipay, E. R., Small, S. G., Pratt, A., Chen, R., et al. (1996). Cognitive profiles of difficult-to-remediate and readily remediated poor readers: Early intervention as a vehicle for distinguishing between cognitive and experiential deficits as basic causes of specific reading disability. *Journal of Educational Psychology, 88*(4), 601.

Vogler, G. P., DeFries, J. C., & Decker, S. N. (1985). Family history as an indicator of risk for reading disability. *Journal of Learning Disabilities, 18*(7), 419–421. doi:10.1177/002221948501800711

Wagner, R. K. (2008). Rediscovering dyslexia: New approaches for identification and classification. In G. Reid, A. J. Fawcett, F. Manis, & L. S. Siegel (Eds.), *The Sage handbook of dyslexia* (pp. 174–191). London: Sage.

Wagner, R. K., & Torgesen, J. K. (1987). The nature of phonological processing and its causal role in the acquisition of reading skills. *Psychological Bulletin, 101*(2), 192–212.

Willcutt, E. G., Pennington, B. F., & DeFries, J. C. (2000). Twin study of the etiology of comorbidity between reading disability and attention-deficit/hyperactivity disorder. *American Journal of Medical Genetics (Neuropsychiatric Genetics), 96*, 293–301.

8

Using Genetic Etiology to Intervene with Students with Intellectual Disabilities

ROBERT M. HODAPP AND MARISA H. FISHER

Translating basic research findings to applied settings can be more difficult than generally supposed. Across research fields, professionals often reject or ignore the latest basic research findings, feeling that their own long-held attitudes and practices are under attack. In addition, the move from basic findings to changes in applied settings sometimes brings about a host of practical – and at times unanticipated – problems.

This has been the case concerning special education and the genetics of intellectual disabilities, particularly the influence of etiology-related information on classroom practice. On one hand, researchers have an increasing understanding of the 750 to 1,000+ specific genetic conditions that result in intellectual disabilities (Dykens, Hodapp, & Finucane, 2000). Knowledge of these conditions – previously of interest mostly to pediatricians, geneticists, psychiatrists, and other biomedical professionals – has now increasingly informed the studies of various developmental, clinical, behavioral, and other psychologists, and other social science researchers. As a result, we now know much more about so-called gene–brain–behavior relationships, as well as about etiology-related cognitive, linguistic, and adaptive trajectories and profiles, maladaptive behavior and psychopathology, and other behavioral characteristics (Hodapp & Dykens, 2012). Such rapidly accumulating information has coalesced into a sub-field referred to as "behavioral phenotypes," with much of that information potentially relating to practices within the special education field.

At the same time, however, special education has been slow to embrace these new advances. To this day, few large-scale datasets of students with intellectual disabilities even report the cause of intellectual disability in their participants. Thus, the National Longitudinal Transition Study – Second Wave (NLTS-2) surveyed 11,000 US students with disabilities as these students were in their final high-school years and for several years thereafter

(Cameto, Levine, & Wagner, 2004). Yet despite all of this information, no questions ask about the cause of disabilities in these teens and young adults. Similarly, the US Department of Education reports only the number of 3 to 21 year-old students who fall under each of its thirteen disability categories (e.g., intellectual disability, emotional disturbance, autism, specific learning disability, speech or language impairment; NICHCY, 2012). No information is provided about the genetic causes of or contributions to intellectual disability (or of any other disability).

In this chapter, we argue for the benefits of an approach that focuses on considering the behavioral phenotype of genetic conditions when intervening with students with intellectual disabilities. After briefly describing the main principles of this approach, we provide examples of how the approach might be incorporated into special education practice. Using both the principles and examples, we then end with more general recommendations for future translations. Although many special education researchers and teachers continue to disregard genetic information, it remains important to highlight advantages of a genetically informed approach to teaching students with intellectual disabilities.

BEHAVIORAL PHENOTYPES: DEFINITION AND PRINCIPLES

At its most basic level, behavioral phenotypes constitute a subset of a more general perspective that examines the movement from one's genotype to one's phenotype. In this more general progression, specifics of one's genetic inheritance influence physical, medical, or other outcomes. In this instance, the outcomes concern behavior. In many genetic conditions that lead to intellectual disabilities, then, a specific genotype (be it a particular deletion, trisomy, or other genetic anomaly) leads to a particular behavioral outcome. This idea, first discussed in the early 1970s by William Nyhan (1972, one of the discoverers of Lesch-Nyhan disease), was explored more fully beginning in the 1990s (Flynt & Yule, 1994; Hodapp & Dykens, 1994). As described over the years, behavioral phenotypes involve "the heightened probability or likelihood that people with a given syndrome will exhibit certain behavioral and developmental sequelae relative to those without the syndrome" (Dykens, 1995, p. 523).

Behavioral phenotypes are characterized by at least three principles, including (1) total specificity and partial specificity; (2) within-syndrome variability; and (3) multiple behavioral and non-behavioral domains that change over development.

Total Specificity and Partial Specificity

The connections of a genetic anomaly and a specific behavioral outcome can occur in several ways. First, a genetic anomaly can result in a one-to-one connection to a particular behavioral outcome that is rarely seen in other genetic conditions. This first pattern, called total specificity (Hodapp, 1997), is most obviously seen in Lesch-Nyhan syndrome itself. More than any other genetic condition, individuals with Lesch-Nyhan syndrome show extreme self-mutilation (Schretlen et al., 2005). No other syndrome shows such behaviors to anywhere near the same extent. Other examples of such 1:1 connections are found in the hyperphagia (overeating) in Prader-Willi syndrome (Dykens et al., 2007), the cat-cry in 5p syndrome (Sigafoos, O'Reilly, & Lancioni, 2009), and the body "self-hugging" and putting objects into bodily orifices in Smith-Magenis syndrome (Finucane & Haas-Givler, 2009). In each case, a particular genetic syndrome leads to a particular behavioral outcome, and that outcome appears in no other genetic syndrome to anywhere near the same degree or frequency.

Although a few instances of total specificity have been noted, the more common pattern involves what has been called partial specificity (Hodapp, 1997). In this case, two or more genetic conditions predispose children to the same behavioral outcome. Consider the distinction in intellectual performance between tasks that involve simultaneous – or holistic – processing and those that involve sequential, step-by-step processing (Das, Kirby, & Jarmin, 1975). On one IQ test (the Kaufman Assessment Battery for Children, or K-ABC) that divides tasks in this way, a pattern of "simultaneous over sequential processing" was found among boys with fragile X syndrome (Dykens, Hodapp, & Leckman, 1987) and among individuals with Prader-Willi syndrome (Dykens, Hodapp, Walsh, & Nash, 1992). Similarly, compared to groups with intellectual disabilities in general, hyperactivity is more frequently found in children with 5p syndrome (Dykens & Clarke, 1997) and in boys with fragile X syndrome (Baumgardner, Reiss, Freund, & Abrams, 1995). In each case, two or more genetic conditions of intellectual disability predispose individuals to a specific profile of strength and weakness or particular maladaptive behavior.

Within-Syndrome Variability

Rarely is it the case that every person with a specific genetic syndrome shows that syndrome's "characteristic" behavioral outcome(s). Instead, behavioral phenotypes are best considered as increasing the probabilities or heightening the risks of showing a particular profile or behavior.

In adopting a perspective that emphasizes heightened risks, we follow the approach of genotype–phenotype connections more generally. Consider the characteristic physical, facial, and medical characteristics found in Down syndrome. Most would agree that children (and adults) with Down syndrome show particular facial features – including epicanthal folds around the eyes, a rounder face, and a short, pug-nose. Upon closer examination, however, not every child with Down syndrome shows any one of these facial features (Pueschel, 1990). Similarly, children with Down syndrome are especially prone to congenital heart disease (roughly 50 percent), respiratory issues, gastrointestinal problems, celiac disease, obesity, and visual and hearing problems (Roizen, 2010). Although percentages of each physical feature or medical problem are markedly higher than those found in either children without disabilities or children with other genetic conditions, none occurs in every child with Down syndrome.

Such within-syndrome variability also occurs when considering etiology-related behaviors. Consider the generally weak linguistic abilities of most children and young adults with Down syndrome, particularly in the areas of grammar and articulation (Abbeduto, Warren, & Conners, 2007). Although this pattern occurs in most youth with Down syndrome, there are some exceptions. Thus, Rondal (1995) reported on the case of Françoise, a thirty-two-year-old woman whose IQ is 64. Although Françoise has trisomy 21, she nevertheless utters long and complex sentences. Rondal (1995, p. 117) reports her saying (translated), "And that does not surprise me because dogs are always too warm when they go outside" ("Et ca m'etonne pas parce que les chiens ont toujours trop chaud quand ils vont a la port"). Although grammatical, articulatory, and expressive language problems may be common in Down syndrome, not every person with the syndrome shows such behavioral characteristics.

Multiple Behavioral and Non-Behavioral Domains That Change Over Development

As the preceding sections illustrate, behavioral phenotypes relate to a variety of behavioral domains. Etiology-related strengths and weaknesses thus occur in cognition and language, as well as in adaptive and maladaptive behavior. Within each of these domains, one sees as well sub-domain strengths and weaknesses – witness, for example, the (relatively high) vocabulary skills as opposed to the (relatively weak) grammatical abilities in the language of children with Down syndrome (Abbeduto, et al., 2007). At the same time, non-normative patterns also occur. Young adults with Williams syndrome thus show high levels of anxiety when faced with performing a difficult cognitive task, even as they show little or no fear of getting up and performing before an audience (Lense & Dykens, 2013). To most people without Williams syndrome, public performances are among the most common and stressful of all life events (Stein, Torgrud, & Walker, 2000).

Such etiology-related ties to specific behaviors or profiles also change with age. For children with Down syndrome, fragile X syndrome, and Williams syndrome, age and developmental factors influence etiology-related profiles of cognitive, linguistic, and adaptive strength and weakness (Fidler, 2011). So-called developmental cascades then serve to make small, earlier occurring strengths and weaknesses more prominent as children develop (Fidler, 2005). Such age-related differences also show themselves in maladaptive behavior-psychopathology. Comparing four age groups of children and adults with Prader-Willi syndrome, Dykens (2004) found a general mellowing as these individuals reached their twenties, thirties, and forties, with externalizing problems decreasing across age groups. For other behaviors – including hoarding – the peak age period was the twenties, with lower average scores before and after that period. Even when a particular behavior or symptom shows itself in most persons with a syndrome, the frequency and intensity of such symptoms often vary by age.

Etiology-related characteristics are also not limited to behaviors per se. Note, for example, hearing problems in children with Down syndrome. These problems, which occur in 35–70 percent of these children (Porter & Tharpe, 2010), are most likely tied to these children's decreased levels of language. Similarly, the articulation problems of children with Down syndrome might be associated with the characteristic physical structure of the tongue, including its thickness, size, and placement within the mouth (Bunton & Leddy, 2011). Even the facial characteristics of some genetic conditions may be important, with the more neonatal or babylike faces of children with Down syndrome (Allanson, O'Hara, Farkas, & Nair, 1993) possibly relating to increasing degrees of nurturant behavior on the part of unfamiliar adults (Fidler & Hodapp, 1999) and of mothers of these children (Fidler, 2003). In short, etiology-related behavioral and non-behavioral characteristics may be important.

Three Examples

We provide next three examples of etiology-related characteristics that might influence educational practice for children with intellectual disabilities.

Cognitive-Linguistic Functioning in Down Syndrome
Down syndrome holds a unique place within studies of behavioral phenotypes. On one hand, there are probably as many behavioral studies of Down
syndrome as of all other genetic intellectual disability conditions together
(Hodapp & Dykens, 2012). And yet, at the same time, many of these studies
are not really about behavior in Down syndrome per se. Many researchers
have used Down syndrome as a control or contrast group to study functioning among children with autism, with Williams syndrome, or with other
conditions. Down syndrome has come to represent all children with intellectual disabilities.

To us, this characterization is incorrect (Hodapp & Dykens, 2001). In
fact, over the years an interesting set of etiology-related behaviors has been
identified in Down syndrome. These include problems in language development – deficits that usually exceed overall mental age levels, especially
in grammar and in expressive language abilities overall. Problems have
also been identified in hippocampal functioning, especially the ability to
recall event knowledge, which has been tentatively linked to specific deficits shown in mouse models of Down syndrome (cf. Fidler & Nadel, 2007).
In addition, many of these children display deficits in short-term working
memory, especially when tasks involve the recall of words and numbers
verbally (for a discussion, see Baddeley & Jarrold, 2007).

Two recent lines of work seem especially interesting relative to educational issues. In the first, Fidler and colleagues examined deficits in executive processing in children with Down syndrome (Fidler, 2005; Fidler,
Philofsky, Hepburn, & Rogers, 2005). Examining working memory using
the Behavior Rating Inventory of Executive Functioning – Preschool version (or BRIEF-P), Lee and colleagues (2011) found that mothers of over
half (50–60 percent) of their sample's 4 ten-year-old children with Down
syndrome rated their children as having particular difficulties (i.e., within
the clinical range) on working memory tasks that involve maintaining disparate pieces of information to complete a task or to generate an appropriate response.

Such problems, in turn, may relate to two other aspects of the Down syndrome phenotype. First, even from the earliest years, these children have
been described as friendly, upbeat, and outgoing; in observational studies
of mother–toddler interactions, these children (vs. MA-matched children
without disabilities) spend greater amounts of time looking at their mothers and smiling (Kasari, Freeman, Mundy, & Sigman, 1995). Second – and
possibly related to these children's difficulties in executive functioning and
other means–ends tasks (Fidler, 2005) – children with Down syndrome

often rely on others for help and avoid attempting to perform difficult cognitive tasks (Kasari & Freeman, 2001; Pitcairn & Wishart, 1994).

Problems in language and executive functioning, along with a general social orientation, lower task persistence, and orientation to ask others for help, all lead to potentially useful educational implications. Fidler and Nadel (2007) suggested, for example, that special education teachers might profitably employ visual supports to help these children process tasks involving verbal commands; be more aware of the stigma associated with deficits in expressive communication and articulation; and employ social techniques such as modeling, peer collaboration, and social groups to increase the social rewards of performing academic tasks. Although such techniques have yet to be formally attempted or evaluated, they do appear promising.

A second line of research pertains to teaching children with Down syndrome to read. This issue is of interest both practically and as it relates to these children's cognitive-linguistic profiles. Until twenty or so years ago, few children with Down syndrome received any literacy instruction. But, as a result of the work of Buckley and her colleagues (Buckley, Bird, & Byrne, 1996), many children with Down syndrome are now being taught to read.

Reading involves phonological processing. As phonological processing is an area of weaknesses in this population, learning to read may be particularly problematic for these children. Still, recent studies have found that: (a) children with Down syndrome can be taught at least the basics of reading (Kay-Raining Bird & Chapman, 2011); (b) these children, much like children without Down syndrome, rely on phonological processing to achieve higher levels of reading proficiency (Lemons & Fuchs, 2010a); and (c) individualized, intensive, and phonologically based reading interventions show some success in teaching reading to school-aged children with Down syndrome (Burgoyne et al., 2013; Lemons & Fuchs, 2010b).

Personality, Vulnerability, and Williams Syndrome

Although important for social and academic success in any educational environment, personality has received only sporadic attention within special education. As an example of the educational influence of personality, consider children with Williams syndrome. From a very early age, these children are described as having social and outgoing personalities (Davies, Udwin, & Howlin, 1998; Doyle, Bellugi, Korenberg, & Graham, 2004). Even as infants and toddlers, individuals with Williams syndrome (compared to typically developing children) spend more time looking at faces and they display a more intense gaze when looking at individuals (Mervis et al., 2003).

The social personality of children with Williams syndrome has also been tied to their language development. For example, compared to typically developing children, children with Williams syndrome are less interested in objects and more interested in faces. This social interest has led to deficits in joint attention, as infants and toddlers with Williams syndrome fail to disengage from the face to follow a gesture (point) to a referent object (Laing et al., 2002). Despite delays in joint attention and the initial development of language, children with Williams syndrome eventually develop relatively good language skills. Older children with Williams syndrome have strong receptive and expressive language skills and they use more affective enhancers and more evaluative devices to engage and maintain their listener's attention during a narrative (Jones et al., 2000; Martens, Wilson, & Reutens, 2008).

Educators should approach these excellent verbal skills and outgoing personalities with caution. In regard to the former, the verbal abilities of children with Williams syndrome could mask other academic difficulties the student might be experiencing. Specifically, children with Williams syndrome have difficulty processing non-verbal information compared to auditory information, they have deficits in visual-spatial processing, and they have difficulty with abstract concepts and abstract reasoning (Mervis & Klein-Tasman, 2000; Mervis, Morris, Bertrand, & Robinson, 1999). Overemphasizing their often impressive vocabularies, classroom teachers sometimes assume that these students are better able to keep up in the general education classroom than they actually are. Teachers should ensure that students with Williams syndrome receive appropriate accommodations to excel in all academic areas.

The social personality of students with Williams syndrome can also affect their school-related social achievements. While individuals with Williams syndrome are extremely friendly and outgoing, many also have poorly developed social skills and often have trouble making or keeping friends (Davies et al., 1998). Lack of a peer network, which usually serves as a protective factor against victimization, then leads to increased risk of bullying and teasing (Fisher, Moskowitz, & Hodapp, 2013). In addition, because of their desire to make friends, students with Williams syndrome may be easily persuaded to commit questionable actions or take the blame for another student's transgressions (Thurman & Fisher, 2015). As a result, students with Williams syndrome often benefit from direct social skills instruction, especially when instruction includes lessons on appropriate interactions with friends and strangers, as well as on ways to maintain healthy relationships (Fisher, 2013). Both inside and outside of the classroom, teachers must also be vigilant regarding bullying and other forms of victimization to which students with Williams syndrome can easily fall prey.

A final personality consideration relates to anxiety. In addition to fears and phobias, individuals with William syndrome show high levels of generalized anxiety, with rates ranging from 8 percent (Leyfer, Woodruff-Borden, & Mervis, 2009) to 24 percent (Kennedy, Kaye, & Sadler, 2006). Educators can help students reduce episodes of anxiety by being aware of sources of anxiety and altering the environment to accommodate such concerns. For example, students with Williams syndrome benefit from a predictable schedule and environment. Thus, students should be provided with a daily schedule, with specific warnings about changes in routine (unexpected changes should be minimized). In addition, due to heightened sensitivity to sounds (Dykens, 2003), individuals with Williams syndrome are often anxious about the presence of loud noises – allowing the use of noise-cancelling headphones before the school bell rings or providing warnings before fire drills might alleviate behavior problems related to anxiety.

Other Etiology-Related Behaviors and "Associated" Characteristics
A host of genetic conditions display characteristic features that also relate to educational issues. Many show specific maladaptive behaviors-psychopathologies. High levels of anxiety disorders have been noted among children and young adults with Williams syndrome (Dykens, 2003; Tonge & Einfeld, 2003); hyperphagia and obsessions and compulsions among children with Prader-Willi syndrome (Dykens, Leckman, & Cassidy, 1996); autism spectrum disorders among children (especially boys) with fragile X syndrome (Rodgers, Wehner, & Hagerman, 2001); and schizophrenia, bipolar disorder, anxiety, and other psychiatric disorders among children with 22q11.2 deletion or VCF syndrome (Schneider et al., 2014). Although within-syndrome variability exists, each genetic condition conveys much higher-than-usual risks for the co-occurring psychiatric disorder.

Beyond the genetic conditions themselves, other issues relate to parents and families of children with certain genetic etiologies. As an inherited condition, fragile X syndrome often runs in families. As such, carrier mothers display high rates of shyness and anxiety (Bourgeois et al., 2009), with recent studies documenting these women's use of psychotropic medication for depression, anxiety, social phobia, and panic (Lachiewicz et al., 2010). These issues likely affect parental interactions with the school system.

In another example, mothers of children with Down syndrome are, on average, four to five years older than mothers of children without Down syndrome when they deliver their newborns (this finding exists even today). Comparing the mother's age at delivery for newborns with Down syndrome versus all mothers who have children without Down syndrome in a mid-size Southeastern state, the Down syndrome group had proportionally fewer offspring delivered during the teen years, proportionally more when mothers were over age thirty-five (Hodapp, Burke, & Urbano, 2012). For all groups (i.e., of newborns with or without Down syndrome), mothers who are older when they give birth also more often have higher levels of education and are married (Hodapp & Urbano, 2008). Such parental demographic characteristics might also influence parents' relationships with school personnel, as well as parents' knowledge of and abilities to access and to benefit from educational, medical, and other services (Hodapp et al., 2012).

Going Beyond the IEP: Why Etiology Matters in Special Education

In considering the role of etiology within the field of special education, we now discuss special education teachers' current knowledge; the potential of such information to help in educational settings; and a helpful concept from the field of medicine.

Special Education's Present State of Affairs vis-à-vis Behavioral Phenotypes

Even if special education's large-scale datasets and federal reporting guidelines pay little attention to etiology, special education teachers may still know and use etiological information in their classrooms. Unfortunately, only a few studies have examined this issue; even the degree to which teachers know about behavioral phenotypes remains largely unexamined. From the existing evidence, however, most special education teachers possess limited knowledge. One study compared pediatrician and special education teachers' knowledge of the physical, cognitive, and behavioral characteristics associated with children with Down syndrome, fragile X syndrome, and VCF (the disorder predisposing adolescents to schizophrenia, bipolar, and other severe psychiatric conditions). Lee and colleagues (2005) found that, while over half (64 percent) of special education teachers were able to identify Down syndrome's physical characteristics, less than half were able to identify etiology-related cognitive (46 percent) or behavioral (32 percent) features. Teachers' levels of knowledge were lower about fragile X syndrome (scores for physical, cognitive, and behavioral characteristics were 26 percent, 24 percent, and 34 percent, respectively), and, for VCF, levels were even lower (21 percent, 8 percent, and 8 percent). Although questions differed slightly across studies, teachers displayed limited knowledge about characteristics when students had specific etiologies (usually fragile X syndrome; Finucane, Haas-Givler, & Simon, 2013; Wilson & Mazzocco, 1993; York et al., 1999).

Potential of Special Education Practice to be Informed by Etiology
While special education teachers may lack knowledge of behavioral phenotypes, what should they know and why will it help? We here provide a few possible answers.

Understanding Etiology-Based Strengths–Weaknesses Might Lead to Targeted Interventions. One hope is that, by knowing etiology-related strengths and weaknesses of different genetic conditions, one might be able to develop etiology-related interventions. Specific educational interventions have now been suggested for children with Prader-Willi syndrome (Chudd, Levine, & Wharton, 2006), Williams syndrome (Levine, 1997), Down syndrome (Buckley & Bird, 2002), and fragile X syndrome (Braden, 2002).

Beyond the need for rigorous tests of these programs' effectiveness, several reviews have questioned the potential usefulness of such interventions. Three issues have been discussed. First, as noted already, behavioral phenotypes are probabilistic. Thus, no single intervention is going to work for all children with, say, Down syndrome, fragile X syndrome, or Prader-Willi syndrome. Second, many different genetic etiologies cause intellectual disabilities, probably over 1,000, and most of these syndromes are relatively rare. As a result, "The fact that many teachers will only teach a small number of children with genetic syndromes through their career would make comprehensive training an inefficient allocation of resources" (Reilly, 2012, p. 937; see also Einfeld, 2005).

The third issue relates to changing the classroom environment. How much can, or should, the classroom be changed to accommodate a single child? Several researchers have suggested that there be a pyramid (similar to that used in Response-to-Intervention) that goes from universal to particular (Braden, 2002; Reilly, 2011). At the bottom of the pyramid, the most universal level, teachers can provide general classroom accommodations that benefit many children with intellectual disabilities. At the second, middle level, teachers might provide accommodations that relate to behaviors or problems that affect children with several genetic etiologies. The best examples here might consist of quiet or distraction-free zones in the classroom, which might benefit children with attentional issues resulting from several genetic conditions. Finally, at the top of the pyramid, teachers might make classroom accommodations for specific phenotypes, such things as controlling access to food in Prader-Willi syndrome. Such a three-step, pyramid approach might allow for accommodations based on behavioral phenotypes within general education or special education classrooms.

Knowledge of Phenotypes Might Lead to Appropriate Referrals. There may also be other school-related uses of genetic etiology. One important, albeit overlooked, use involves diagnosis. As teachers constitute the professional group with which parents potentially have the closest and longest-lasting interactions, teachers might serve to alert parents as to the need for genetic diagnoses.

One clear instance of this role occurs in fragile X syndrome. Considered the most common hereditary genetic condition, occurring in one per 4,000 boys and one per 8,000 girls (Lachiewicz et al., 2010), fragile X syndrome also has a characteristic physical appearance, particularly among affected males. Specifically, these boys often display a long, thin face and large ears. Yet, despite this appearance, many children with fragile X syndrome remain undiagnosed until the early school years. As Lee et al., (2005) noted, "late diagnosis is a problem. A study of fragile X diagnosis concluded that affected children typically were not identified until a mean age of 35 months, thus missing 2 to 3 years of intervention (Bailey, Skinner, Hatton & Roberts, 2000)" (p. 104). This diagnostic function is also highly valued by parents. In one recent study, several parents lauded school personnel for their awareness of etiology-related information. As one parent noted, "We have had a very positive experience with our school district as our daughter has been in the district since pre-k where they actually diagnosed her with Williams Syndrome" (Burke & Hodapp, 2016, p. 147).

Etiological Knowledge Might Predict, Ameliorate, or Circumvent Later Individual or Interpersonal Problems. As noted earlier in reference to the development of cognitive and linguistic phenotypes, knowledge of emerging etiology-related patterns can help to identify emerging patterns. School personnel should find it important to know that children with Down syndrome are apt to avoid attempting difficult cognitive tasks (Kasari & Freeman, 2001; Pitcairn & Wishart, 1994).

Other, non-cognitive changes can also be predicted in certain etiological groups. One can thus predict that many children with Prader-Willi syndrome will show hyperphagia (as well as obsessions and compulsions) starting in the two- to five-year period (Dykens, Cassidy, & DeVries, 2011). The beginnings of other maladaptive behaviors/psychopathology also often begin at a particular age period: note here the rise of anxiety among children with Williams syndrome (Dykens, 2003), and of schizophrenia and other psychiatric problems among adolescents with VCF syndrome (Schneider et al., 2014).

Finally, it is important to emphasize the ways that changing cognitive, linguistic, academic profiles – as well as overfriendliness and other personality characteristics – relate to the reactions of peers and adults within the child's surrounding environment. Having friends, not being bullied or taken advantage of, being a rewarding student to a teacher – all are important non-academic issues to children and adolescents with different genetic conditions. While such child–environment interactions are important and change over time, few have received much research attention.

Anticipatory Guidance in Special Education
In considering behavioral phenotypes, there exist many potential issues and domains of information; any individual child will not necessarily need any particular change or accommodation; and there are many different genetic etiologies of intellectual disability. It, therefore, becomes unclear how to advise teachers and other professionals to best employ etiology-specific information.

Borrowing from pediatrics, we adopt the idea of anticipatory guidance. Although described in multiple ways, anticipatory guidance involves the idea that physicians are able to provide parents a set of predictions, warnings, and advice about potential upcoming issues. As Nelson, Wissow, and Cheng (2003) noted, "Anticipatory guidance consists of information that clinicians give families about what they should expect in their child's development, what they should do to promote this development, and the benefits of these healthy lifestyles and practices. It is distinct from counseling, which is advice given in response to specific problems" (p. 630). Within pediatrics, articles have appeared on anticipatory guidance in connection to specific topics (e.g., breastfeeding, healthy lifestyles, teeth brushing, toileting, firearms); to the best ways of presenting information (during well-child visits, by DVD, or through written materials disseminated via the web, baby books or pamphlets); and to different audiences (teen mothers, fathers, Latinos, parents of children with genetic syndromes).

As adapted for teachers, anticipatory guidance involves providing to teachers etiology-relevant information whenever they are teaching a child with one or another specific genetic etiology. For teachers, anticipatory guidance begins with an appreciation that etiology might matter and that, for many etiological conditions, there exist parent and professional groups with wide knowledge and experience in dealing with children with the condition. Reached through phone calls, e-mails, chat rooms, or web or printed materials, such groups can then help teachers to anticipate those behaviors or profiles that are likely to occur. Although not every child with a particular genetic etiology will ultimately display any or all of these issues and problems, many will. Teachers and other educational personnel will benefit from the wisdom of those with longer, more in-depth experience in dealing with children with a particular etiology. Fidler and Nadel (2007), who also invoke the idea of anticipatory guidance, refer to this orientation as a "ready stance" that teachers and other school personnel can take whenever they teach a child with a specific genetic etiology.

CONCLUSION

Given the incredible advances in finding and describing the genetic causes of intellectual disability, one might assume that this information is now being used to improve the teaching of these children. At the current time, however, we continue to see two parallel tracks. On one hand, the field of behavioral phenotypes is providing increased information about cognitive, linguistic, and adaptive profiles, maladaptive behavior and psychopathology, and even those non-behavioral aspects of phenotypes that affect others (facial appearance, sickness, among others). On the other hand, much of this information does not seem to have yet affected the everyday classroom experiences of these students. Gradually, however, genetic information is becoming more available and, increasingly, teachers seem to be open to helping children with various genetic etiologies. As we increasingly encourage students with intellectual disabilities to succeed in inclusive classrooms, we are going to need to pay attention to all of these children's personal characteristics, including behavioral and personality aspects of their causes of intellectual disability.

REFERENCES

Abbeduto, L., Warren, S. F., & Conners, F. A. (2007). Language development in Down syndrome: From the prelinguistic period to the acquisition of literacy. *Mental Retardation and Developmental Disabilities Research Reviews, 13*, 247–261.

Allanson, J. E., O'Hara, P., Farkas, G., & Nair, R. C. (1993). Anthropometric craniofacial pattern profiles in Down syndrome. *American Journal of Medical Genetics, 47*, 748–752.

Baddeley, A., & Jarrold, C. (2007). Working memory and Down syndrome. *Journal of Intellectual Disability Research, 51*(12), 925–931.

Bailey, D. B., Skinner, D., Hatton, D. D., & Roberts, J. (2000). Family experiences and factors associated with diagnosis of fragile X syndrome. *Journal of Developmental and Behavioral Pediatrics, 21*, 315–321.

Baumgardner, T. L., Reiss, A. L., Freund, L. S., & Abrams, M. T. (1995). Specification of the neurobehavioral phenotype in males with fragile X syndrome. *Pediatrics, 95*, 744–752.

Bourgeois, J. A., Coffey, S. M., Rivera, S. M., Hessl, D., Gane, L. W., Tassone, F., et al. (2009). A review of fragile X permutation disorders: Expanding psychiatric perspectives. *Journal of Clinical Psychiatry, 70*, 852–862.

Braden, M. L. (2002). Academic interventions. In R. J. Hagerman & P. J. Hagerman (Eds.), *Fragile X syndrome: Diagnosis, treatment, and Research* (3rd ed.; pp. 428–464). Baltimore, MD: Johns Hopkins University Press.

Buckley, S., & Bird, G. (2002). Cognitive development and education: Perspectives on Down syndrome from a twenty-year research programme. In M. Cuskally, A. Jobling, & S. Buckley (Eds.), *Down syndrome across the life span* (pp. 66–80). London, UK: Whurr.

Buckley, S. J., Bird, G., & Byrne, A. (1996). Reading acquisition by young children with Down syndrome. In B. Stratford & P. Gunn (Eds.), *New approaches in Down syndrome* (pp. 268–279). London, UK: Cassell.

Bunton, K., & Leddy, M. (2011). An evaluation of articulatory working space area in vowel production of adults with Down syndrome. *Clinical Linguistics and Phonetics, 25*, 321–334.

Burgoyne, K., Duff, F., Snowling, M., Clarke, P. J., Buckley, S., & Holme, C. (2013). Efficacy of a reading and language intervention for children with Down syndrome: A controlled study. *Journal of Psychology & Psychiatry, 53*, 1044–1053.

Burke, M. M., & Hodapp, R. M. (2016). The nature, correlates, and conditions of parental advocacy in special education. *Exceptionality, 24*, 137–150.

Cameto, R., Levine, P., & Wagner, M. (2004). *Transition planning for students with disabilities. A special topic report from the National Longitudinal Transition Study-2 (NLTS-2).* Menlo Park, CA: SRI International.

Chudd, N., Levine, K., & Wharton, R. H. (2006). Educational considerations for children with Prader-Willi syndrome. In M. G. Butler, P. D. K. Lee, & B. Y. Whitman (Eds.), *Management of Prader-Willi syndrome* (pp. 302–316). New York: Springer.

Das, J. P., Kirby, J., & Jarman, R. F. (1975). Simultaneous and successive synthesis: An alternative model for cognitive abilities. *Psychological Bulletin, 82*, 87–103.

Davies, M., Udwin, O., & Howlin, P. (1998). Adults with Williams syndrome: Preliminary study of social, emotional and behavioural difficulties. *British Journal of Psychiatry, 172*, 273–276.

Doyle, T. F., Bellugi, U., Korenberg, J. R., & Graham, J. (2004). "Everybody in the world is my friend." Hypersociability in young children with Williams syndrome. *American Journal of Medical Genetics, 124A*, 263–273.

Dykens, E. M. (1995). Measuring behavioral phenotypes: Provocations from the "new genetics." *American Journal on Mental Retardation, 99*, 522–532.

(2003). Anxiety, fears, and phobias in persons with Williams syndrome. *Developmental Neuropsychology, 23*, 291–316.

(2004). Maladaptive and compulsive behavior in Prader-Willi syndrome: New insights from older adults. *American Journal on Mental Retardation, 109*, 142–153.

Dykens, E. M., Cassidy, S. B., & DeVries, M. L. (2011). Prader-Willi syndrome. In S. Goldstein & C. R. Reynolds (Eds), *Handbook of neurodevelopmental and genetic disorders* (2nd ed., pp. 484–511). New York, NY: Guilford.

Dykens, E. M., & Clarke, D. J. (1997). Correlates of maladaptive behavior in individuals with 5p-(cri-du-chat) syndrome. *Developmental Medicine and Child Neurology, 39*, 752–756.

Dykens, E. M., Hodapp, R. M., & Finucane, B. (2000). *Genetics and mental retardation syndromes: A new look at behavior and treatments.* Baltimore, MA: Paul H. Brookes Publishing.

Dykens, E. M., & Leckman, J. F. (1987). Strengths and weaknesses in intellectual functioning of males with fragile X syndrome. *American Journal of Mental Deficiency, 92*, 234–236.

Dykens, E. M., Hodapp, R. M., Walsh, K. K., & Nash, L. (1992). Profiles, correlates, and trajectories of intelligence in Prader-Willi syndrome. *Journal of the Academy of Child and Adolescent Psychiatry, 31*, 1125–1130.

Dykens, E. M., Leckman, J. F., & Cassidy, S. B. (1996). Obsessions and compulsions in Prader-Willi syndrome. *Journal of Child Psychology and Psychiatry, 37*, 995–1002.

Dykens, E. M., Maxwell, M. A., Pantino, E., Kossler, R., & Roof, E. (2007). Assessment of hyperphagia in Prader-Willi syndrome. *Obesity, 15*, 1816–1826.

Einfeld, S. (2005). Behaviour problems in children with genetic disorders causing intellectual disabilities. *Educational Psychology, 25*, 341–346.

Fidler, D. J. (2003). Parental vocalization patterns and perceived immaturity in Down syndrome. *American Journal on Mental Retardation, 108*, 425–434.

(2005). The emerging Down syndrome behavioral phenotype in early childhood: Implications for practice. *Infants & Young Children, 18*(2), 86–103.

Fidler, D. J. (Ed.) (2011). Early development in neurogenetic disorders. *International Review of Research in Developmental Disorders, 40*, 1–318.

Fidler, D. J., & Hodapp, R. M. (1999). Craniofacial maturity and perceived personality in children with Down syndrome. *American Journal on Mental Retardation, 104*, 410–421.

Fidler, D. J., & Nadel, L. (2007). Education and children with Down syndrome: Neuroscience, development, and intervention. *Mental Retardation and Developmental Disabilities Research Reviews, 13*, 262–271.

Fidler, D. J., Philofsky, A., Hepburn, S., & Rogers, S. (2005). Nonverbal requesting and problem solving by toddlers with Down syndrome. *American Journal on Mental Retardation, 110*(4), 312–322.

Finucane, B. R., & Haas-Givler, B. (2009). Smith-Magenis syndrome: Genetic basis and clinical implications. *Journal of Mental Health Research in Intellectual Disabilities, 2*, 134–148.

Finucane, B. R., Haas-Givler, B., & Simon, E. W. (2013). Knowledge and perceptions about fragile X syndrome: Implications for diagnosis, intervention, and research. *Intellectual and Developmental Disabilities, 51,* 226–236.

Fisher, M. H. (2013). Evaluation of a stranger safety training program for adults with Williams syndrome. *Journal of Intellectual Disability Research.* doi:10.1111/jir.12108

Fisher, M. H., Moskowitz, A. L., & Hodapp, R. M. (2013). Differences in social vulnerability among individuals with autism spectrum disorder, Williams syndrome, and Down syndrome. *Research in Autism Spectrum Disorders, 7,* 931–937.

Flynt, J., & Yule, W. (1994). Behavioural phenotypes. In M. Rutter, E. Taylor, & L. Hersov (Eds.), *Child and adolescent psychiatry: Modern approaches* (3rd ed., pp. 666–687). London: Blackwell Scientific.

Hodapp, R. M. (1997). Direct and indirect behavioral effects of different genetic disorders of mental retardation. *American Journal on Mental Retardation, 102,* 67–79.

Hodapp, R. M., Burke, M. M., & Urbano, R. C. (2012). What's age got to do with it? Implications of maternal age on families of offspring with Down syndrome. *International Review of Research in Developmental Disabilities, 42,* 109–145.

Hodapp, R. M., & Dykens, E. M. (1994). Mental retardation's two cultures of behavioral research. *American Journal on Mental Retardation, 98,* 675–687.

(2001). Strengthening behavioral research on genetic mental retardation syndromes. *American Journal on Mental Retardation, 106,* 4–15.

(2012). Genetic disorders of intellectual disability: Expanding our concepts of phenotypes and of family outcomes. *Journal of Genetic Counseling, 21,* 761–769.

Hodapp, R. M., & Urbano, R. C. (2008). Demographics of African-American and European-American mothers of newborns with Down syndrome. *Journal of Policy and Practice in Intellectual Disabilities, 5,* 187–193.

Jones, W., Bellugi, U., Lai, Z., Chiles, M., Reilly, J., Lincoln, A., & Adolphs, R. (2000). II. Hypersociability in Williams syndrome. *Journal of Cognitive Neuroscience, 12* (Supplement 1), 30–46.

Kasari, C., & Freeman, S. F. N. (2001). Task-related social behavior in children with Down syndrome. *American Journal on Mental Retardation, 106,* 253–264.

Kasari, C., Freeman, S. F. N., Mundy, P., & Sigman, M. (1995). Affective regulation by children with Down syndrome: Coordinated joint attention and social referencing. *American Journal on Mental Retardation, 100,* 128–136.

Kay-Raining Bird, E., & Chapman, R. S. (2011). Literacy development in childhood, adolescence, and young adulthood in persons with Down syndrome. In J. A. Burack, R. M. Hodapp, G. Iarocci, & E. Zigler (Eds.), *The Oxford handbook of intellectual disabilities and development* (2nd ed., pp. 184–189). Oxford, UK: Oxford University Press.

Kennedy, J. C., Kaye, D. L., & Sadler, L. S. (2006). Psychiatric diagnoses in patients with Williams syndrome and their families. *Jefferson Journal of Psychiatry, 20,* 22–31.

Lachiewicz, A., Dawson, D., Spiridgliozzi, G., Cuccaro, M., Lachiewicz, M., & McConkie-Rossell, A. (2010). Indicators of anxiety and depression in women with fragile X permutation: Assessment of a clinical sample. *Journal of Intellectual Disability Research, 54,* 597–610.

Laing, E., Butterworth, G., Ansari, D., Gsödl, M., Longhi, E., & Panagiotaki, G. (2002). Atypical development of language and social communication in toddlers with Williams syndrome. *Developmental Science, 5,* 233–246.

Lee, N. R., Fidler, D. J., Blakeley-Smith, A., Daunhauer, L., Robinson, C., & Hepburn, S. L. (2011). Caregiver report of executive functioning in a population-based sample of young children with Down syndrome. *American Journal on Intellectual and Developmental Disabilities, 116,* 290–304.

Lee, T. H., Blasey, C. M., Dyer-Friedman, J., Glaser, B., Reiss, A. L., & Eliez, S. (2005). From research to practice: Teacher and pediatrician awareness of phenotypic traits in neurogenetic syndromes. *American Journal on Mental Retardation, 110,* 100–106.

Lemons, C. J., & Fuchs, D. (2010a). Modeling response to reading intervention in children with Down syndrome: An examination of predictors of differential growth. *Reading Research Quarterly, 45,* 134–168.

  (2010b). Phonological awareness of children with Down syndrome: Its role in learning to read and the effectiveness of related interventions. *Research in Developmental Disabilities, 31,* 316–330.

Lense, M. D., & Dykens, E. M. (2013). Cortisol reactivity and performance abilities in social situations in adults with Williams syndrome. *American Journal on Intellectual and Developmental Disabilities, 118,* 381–393.

Levine, K. (1997). *Williams syndrome: Information for teachers.* Clawson, MI: Williams Syndrome Association.

Leyfer, O., Woodruff-Borden, J., & Mervis, C. B. (2009). Anxiety disorders in children with Williams syndrome, their mothers, and their siblings: Implications for the etiology of anxiety disorders. *Journal of Neurodevelopmental Disorders, 1,* 4–14.

Martens, M. A., Wilson, S. J., & Reutens, D. C. (2008). Research review: Williams syndrome: A critical review of the cognitive, behavior, and neuroanatomical phenotype. *Journal of Child Psychology and Psychiatry, 49,* 576–608.

Mervis, C. B., & Klein-Tasman, B. P. (2000). Williams syndrome: cognition, personality, and adaptive behavior. Mental Retardation and Developmental Disabilities Research Reviews, 6(2), 148–158.

Mervis, C. B., Morris, C. A., Bertrand, J., & Robinson, B. F. (1999). Williams syndrome: Findings from an integrated program of research. In H. Tager-Flusberg (Ed.), *Neurodevelopmental disorders* (pp. 65–110). Cambridge, MA: MIT Press.

Mervis, C. B., Morris, C. A., Klein-Tasman, B. P., Bertrand, J., Kwitny, S., Appelbaum, L. G., et al. (2003). Attentional characteristics of infants and toddlers with Williams syndrome during triadic interactions. *Developmental Neuropsychology, 23,* 243–268.

Nelson, C. S., Wissow, L. S., & Cheng, T. L. (2003). Effectiveness of anticipatory guidance: Recent developments. *Current Opinion in Pediatrics, 15,* 630–635.

NICHCY. (2012). *Categories of disability under IDEA.* Washington, DC: National Dissemination Center for Children with Disabilities.

Nyhan, W. L. (1972). Behavioral phenotypes in organic genetic diseases: Presidential address to the Society of Pediatric Research (May 1, 1971). *Pediatric Research, 6,* 1–9.

Pitcairn, T. K., & Wishart, J. G. (1994). Reactions of young children with Down's syndrome to an impossible task. *British Journal of Developmental Psychology*, *12*, 485–489.

Porter, H., & Tharpe, A. M. (2010). Hearing loss among persons with Down syndrome. *International Review of Research in Mental Retardation*, *39*, 195–220.

Pueschel, S. M. (1990). Clinical aspects of Down syndrome from infancy to adulthood. *American Journal of Medical Genetics* (Suppl. 7), 52–56.

Reilly, C. (2012). Behavioural phenotypes and special education needs: Is aetiology important in the classroom? *Journal of Intellectual Disability Research*, *56*, 929–946.

Riley, K. (2011). Infusing etiology into interventions: A model for translational research in education. *International Review of Research in Developmental Disabilities (Early Development in Neurogenetic Disorders; D.J. Fidler, Ed.)*, *40*, 261–292.

Rodgers, S. J., Wehner, E. A., & Hagerman, R. J. (2001). The behavioral phenotype in fragile X: Symptoms of autism in very young children with fragile X syndrome, idiopathic autism, and other developmental disorders. *Journal of Developmental and Behavioral Pediatrics*, *22*, 409–417.

Rondal, J. (1995). *Exceptional language development in Down syndrome*. Cambridge, UK: Cambridge University Press.

Roizen, N. (2010). Overview of health issues among persons with Down syndrome. *International Review of Research in Mental Retardation*, *39*, 3–33.

Schneider, M., Debbane, M., Bassett, A. S., Chow, E. W. C., Fung, W. L. A., van den Bree, M. B. M. et al. (2014). Psychiatric disorders from childhood to adulthood in 22q11.2 deletion syndrome: Results from the International Consortium on Brain and Behavior in 22q11.2 Deletion Syndrome. *American Journal of Psychiatry*, *171*, 627–639.

Schretlen, D. J., Ward, J., Meyer, S. M., Yun, J., Puig, J. G., Nyhan, W. L., et al. (2005). Behavioral aspects of Lesch-Nyhan disease and its variants. *Developmental Medicine and Child Neurology*, *47*, 673–677.

Sigafoos, J., O'Reilly, M. F., & Lancioni, G. E. (2009). Editorial: Cri-du-chat. *Developmental Neurorehabilitation*, *12*, 119–121.

Stein, M. B., Torgrud, L. J., & Walker, J. R. (2000). Social phobia symptoms, subtypes, and severity: Findings from a community survey. *Archives of General Psychiatry*, *57*, 1046–1052.

Thurman, A. J., & Fisher, M. H. (2015). The Williams syndrome social phenotype: Disentangling the contributions of social interest and social difficulties. *International Review of Research in Developmental Disabilities*, *49*, 191–227.

Tonge, B. J., & Einfeld, S. L. (2003). Psychopathology and intellectual disability: The Australian child to adult longitudinal study. *International Review of Research in Mental Retardation*, *26*, 61–91.

Wilson, P. G., & Mazzocco, M. M. M. (1993). Awareness and knowledge of fragile X syndrome among special educators. *Mental Retardation*, *31*, 221–227.

York, A., Von Fraunholder, N., Turk, J., & Sedgwick, P. (1999). Fragile X, Down's syndrome, and autism: Awareness and knowledge amongst special educators. *Journal of Intellectual Disability Research*, *43*, 314–324.

9

Ethical Implications of Behavioral Genetics on Education

VICTORIA J. SCHENKER AND STEPHEN A. PETRILL

INTRODUCTION

Numerous aspects of children's lives contribute to or inhibit academic skills and successes. Deservedly, much of the focus traditionally has been on children's environments. Children's homes and schools especially have a large influence on reading and math skills. Academic skills are, after all, acquired through many years of structured and unstructured support. On the other hand, not all children develop academic skills at the same rate, or to the same degree. Some children continue to struggle despite adequate support, whereas others thrive in the face of significant environmental privation. This has led many to consider the role that genetics may also play in children's academic abilities, resulting in several important findings but also some controversy.

The purpose of this chapter is to examine the ethical implications of the role of genetics and environment on education from the perspective of behavioral genetics. In particular, this chapter will provide examples of behavioral genetic studies to examine some of the promises and barriers to using genetic information in educational settings. It is our contention that the issue of genetics in education, although controversial, is simultaneously necessary and problematic due to (1) difficulty in translating genetic studies into educational practice, (2) misconceptions concerning how genetic effects operate in individuals versus populations, and (3) misapplication and misinterpretation of genetic information. However, we also assert that genetically sensitive information may provide schools, teachers, and parents with more individualized information on children's academic achievement in order to give them the best chances at success.

ENVIRONMENTAL EFFECTS ON ACADEMIC
ACHIEVEMENT

There are many aspects of the environment that contribute to academic achievement. Clearly, children are taught how to read, solve math problems, and succeed in other academic subjects. The home literacy environment, for example, describes the aspects of the home that are most relevant to reading development, including how often a parent reads to a child and the number of books a child possesses. Several studies have demonstrated ways in which the home literacy environment relates to children's early reading abilities, such as the development of oral language ability (e.g., Burgess, Hecht, & Lonigan, 2002; Payne, Whitehurst, & Angell, 1994; Rodriguez et al., 2009). Children's oral language skills relate to their early reading achievement, and aspects of the home literacy environment, including frequency of shared picture book reading and the age of onset of picture book reading, explain 12–18.5 percent of the variance in children's language scores (Payne et al., 1994). Furthermore, studies of the home literacy environment are especially relevant to children from low-income families, because these children tend to begin school with lower average scores in measures such as vocabulary. For example, Rodriguez and colleagues (2009) examined language and cognitive abilities in 1,046 children from low-income families at fourteen, twenty-four, and thirty-six months of age. The researchers measured home literacy environments by examining the frequency of children's participation in literacy activities, including shared book reading and storytelling, the quality of mothers' engagements with their children, such as cognitive stimulation and literacy activities, and the provision of age-appropriate learning materials. These aspects of the home literacy environment uniquely contributed to the prediction of children's language and cognitive skills at all three ages, and the home literacy environment at all three ages explained unique variance in the children's thirty-six-month language and cognitive skills. Children from low-income families who had consistently high home literacy environments performed similar to the general population, whereas children who had lower home literacy experiences performed worse, possibly placing them at risk for later learning difficulties. Overall these findings indicate that children from low-income populations could benefit from early interventions targeting all aspects of the home literacy environment. Fortunately, there is evidence to suggest that parents can be taught to promote literacy in their homes, such as by learning how to read books to their children, how to engage their children in discussion about book reading, and how to encourage children to read more books (Saracho, 1997).

Similarly, studies from various countries have demonstrated that the home learning environment is related to children's numeracy skills, which predict later success in mathematics. One study on German children revealed that both home activities associated with numeracy and home activities associated with literacy were related to children's numeracy skills in the first year of preschool, and the advantages gained remained present in later years (Anders et al., 2012). A study on English students had similar findings when studying over- and underachievement (Melhuish et al., 2008). The researchers considered the performances expected of children based on their family and background characteristics, such as socioeconomic status, and compared their expectations with children's actual performances. They found that at five years of age, children with higher quality home learning environments, related to both literacy and numeracy, were more likely to overachieve, whereas children with lower quality home learning environments were more likely to underachieve. At seven years of age, children who were underachievers were more likely to have lower quality home learning environments, but children who were average or who were overachieving showed little difference in the quality of their home learning environments. Finally, a study examining five- to seven-year-old Dutch students examined parent–child numeracy activities and parents' numeracy expectations for their children as well as children's cognitive and linguistic skills (Kleemans, Peeters, Segers, & Verhoeven, 2012). Children's early literacy skills and grammatical ability were associated with their numeracy skills, and both the numeracy activities and parents' numeracy expectations explained variance in early numeracy, after controlling for cognitive and linguistic skills. Interestingly, across the three studies, not only were home activities related to numeracy associated with children's numeracy abilities, but also home activities related to literacy were significant contributors to early numeracy ability. Anders et al. (2012) offer several explanations, including the higher number of literacy activities taking place in the home in comparison to lower amounts of numeracy-related activities. Melhuish et al. (2008) also suggest that the home learning environment may be a more generalized and motivation-related aspect of development, as opposed to the specific teaching and learning of different skills. Kleemans et al. (2012) interpret their findings by suggesting the integration of language and numeracy practices in order to develop numeracy skills because of the relationship between early numeracy skills and linguistic skills. Although the possible explanations and applications for the association between home literacy activities and numeracy skills vary across studies, they do agree on an important conclusion: home learning environment is an important component of the development of numeracy skills.

Schools are also clearly important to reading and math development. Many studies have examined the efficacy of certain teaching methods and interventions in schools, such as the eighteen studies published between 1995 and 2005 reviewed by Wanzek and Vaughn (2007). All eighteen studies demonstrated positive outcomes for students who participated in early reading interventions. This was especially true when the students participated in the interventions in earlier grades and when phonics instruction and text reading were emphasized. A more recent study examined 315 students in twenty-seven third-grade classrooms to examine the effects of content and the amount of time the students spent in literacy instruction as well as the global quality of the classroom learning environment (Connor et al., 2014). Interestingly, the quality of the classroom on its own did not predict children's vocabulary and comprehension gains, nor did the literacy instruction. However, an interaction of the two variables was associated with students' vocabulary and comprehension gains, suggesting a more complex influence of the classroom on children's reading abilities. For example, teachers who created high-quality classrooms but in which students received little teacher/child-managed meaning-focused instruction (teacher and students working together) were no more effective than teachers with low-quality classroom environments. The students who made the most gains were in high-quality classrooms and spent greater amounts of time in teacher/child-managed meaning-focused instruction, and were especially successful when this instruction was provided in small groups.

Home learning practices and classrooms do not contribute separately to children's reading and math development. On the contrary, Anders et al. (2012) found that children from homes with low-quality learning environments did not benefit from even the highest quality of preschools in their study. The authors concluded that a threshold of support at home must exist in order for children to gain from high-quality classrooms. Furthermore, the home learning environments and classroom environments are likely combined with other environmental variables to influence reading and math development. These home learning and classroom environment examples are just several of many demonstrating that certain home and school practices lead to better reading and math outcomes in students, thus supporting the notion that environments can play a large role in a child's academic success.

BEHAVIORAL GENETIC STUDIES

Despite clear and replicable effects on academic outcomes, the extensive research on children's environments does not completely explain children's academic achievements. There is substantial variability in reading and math outcomes, in the same home, in the same classroom, even in the face of similar levels of home and school support. Anecdotally, parents and teachers observe that children are different from one another, even in the same environments. Some children learn more quickly or slowly, or differently. Sometimes it takes different strategies to help different children come to the same point in their academic achievement. Sometimes teachers and parents will use strengths a child has to overcome weaknesses.

One possibility is that genetic factors account for some of these individual differences. Genetics in its entirety is a very large and fast-growing field that is beyond the scope of this chapter (see Kornilov, this volume, however). Instead, we will focus on the behavioral genetic perspective and the ethical implications of behavioral genetic findings. Behavioral genetic approaches (also called Quantitative Genetics; for a more general introduction, see Tan, this volume) compare special populations; for example, identical twins versus fraternal twins, adoptive siblings versus biological siblings, or more recently children from assisted reproductive procedures that may or may not be biologically related to their birth parents (Harold et al., 2013). The key assumption is that if genetic influences are important, then individuals who are more genetically related will be more similar in their academic achievement than will those who are less genetically related. This comparison is examined statistically to yield heritability, or h^2. Heritability refers to the proportion of differences in a group that can be attributable to genetic differences in that group. A heritability of $h^2 = .5$ means that 50 percent of the differences in that trait in a group of people are due to genetic differences in that group.

The quantitative genetic method also can be used to examine the environment. If adoptive siblings are just as similar on a trait as biological siblings, or fraternal twins are just as similar as identical twins, then shared family environment, or c^2, is important. These environmental influences refer to *anything* that is non-genetic that makes family members who live together similar: same home environment, same schools, but also same pre- and peri-natal environments, same neighborhoods, and so on. Finally, identical twins, although often highly similar, are not the same despite their genetic and environmental similarities. This is also the case in other siblings. The environments they experience, although similar, are not always the same. These differences can be used to estimate the non-shared environment, or e^2.

Looking across the thousands of genetic studies conducted in reading and mathematics from across the world, genetic, shared environment, and non-shared environmental influences on reading and math abilities are not only substantial but also highly similar across cultures. Genetic variance accounts for 50–80 percent of individual differences in reading at the end of first grade in the United States, Australia, and Scandinavia (Byrne et al., 2006, 2007; Petrill, Deater-Deckard, Thompson, DeThorne, & Schatschneider, 2006; Petrill et al., 2007). In the twin sample of the Byrne et al. (2006) project, there was a strong genetic influence on preschool phonological awareness ($h^2 = .61$), rapid naming ($h^2 = .64$), and verbal memory ($h^2 = .57$). The shared environment had a strong influence on the variance in print awareness ($c^2 = .68$), vocabulary ($c^2 = .60$), and grammar/morphology ($c^2 = .59$). In kindergarten, reading, phonological awareness, and rapid naming were mostly influenced by genetic factors ($h^2 = .70, .63, .60$, respectively), and spelling was equally affected by genes and shared environment ($h^2 = .39, c^2 = .40$).

Although a large portion of the behavioral genetics research on academic abilities has focused on reading skills, several studies have examined math and other disciplines as well. Kovas et al. (2005) considered both reading and math and found that both were highly heritable, with a reading heritability estimate of $h^2 = .69$ and a math heritability estimate of $h^2 = .67$. In addition, there were smaller non-shared environmental influences on reading and math ($e^2 = .15$ and $.24$, respectively) and a shared environmental influence on reading ($c^2 = .17$). There are even fewer studies on science skills. One study using the Twins Early Development Study (Haworth, Dale, & Plomin, 2009) found that science achievement in nine-year-olds was highly heritable ($h^2 = .64$) and also influenced by shared environment ($c^2 = .16$) and non-shared environment ($e^2 = .20$). For twelve-year-olds, science achievement was heritable, although less so than when the children were younger ($h^2 = .47$), and was also influenced by shared environment ($c^2 = .32$) and non-shared environment ($e^2 = .21$).

Finally, several studies examine traits related to academic skills such as attention and motivation. For example, researchers used children from The Netherlands Twin Registry to examine overactivity and attention problems and found that they were highly heritable, with estimates of about $h^2 = .75$ across different ages (Rietveld, Hudziak, Bartels, van Beijsterveldt, & Boomsma, 2004). Similarly, research on academic motivation across six countries demonstrated that enjoyment and self-perceived ability of academic subjects are heritable, with estimates of about h^2 = .40 (Kovas et al., 2015). Academic motivation, including motivation to read, has also been found to be largely influenced by non-shared environment (Kovas et al., 2015; Schenker & Petrill, 2015).

These studies suggest strongly that genetic *and* environmental influences are important to academic outcomes. More importantly, beyond showing that academic outcomes are influenced by both genetics and environment, these methods can also be used to examine the etiology of the relationships among academics. For example, it is possible to examine one twin's reading score at one time point and the other twin's reading score at a later time point. Using the same logic as described above, it is possible to examine the extent to which genetic and environment influences affect the longitudinal stability and instability of academic achievement over time. Using this approach, studies of reading and math skills have shown a high degree of genetic stability; the genetic variance that affects reading at one time point is correlated with the genetic variance affecting reading at later time points. For example, several longitudinal twin studies have found that genetic factors account for significant portions of the stability of reading outcomes in early elementary school (Byrne et al., 2005; Harlaar, Dale, & Plomin, 2007; Petrill et al., 2007). Participants in the International Longitudinal Twin Study demonstrated genetic and environmental influences on the stability of their reading outcomes between preschool and first grade (Byrne et al., 2005). Petrill and colleagues (2007) examined the stability of reading outcomes of slightly older children from the Western Reserve Reading and Math Project, beginning at kindergarten and first grade and then tested again a year later, and found similar results of genetic stability. In this study, genetic influences accounted for a significant portion of the stability of the following reading measures: phonological awareness, expressive vocabulary, letter knowledge, word knowledge, phonological decoding, and passage comprehension. This pattern of genetic stability continues into later years, as found by Wadsworth, Corley, Hewitt, Plomin, and DeFries (2002). The researchers found genetic correlations for a single measure of reading at ages seven, twelve, and sixteen years. There is also evidence of the genetic stability of reading from the Florida Twin Project on Reading, which demonstrated this stability in grades one through five (Hart et al., 2013).

Similarly, evidence suggests that math abilities demonstrate genetic stability, although perhaps less clearly than with reading. For example, three articles on the Twins Early Development Study have produced high estimates of genetic influences on math achievement across different age groups. When the twin participants were seven years of age, the heritability estimate of their math abilities was $h^2 = .66$ (Oliver et al., 2004). When participants from the same sample were nine years of age, the heritability of their math skills remained high and significant, with an estimate of $h^2 = .68$ (Haworth, Kovas, Petrill, & Plomin, 2007). At ten years of age, the heritability of their math abilities was more moderate, but still significant ($h^2 = .42–.45$; Kovas et al., 2005). Furthermore, there is evidence to suggest overlap in the genetic and environmental influences on reading and math abilities. Kovas et al. (2005) found a genetic correlation of .74 between reading and mathematics, suggesting that most of the genes contributing to differences in reading ability are the same genes that influence math ability. Furthermore, Hart, Petrill, Thompson, and Plomin (2009) examined various factors of reading and math ability and also found genetic and environmental overlap between reading, math, and general cognitive ability. This suggests that reading is vital to solving math problems and possibly that the specific skills required in reading ability and math ability involving reading overlap with general cognitive ability.

These and other findings suggest strongly that genetic and environmental influences are important to understanding academic outcomes, their growth, and their inter-relationships. However, these data come with important caveats with respect to their application in educational settings. First, h^2, c^2, and e^2 estimate the total effect of genetic, shared environmental, and non-shared environmental differences in a group of people. These estimates do not imply how much genes and environments impact the development of an individual. Moreover, estimates of h^2, c^2, and e^2 are a snapshot of the etiology of differences in a population in a particular context, and can vary if the context changes. One common fallacy of genetics that often leads to distrust of using genetic information is that traits high in heritability, such as reading and math skills, are unalterable by the environment (Berryessa, 2013). In reality, estimates of h^2, c^2, and e^2 reflect a certain context and can vary over time and in different situations. If the environment were more varied, then heritability, because it is a proportion of variance, would necessarily go down.

A second caveat is that genetic estimates include both direct and indirect effects of genes. For example, genetic mechanisms may impact the development of brain structures associated with reading (Eicher & Gruen, 2013). However, the heritability of reading may also reflect the accumulation of environmental exposure. This is called *gene-environment correlation*. For example, twins who are more similar in their reading skills have been shown to be more similar in how much they tend to read for pleasure, which in turn may amplify their reading similarity (Harlaar, Deater-Deckard, Thompson, DeThorne, & Petrill, 2011). This study is an example of active or evocative gene–environment correlation, where the probability of the environment that a child selects or has selected for him/her is influenced by genetic factors. Beyond these potential correlations between genes and environments, the relative importance of genetics and environment may vary in other significant ways. Gene x environment interaction occurs when environments trigger genetic risks and/or positive environments allow genetically influenced skills to emerge. For example, a series of studies (Hart, Soden, Johnson, Schatschneider, & Taylor, 2013b; Taylor & Schatschneider, 2010; Taylor, Roehrig, Soden Hensler, Connor, & Schatschneider, 2010) have suggested that genetic influences on reading outcomes may be lower at the lower end of socioeconomic status in the home, and in more impoverished school environments. This is consistent with the environmental studies described earlier which suggest that the home literacy environment has an especially large impact on the reading skills of children from low-income homes, such that the children who had consistently high home literacy environments performed similarly to the general population, whereas children who had lower home literacy experiences performed worse (Rodriguez et al., 2009).

Recent research using the genome-wide complex trait analysis, which is based on similarities between unrelated individuals rather than familial similarities, has found molecular support for the genetic influences on educational achievement (Krapohl & Plomin, 2016). The researchers found that genome-wide single nucleotide polymorphisms (SNPs) explained about a third of the variance in scores on an educational examination. SNPs also explained half of the correlation between scores on the examination and family socioeconomic status, providing further support for the importance of both genes and environment in influencing educational achievement. This method provides an exciting new direction for examining genetic and environmental influences on education outcomes.

POSITIVES OF USING GENETIC INFORMATION
IN EDUCATION

In sum, behavioral genetic studies suggest that ignoring genes may be leaving aside an important contributor to the development of academic abilities. Moreover, potential gene–environment correlation and gene × environment interaction have several ethical implications. First, negative environments may be particularly hazardous to some children. Furthermore, children from families of lower socioeconomic status or schools in low-income areas tend to have lower estimates of the heritability of their academic abilities. It is, therefore, possible that these children's environments are artificially truncating their individual differences that would otherwise be present in different socioeconomic contexts. In other words, in the current educational and socioeconomic context, lower heritability is associated with poorer environments.

Taking this research on board, as research on genetics advances, scientists have begun to consider the possibilities of using this genetic information to improve education. One possibility has been to consider using genetic information to identify children from an early age who are at the greatest risk for reading and math struggles and proactively develop a solution or treatment plan to help them. School instruction often begins in kindergarten when children are around the age of five, but the first five years of a child's life is an important time to set a foundation for further learning. One thought has been to use genetic information to screen children from an early age for risk of reading and math difficulties and to create early interventions so that they have the best chance of later academic success.

Moreover, genetics may also help us understand the underlying biological mechanisms that contribute to the multiple pathways to academic achievement. Optimistic researchers believe we can use genetic information to develop teaching methods that are most suitable to a child's genetic background. In this sense, educators could develop strategies for more personalized learning and more appropriate targeted help for each child's individual needs. In their book *G Is for Genes*, Asbury and Plomin (2014) develop a hypothetical educational system built on the principles of personalized learning. This type of learning emphasizes choice in children's education, allowing children to study what interests them most (along with studying required basics), to partake in activities in which they show the most talent and interest, and to choose with whom they are working. Furthermore, the authors suggest that schools have a profile of each child's strengths and weaknesses, including genetic information, so that when a problem arises, they have the best and most relevant information in order to help the child work through his or her problem. Asbury and Plomin suggest computers as a method for personalized learning. Software such as Carnegie Learning's math program respond to individual development and would allow teachers and teaching assistants more free time to work closely with children who need extra support. The researchers believe that such software will one day be able to interact with genetic information on children's academic abilities to create an optimal program for each child. They do note, however, that at this time there is no empirical support for the benefits of using computers in classroom instruction.

Furthermore, it may one day be possible to compare children's projected levels of achievement, based on their genetic makeup, with their observed academic performances and find children who are underachieving. Realizing the genetic and environmental influences on underachievement and overachievement could be used to individualize solutions for underachieving children. With the many possible benefits of using genetic information in education, it seems unethical to ignore genetics, especially considering the consistent and substantial literature on the importance of genetic variance in academic achievement.

DIFFICULTIES OF INCLUDING GENETICS IN EDUCATION

Despite the likelihood that academic achievement is the result of both genetic and environmental influences, there are, on the other hand, many obstacles to bringing genetically sensitive designs to bear on education. First, from a scientific perspective, there is a fundamental and ethically dangerous misconception about the causality of genetics. In particular, unlike the quantitative genetic studies that estimate the effect of genetics and environment, molecular genetic studies look for DNA markers that affect traits (Strachan & Read, 2004). These methods examine whether differences in DNA are associated with differences we see in measured outcomes. Interestingly, this approach suggests that genes can be considered across two dimensions (see Figure 9.1): (1) effect size – whether the gene by itself causes an outcome or is, instead, a small contributor of risk; and (2) frequency – how often or how rarely the risk or protective version of the gene can be found in the population.

For several decades, the genetics literature has understood that some genes, although rare, are highly predictive of learning problems. This is called the One Gene One Disorder (OGOD) hypothesis, where a gene or small set of genes is both necessary and sufficient to cause a disorder. There are hundreds of examples of these genes that impact learning disabilities, and are most predictive in certain severe and/or identifiable cases of learning impairment, such as Lesch-Lyan syndrome, Down syndrome, or Fragile X. However, because they are relatively rare to very rare, these DNA markers by themselves do not account for a majority of the cases of learning impairments (van Bokhoven, 2011).

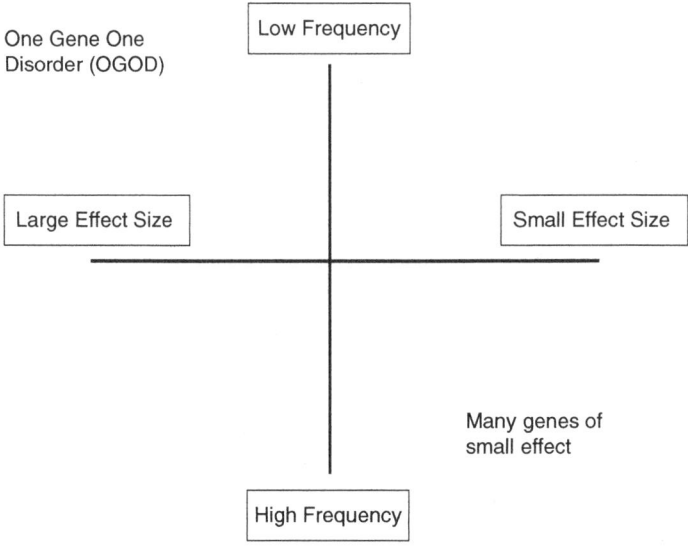

FIGURE 9.1 Genetic effects on academic achievement.

In fact, most children with academic or learning difficulties show no identifiable genetic risk factors at the level of DNA. Thus, the empirical and ethical danger is that the OGOD model does not explain many important learning-related outcomes, including most autism spectrum disorders, language impairment, reading disability, and math difficulties (Plomin, 2013). Instead, the identifiable DNA markers that influence these traits appear to be much, much smaller in effect, individually accounting for an exceedingly small proportion of the differences we see in the entire population (Asbury & Plomin, 2014). As a result, the number of genes that can influence a trait may be in the hundreds or thousands (Plomin, 2013). Although these genes are beginning to be identified and synthesized into plausible biological models (Poelmans, Buitelaar, Pauls, & Franke, 2011), the genes themselves do not explain enough of academic achievement to be clinically meaningful. In the scientific realm, this is called the "Missing Heritability Paradox" (see Maher, 2008; Manolio et al., 2009). This paradox means that, although genetic effects appear to be significant in quantitative genetic studies, finding these genes and combining them into a way to explain this heritability has been very elusive. This is true not only for education outcomes, but also for complex diseases.

From an ethical perspective, this raises several issues. On one hand, genetic effects are important in terms of heritability, but identifying profiles of risk using current DNA technology is likely to result in very small effects. This is very different from the commonly held assumption that human traits stem from an individual gene (i.e., a "reading gene") when in reality many genes contribute small amounts to academic abilities. Therefore, it is potentially problematic to base expectations of students on a genetic model that does not appear to be true for most academic outcomes. This is especially true when considering the possibility of self-fulfilling prophecies, which is the idea that people will behave a certain way as a result of other people's expectations. In a classic experiment from the 1960s, two researchers, Rosenthal and Jacobson (1968), told teachers at a public elementary school that certain students in their classroom would be likely to have a spurt of academic progress based on an assessment given in the classroom. The assessment was actually a test of intelligence, not of the likelihood of an academic spurt, and the children named were randomly chosen and, therefore, were not truly more likely to show an academic spurt. Eight months later the children were again assessed on intelligence. The "special" children, or those whom the teachers had been told were likely to have an academic spurt, gained more points on the IQ test compared to the other children. Overall, these supposedly special children gained an average of 12.22 points, while the other children gained 8.42 points, which is a 3.80 point difference. In particular, the younger grades demonstrated a large amount of variation between the supposedly special children and the other children, with supposedly special children in the first grade gaining an average of 15.4 more points than the other children and supposedly special children in the second grade gaining an average of 9.5 more points than the other children. Why might this have happened? Perhaps teachers were more encouraging to those children from whom they expected academic spurts, or they were more attentive to those children, or they were more reflective in their evaluations of those children. Whatever the method, teachers' higher expectations of children who might have an academic spurt translated into higher average IQ gains by those children, thus carrying out a self-fulfilling prophecy.

Unfortunately, this effect may also work in the opposite direction. If teachers (and parents, administrators, and other relevant professionals) have low expectations of children, then those children may not perform as well (Babad, Inbar, & Rosenthal, 1982). This becomes an ethical concern when considering using a child's genetic makeup to structure his education. If, for example, a child's teacher is told that the child's genetic makeup shows she may be low achieving in reading, then it is possible that the teacher will develop low expectations for that child and that those low expectations will translate into less encouragement and attentiveness, even if it is done subconsciously by the teacher. The child may then only perform to meet those low expectations and demonstrate a self-fulfilling prophecy, when perhaps if she had received higher teacher expectations and more encouragement, she would have performed better.

Along these same lines, genetic information that predicts low achievement may lead to labeling of these children, which in turn could result in stigmatization. In fact, in the 1980s a Massachusetts school system decided to replace labels of children who had learning or intellectual disabilities with a number system that indicated children who needed the most and least restrictive support (Siperstein, Pociask, & Collins, 2010). In this system, a child who needed to be placed in a special education program was given the number .5. However, even these seemingly innocent labels resulted in stigmatization, as children began to use the number systems in a cruel manner, such as saying to a peer, "What are you, a .5?" From this example and many others in our daily interactions, we must be aware that labeling leads to prejudice and can be especially stigmatizing in schools. These concerns arise when considering what effect using genes in schools could have on people with disabilities, many of whom may already face prejudice from being labeled with a disability. Having genetic support of these labels may only further the stigmatization. Furthermore, children who would not otherwise be faced with a label of disability may feel some of these same effects from being targeted as someone with genes that indicate low reading and math abilities. Thus, self-fulfilling prophecies and stigmatizing labels could stem from a misunderstanding of genetics if educators assume a deterministic OGOD model, when the literature suggests that this is not likely for academic achievement.

CONCLUSION: WEIGHING THE POSITIVES AND NEGATIVES

Overall, there are a number of reasonable concerns regarding the use of genetics in education that question whether the possible costs of using this information outweigh the benefits. On the other hand, there is also considerable evidence that genetic influences are important to the development of academic skills, and that these genetic effects may also permeate the environment. How do we balance these concerns with the positive possibilities? We assert the following guiding principles:

1. Genetic information should only be used to enhance opportunities rather than limit them. In a discussion on children who may be

predisposed to committing violent acts, Stone (2003) argues that the child welfare system should not be allowed to reallocate resources or services away from predisposed children because they believe them to be lost causes. This argument is equally applicable to schooling practices; we cannot allow schools to take services away from children who may be at risk for low reading and math abilities because they are believed to be lost causes or less deserving of support. If genetic information is used at all for individual children, it should be used to find the best possible ways to help and encourage the academic progress of all students.

2. The benefits of understanding the genetics behind academic success at the level of the population are much clearer than at the individual level. If parents, educators, and policymakers better understood the heritability of reading and math abilities, our society could create optimal education goals and strive toward the achievement of those goals. Policymakers often discuss closing the "achievement gap," which is the difference between high and low achieving students. The idea of closing this gap is a noble one. As a society, we recognize the need for equal opportunity, and to many a gap in achievement signifies that this need is not being met. However, an understanding of the heritability of reading and math allows us to see that closing the gap to zero is a very difficult goal. We do not wish to downplay the good intentions behind closing the gap nor the work that has been done to increase opportunities for disadvantaged children, but what do we say to the children and their parents who have tried everything but still struggle with academic problems?

3. It is important to think about genetic effects on academic achievement as non-deterministic risk and protective factors, as opposed to causal determinants of outcome. All of the available data suggest that the majority of variance in complex human traits, including academic achievement, is influenced by a multitude of genetic and environmental factors that act in non-deterministic ways. A reasonable rule-of-thumb is to think of genes the way we think of environments. Some genetic effects are so severe that they may have strong and long lasting effects, but their occurrence is rare. Instead, most genetic factors are small in effect, that act cumulatively, over time, in concert with other factors. Equality of opportunity is not the same as equality in outcome. In an education system where every child receives equal opportunities, equally excellent teaching, and equally dispersed resources, we would likely still have a distribution of ability

based on genetics. We can and should take measures to increase performance. Hopefully, we are able to move the distribution of academic achievement so even the lowest performing student is able to thrive. However, as we are working to increase performance, we may expect that the variance in performance will have a significant genetic component, even if we are able to move the average level of performance. Consider that height is highly heritable even though the average height has increased dramatically in the population over the past 200 hundred years due to better nutrition and health care.

4. Genetic testing for determining the proper form of instruction or intervention across the entire population of students is not feasible at this time. This is due to the fact that academic traits are influenced by many genes of small effect. Given this situation, we must be aware of the negative impacts that could come from lowered expectations of children genetically predisposed to have poor academic abilities so we do not create self-fulfilling prophecies, as well as the possibility of stigmatization coming from knowing a child's genetic predisposition for academic success and labeling them in such a way.

That said, there may be some benefits of using genetics in individual children's education plans, particularly in children who have a strong family history of learning problems, to ascertain whether these children have genetic risks on top of any environmental risk or protective factors. Moreover, genetic information, particularly if it is relatable to a meaningful biological or brain-based mechanism, may be useful when children who are identified as at risk do not learn from typical instructional methods. Furthermore, if as a society we determine that personalized learning is a priority in our education systems, we can achieve many of the goals laid out by Asbury and Plomin by focusing first on children who are most at risk for learning difficulties. In this way, genetic information would be brought to bear on early identification, treatment, and prevention, as it is generally used in other public health settings.

Research suggests genetic effects are significant and substantial, and thus reflect potential untapped means for understanding the development and course of academic achievement.

Having a more complete understanding of the role of genetics in academic abilities, *if properly understood and used,* may be beneficial to everyone in the education system. Children are not blank slates, and differences between learners are real. It is important that the education system and we as a society recognize the need to respect these differences as an expression of human individuality, and create the best learning environments for all children.

REFERENCES

Anders, Y., Rossbach, H.-G., Weinert, S., Ebert, S., Kuger, S., Lehrl, S., et al. (2012). Home and preschool learning environments and their relations to the development of early numeracy skills. *Early Childhood Research Quarterly, 27*, 231–244.

Asbury, K., & Plomin, R. (2014). *G is for genes: The impact of genetics on education and achievement.* West Sussex, UK: Wiley-Blackwell.

Babad, E. Y., Inbar, J., & Rosenthal, R. (1982). Pygmalion, Galatea, and the Golem: Investigations of biased and unbiased teachers. *Journal of Educational Psychology, 74*(4), 459–474.

Berryessa, C. M. (2013). Ethical, legal, social, and policy implications of behavioral genetics. *Annual Review of Genomics and Human Genetics, 14*, 515–534.

Burgess, S. R., Hecht, S. A., & Lonigan, C. J. (2002). Relations of the home literacy environment (HLE) to the development of reading-related abilities: A one-year longitudinal study. *Reading Research Quarterly, 37*(4), 408–426.

Byrne, B., Olson, R. K., Samuelsson, S., Wadsworth, S., Corley, R., DeFries, J. C., et al. (2006). Genetic and environmental influences on early literacy. *Journal of Research in Reading, 29*(1), 33–49.

Byrne, B., Wadsworth, S., Corley, R., Samuelsson, S., Quain, P., DeFries, J.C., et al. (2005). Longitudinal twin study of early literacy development: Preschool and kindergarten phases. *Scientific Studies of Reading, 9*(3), 219–235.

Byrne, B., Samuelsson, S., Wadsworth, S., Hulslander, J., Corley, R., DeFries, J. C., et al. (2007). Longitudinal twin study of early literacy development: Preschool through grade 1. *Reading and Writing, 20* (1–2), 77–102.

Connor, C. M., Spencer, M., Day, S. L., Giuliani, S., Ingebrand, S. W., McLean, L., et al. (2014). Capturing the complexity: Content, type, and amount of instruction and quality of the classroom learning environment synergistically predict third graders' vocabulary and reading comprehension outcomes. *Journal of Educational Psychology.* doi:10.1037/a0035921

Eicher, J. D., & Gruen, J. R. (2013). Imaging-genetics in dyslexia: Connecting risk genetic variants to brain neuroimaging and ultimately to reading impairments. *Molecular Genetics and Metabolism, 110*(3), 201–212.

Harlaar, N., Dale, P. S., & Plomin, R. (2007). From learning to read to reading to learn: Substantial and stable genetic influence. *Child Development, 78*(1), 116–131.

Harlaar, N., Deater-Deckard, K., Thompson, L. A., DeThorne, L. S., & Petrill, S. A. (2011). Associations between reading achievement and independent reading in early elementary school: A genetically informative cross-lagged study. *Child Development, 82*(6), 2123–2137.

Harold, G. T., Leve, L. D., Elam, K. K., Thapar, A., Neiderhiser, J. M., Natsuaki, M. N., et al. (2013). The nature of nurture: Disentangling passive genotype-environment correlation from family relationship influences on children's externalizing problems. *Journal of Family Psychology, 27*(1), 12–21.

Hart, S. A., Logan, J. A. R., Soden-Hensler, B., Kershaw, S., Taylor, J., & Schatschneider, C. (2013a). Exploring how nature and nurture affect the development of reading: An analysis of the Florida Twin Project on Reading. *Developmental Psychology, 49*(10), 1971–1981.

Hart, S. A., Soden, B., Johnson, W., Schatschneider, C., & Taylor, J. (2013b). Expanding the environment: Gene x school-level SES interaction on reading comprehension. *Journal of Child Psychology and Psychiatry, 54*(10), 1047–1055.

Hart, S. A., Petrill, S. A., Thompson, L. A., & Plomin, R. (2009). The ABCs of math: A genetic analysis of mathematics and its links with reading ability and general cognitive ability. *Journal of Educational Psychology, 101*(2), 388–402.

Haworth, C. M. A., Dale, P. S., & Plomin, R. (2009). The etiology of science performance: Decreasing heritability and increasing importance of the shared environment from 9 to 12 years of age. *Child Development, 80*(3), 662–673.

Haworth, C. M. A., Kovas, Y., Petrill, S. A., & Plomin, R. (2007). Developmental origins of low mathematics performance and normal variation in twins from 7 to 9 years. *Twin Research and Human Genetics, 10*(1), 106–117.

Kleemans, T., Peeters, M., Segers, E., & Verhoeven, L. (2012). Child and home predictors of early numeracy skills in kindergarten. *Early Childhood Research Quarterly, 27*, 471–477.

Kovas, Y., Garon-Carrier, G., Boivin, M, Petrill, S. A., Plomin, R., Malykh, S. B., et al. (2015). Why children differ in motivation to learn: Insights from over 13,000 twins from 6 countries. *Personality and Individual Differences, 80*, 51–63.

Kovas, Y., Harlaar, N., Petrill, S. A., & Plomin, R. (2005). "Generalist genes" and mathematics in 7-year-old twins. *Intelligence, 33*(5), 473–489.

Krapohl, E., & Plomin, R. (2016). Genetic link between family socioeconomic status and children's educational achievement estimated from genome-wide SNPs. *Molecular Psychiatry, 21*, 437–443.

Maher, B. (2008). Personal genomes: The case of the missing heritability. *Nature, 456*(7218), 18–21.

Manolio, T. A., Collins, F. S., Cox, N. J., Goldstein, D. B., Hindorff, L. A., Hunter, D. J., et al. (2009). Finding the missing heritability of complex diseases. *Nature, 461*(7265), 747–753.

Melhuish, E. C., Phan, M. B., Sylva, K., Sammons, P., Siraj-Blatchford, I., & Taggart, B. (2008). Effects of the home learning environment and preschool center experience upon literacy and numeracy development in early primary school. *Journal of Social Issues, 64*(1), 95–114.

Oliver, B., Harlaar, N., Hayiou-Thomas, M. E., Kovas, Y., Walker, S. O., Petrill, S. A., et al. (2004). A twin study of teacher-reported mathematics performance and low performance in 7-year-olds. *Journal of Educational Psychology, 96*(3), 504–517.

Payne, A. C., Whitehurst, G. J., & Angell, A. L. (1994). The role of home literacy environment in the development of language ability in preschool children from low-income families. *Early Childhood Research Quarterly, 9*(3–4), 427–440.

Petrill, S. A., Deater-Deckard, K., Thompson, L. A., DeThorne, L. S., & Schatschneider, C. S. (2006). Genetic and environmental effects of serial naming and phonological awareness on early reading outcomes. *Journal of Educational Psychology, 98*(1), 112–121.

Petrill, S. A., Deater-Deckard, K., Thompson, L. A., Schatschneider, C., DeThorne, L. S., & Vandenbergh, D. J. (2007). Longitudinal genetic analysis of early reading: The Western Reserve Reading Project. *Reading and Writing, 20*(1–2), 127–146.

Plomin, R. (2013). Child development and molecular genetics: 14 years later. *Child Development, 84*(1), 104–120.

Poelmans, G., Buitelaar, J. K., Pauls, D. L., & Franke, B. (2011). A theoretical molecular network for dyslexia: Integrating available genetic findings. *Mol Psychiatry, 16*(4), 365–382.

Rietveld, M. J. H., Hudziak, J. J., Bartels, M., van Beijsterveldt, C. E. M., & Boomsma, D. I. (2004). Heritability of attention problems in children: Longitudinal results from a study of twins, age 3 to 12. *Journal of Child Psychology and Psychiatry, 45*(3), 577–588.

Rodriguez, E. T., Tamis-LeMonda, C. S., Spellmann, M. E., Pan, B. A., Raikes, H., Lugo-Gil, J., et al. (2009). The formative role of home literacy experiences across the first three years of life in children from low-income families. *Journal of Applied Developmental Psychology, 30*(6), 677–694.

Rosenthal, R., & Jacobson, L. (1968). *Pygmalion in the classroom: Teacher expectation and pupils' intellectual development.* New York: Holt, Rinehart and Winston.

Saracho, O. N. (1997). Using the home environment to support emergent literacy. *Early Child Development and Care, 127*(1), 201–216.

Schenker, V. J., & Petrill, S. A. (2015). Overlapping genetic and child-specific non-shared environmental influences on listening comprehension, reading motivation, and reading comprehension. *Journal of Communication Disorders, 57,* 94–105.

Siperstein, G. N., Pociask, S. E., & Collins, M. A. (2010). Sticks, stones, and stigma: A study of students' use of the derogatory term "retard." *Intellectual and Developmental Disabilities, 48*(2), 126–143.

Stone, R. D. (2003). The cloudy crystal ball: Genetics, child abuse, and the perils of predicting behavior. *Vanderbilt Law Review, 56*(5), 1557–1590.

Strachan, T. & Read, A. P. (2004). *Human molecular genetics 3.* London: Garland Press.

Taylor, J., Roehrig, A. D., Soden Hensler, B., Connor, C. M., & Schatschneider, C. (2010). Teacher quality moderates the genetic effects on early reading. *Science, 328*(5977), 512–514.

Taylor, J., & Schatschneider, C. (2010). Genetic influence on literacy constructs in kindergarten and first grade: Evidence from a diverse twin sample. *Behavioral Genetics, 40*(5), 591–602.

van Bokhoven, H. (2011). Genetic and epigenetic networks in intellectual disabilities. *Annual Review of Genetics, 45,* 81–104.

Wadsworth, S. J., Corley, R. P., Hewitt, J. K., Plomin, R., & DeFries, J. C. (2002). Parent-offspring resemblance for reading performance at 7, 12, and 16 years of age in the Colorado Adoption Project. *Journal of Child Psychology and Psychiatry, 43*(6), 769–774.

Wanzek, J., & Vaughn, S. (2007). Research-based implications from extensive early reading interventions. *School Psychology Review, 36*(4), 541–561.

10

Genomic Literacy and the Communication of Genetic and Genomic Information

KIMBERLY A. KAPHINGST

INTRODUCTION

Advances in genomics and sequencing technologies have the potential to greatly improve patient health through applications of clinical care based on genomic information (i.e., genomic medicine) (Biesecker et al., 2009; Green, Guyer, & National Human Genome Research Institute, 2011; Hall, Forman, Montgomery, Rainey, & Daly, 2014). While research is needed to develop medical applications of genomics (Biesecker et al., 2009), genomic information has already had an impact in diagnostics and treatment for a few clinical circumstances (Green et al., 2011). For example, pilot projects in implementing genomic medicine have focused on genotyping of somatic mutations in tumors to inform treatment decisions, genome sequencing to diagnose genetic causes of rare conditions, and pharmacogenomics (Manolio et al., 2013). Building on such efforts, genomic information is expected to have an increasing role in clinical care and public health initiatives (Guttmacher & Collins, 2002; Hall et al., 2014; McBride et al., 2008).

In order for the translation of genomic information to improve medicine and public health, individuals will need to be able to understand the information provided to them and use that information to make health decisions and participate in public policy discussions about genomics (Green et al., 2011; Hurle et al., 2013). In applications of genomic medicine, patients and those at risk (e.g., patients' family members) will likely need to be able to comprehend genomic test results and make informed decisions about treatment, disease management and prevention, and use of genetic services (Christensen, Jayaratne, Roberts, Kardia, & Petty,

Dr. Kaphingst was supported by R01CA168608, U54CA153460, and the Huntsman Cancer Institute.

2010; Condit, 2010; Hall et al., 2014; Manolio et al., 2013; Pearson & Liu-Thompkins, 2012), which will require knowledge and skills related to genetics and genomics (Catz et al., 2005; Jallinoja & Aro, 1999). Being able to understand and use genomic information will require knowledge about genetic and genomic concepts, familiarity with genetic and genomic terminology (Kessler, Collier, & Halbert, 2007; McInerney, 2002; Richards & Ponder, 1996; Wang, Bowen, & Kardia, 2005), and skills in understanding and using print, oral, and numeric information (Lea, Kaphingst, Bowen, Lipkus, & Hadley, 2011). Levels of such knowledge and skills among patients and the general public will, therefore, likely have a critical impact on efforts to translate genomic discoveries into individual and population health benefits (Dougherty, Lontok, Donigan, & McInerney, 2014; Khoury, 2003; Khoury et al., 2007; Smerecnik, Mesters, de Vries, & de Vries, 2008; Wang et al., 2005).

However, knowledge about genetics and genomics among the public has been identified as an important barrier to the implementation of genomic medicine (Manolio et al., 2013). Prior research has demonstrated substantial gaps in public understanding of genetics (Condit, 2010). Although individuals may have some familiarity with genetic and genomic terms, they have gaps in their understanding of underlying concepts. In addition, limited literacy or numeracy skills among individuals may impact communication of genetic and genomic information and the integration of genomic medicine into practice (Hall et al., 2014; Lea et al., 2011). Studies have shown that individuals' literacy skills affect their understanding of print and oral communications about genetic and genomic information, and that numeracy is likely to be an important predictor of being able to understand and apply quantitative information about genomic risk. It is, therefore, critical to understand individuals' knowledge and skills related to genetics and genomics, or their genomic literacy, in planning for the increasing integration of genomic information into healthcare and the public sector (Green et al., 2011; Hurle et al., 2013). Limited genomic literacy may adversely impact the public's understanding and use of genetic and genomic information in healthcare and public health (Hurle et al., 2013).

This chapter will, therefore, focus on the issue of genomic literacy. The chapter will first describe various definitions of genetic and genomic literacy, and then present prior research regarding knowledge about genetics and genomics and the effects of literacy and numeracy skills on responses to genetic and genomic information. Priority areas for research on genomic literacy and educational practice will also be described.

DEFINITIONS OF GENETIC AND GENOMIC
LITERACY

With the increasing recognition of the importance of individuals' knowledge and skills to being able to understand and use genetic and genomic information, various definitions have been developed of the concepts of genetic literacy and genomic literacy. A number of definitions of genetic literacy have focused on individuals' knowledge of basic genetic and genomic concepts (Pearson & Liu-Thompkins, 2012), often in the functional context of having sufficient knowledge to inform health decision making and participate in public policy discussions on genetic issues (Bowling et al., 2008; Haga et al., 2013). Other definitions of genetic literacy incorporate both ability to decode genetic terminology and comprehension of genetic concepts, recognizing that unfamiliar language and concepts related to genetics may present difficulty for many patients in understanding and using genetic and genomic information (Hooker et al., 2014).

A more expanded definition of genomic literacy was developed by an expert working group convened by the National Human Genome Research Institute (NHGRI) of the US National Institutes of Health (Hurle et al., 2013), building upon conceptual work in the related area of health literacy. Health literacy has been defined as the degree to which individuals can obtain, process, and understand the basic health information and services needed to make appropriate health decisions (Nielsen-Bohlman, Panzer, & Kindig, 2004). Within this framework, the Institute of Medicine has operationalized health literacy as having the following components: oral literacy (listening and speaking skills); print literacy (reading and writing skills); numeracy (basic quantitative skills); and cultural and conceptual knowledge (Nielsen-Bohlman et al., 2004). About 36 percent of US adults have limited health literacy (Kutner, Greenberg, Jin, Paulsen, & White, 2006), and these individuals have, on average, lower health knowledge, increased incidence of chronic illness, lower utilization of preventive health services, and poorer self-reported health (Berkman et al., 2011; Nielsen-Bohlman et al., 2004). Based on this conceptualization of health literacy, the specific domain of genomic health literacy was defined as the capacity to obtain, process, understand, and use genomic information for health-related decision making (Hurle et al., 2013). This definition of genomic literacy, organized into the domains of conceptual knowledge, print and oral literacy skills, and numeracy skills, will be the organizing framework for this chapter.

KNOWLEDGE ABOUT GENETICS AND GENOMICS

Of the domains of genomic literacy, much of the prior research has focused on individuals' conceptual knowledge about genetics and genomics (Lea et al., 2011). Early research into public knowledge about genetics focused on areas such as beliefs about the causation of birth defects and carrier risk. In one study, Cohen, Fine, and Pergament (1998) observed variability in beliefs about causation of birth defects based on racial/ethnic group and family history of a birth defect (Cohen, Fine, & Pergament, 1998). In another study, Lafayette and colleagues (1999) found that most relatives of individuals with cystic fibrosis underestimated their carrier risk and overestimated their risk of having a child with the disease. They also found lower knowledge about cystic fibrosis among participants with lower educational attainment (Lafayette, Abuelo, & Passero, 1999), although another study suggested that adults who had had genetic counseling and testing had fairly high knowledge about cystic fibrosis (Callanan, Bloom, Sorenson, DeVellis, & Cheuvront, 1995). These early studies, therefore, indicated that understanding of the role of genes in inherited conditions may differ across sociodemographic groups.

More recent research on public understanding of genetics and genomics has examined individuals' knowledge about genetic causation for common, complex health conditions. In a focus group study, Parrott, Silk, and Condit (2003) found that individuals' perceptions of the relative role of genes in causation varied across conditions; for example, genes were seen as playing a greater role in the causation of breast and prostate cancer compared with lung cancer. Differences in attributions about the role of genes in disease causation were observed by race, gender, income, and having had a college biology course (Parrott, Silk, & Condit, 2003). In another focus group study, Bates and colleagues (2003) found that only about a quarter of participants believed that those who had a genetic predisposition for heart disease would certainly get the disease, although risk perceptions did differ somewhat by racial group (Bates, Templeton, Achter, Harris, & Condit, 2003). These and other studies suggest that the general public does have an understanding that genes are not the sole cause of common, complex diseases, although differences have been observed across population subgroups.

Other prior qualitative interview and focus group research has shown important gaps in individuals' knowledge about genetic and genomic concepts. Lanie and colleagues (2004) found that participants had a limited understanding of genes (e.g., what they are, how they are inherited, where they are located in the body) (Lanie et al., 2004). In a focus group study,

Mesters, Ausem, and DeVries (2005) observed that although participants were familiar with the terms "genes" and "DNA" in association with cancer, they did not have a good understanding of these concepts (Mesters, Ausems, & De Vries, 2005). Focus groups conducted with fifty-five patients from diverse, underserved communities in the United States also showed that participants had limited understanding of genetics or genetic testing (Catz et al., 2005).

Findings from larger quantitative studies have been generally similar. In a telephone survey conducted with 1,009 Australian adults, Molster and colleagues (2009) showed that although most respondents were aware of the connection between genes, inheritance, and disease risk, significantly fewer understood the biological mechanisms underlying these connections. This study found that genetic knowledge was associated with education, age, gender, and income (Molster, Charles, Samanek, & O'Leary, 2009). Haga and colleagues (2013) found that 300 participants from a general public sample had significantly higher knowledge about the inheritance and causes of disease compared with biological knowledge about genes, chromosomes, and cells. In this study, age, racial group, and education were significantly associated with genetic knowledge (Haga et al., 2013). Higher educational attainment has consistently been shown to be significantly associated with higher genetic knowledge across multiple studies conducted in different countries (Carlsbeek, Morren, Bensing, & Rijken, 2007; Henneman, Timmermans, & van der Wal, 2004; Ishiyama et al., 2008; Jallinoja & Aro, 1999; Molster et al., 2009; Rose, Peters, Shea, & Armstrong, 2005), although low levels of genetic knowledge have been observed even among college students (Bowling et al., 2008) and limited comprehension of genetics terms has been found among a well-educated clinical population (Hooker et al., 2014). Limited health literacy has also been shown to be associated with lower genetic knowledge (Kaphingst et al., 2016). In a number of prior studies, age has also been shown to be significantly and negatively related to genetic knowledge (Carlsbeek et al., 2007; Henneman et al., 2004; Ishiyama et al., 2008; Jallinoja & Aro, 1999; Molster et al., 2009; Rose et al., 2005). For example, Ashida and colleagues found that, among 971 medically underserved individuals, genetic knowledge was lower among older patients than younger patients (Ashida et al., 2011).

Race and ethnicity may also be associated with genetic knowledge, although research is needed on this question. In a national telephone survey of US adults, Christensen and colleagues (2010) found that although most respondents answered items about Mendelian inheritance correctly, less than half answered any other item correctly, with misunderstandings about basic genetic concepts. The authors found some differences in genetic knowledge by racial and ethnic group, although they observed that many misunderstandings were shared across groups (Christensen et al., 2010). Taken together, the findings of these studies, therefore, suggest that members of the general public have some familiarity with genetic and genomic terms, but substantial gaps exist in their understanding of the underlying concepts and there may be some differences in genetic knowledge across sociodemographic groups.

A number of different mechanisms have been proposed to explain possible differences in genetic knowledge across population subgroups. For example, observed differences in genetic knowledge might be due to differences in exposure to genetic information (Hughes et al., 1997), or differences in information sources (Catz et al., 2005). Differences in genetic knowledge could also be due to differential formal education about genetics (Furr & Kelly, 1999), although most people have not received extensive formal education about this topic (Condit, 2010). Health literacy research also suggests mechanisms that could underlie some differences in genetic knowledge across population subgroups. For example, prior research has shown that exposure to newspaper articles about genes and health was associated with higher genetic knowledge (Parrott, Silk, Krieger, Harris, & Condit, 2004). Individuals with limited print literacy skills may be less likely to use print sources of health information (Kutner et al., 2006), and may also be less likely to comprehend print information (Nielsen-Bohlman et al., 2004), thereby perhaps leading to lower genetic knowledge. Prior research in this area to date is limited, however.

Smerecnik and colleagues (2008) have added to the theoretical basis of research into genetic and genomic knowledge. They have proposed a theory-based framework that distinguishes between three different types of genetic knowledge: awareness knowledge (i.e., knowing that there are genetic risk factors for a disease); how-to knowledge (i.e., practical knowledge of genetic risk factors); and principles knowledge (i.e., theoretical knowledge of genetic risk factors). The authors place these forms of knowledge along a continuum, representing increasing complexity from awareness knowledge to principles knowledge. Most studies have been done on awareness knowledge; the limited existing research suggests low how-to knowledge and principles knowledge in the general public (Smerecnik et al., 2008). Although the public may primarily be interested in how-to knowledge (Emery, Kumar, & Smith, 1998; Smerecnik et al., 2008), they may also be interested in principles knowledge that is personally relevant (Mesters et al., 2005). How-to knowledge about genetics may also require knowledge of at least some underlying principles (Condit, 2010). Future research is needed to examine what types of genetic knowledge are relevant for an individual at different times during the lifespan (Condit, 2010; Syurina, Brankovic, Probst-Hensch, & Brand, 2011). While this theory-based framework could allow researchers to examine how deeply the general public understands genetic and genomic concepts (Smerecnik et al., 2008), empirical work based on the framework is as yet limited.

An additional important line of previous research has been to identify educational approaches that may improve knowledge about genetics and genomics, including education during genetic counseling (Edwards et al., 2008; Kelly et al., 2004; Lerman et al., 1997), culturally tailored education sessions led by lay health advisors (Kaphingst, Lachance, Gepp, D'Anna, & Rios-Ellis, 2011), and computer-based educational approaches (Green et al., 2004; Meilleur & Littleton-Kearney, 2009). However, more research is needed to develop and test educational approaches targeting relevant and theory-based components of genetic knowledge that inform health decision making (Decruyenaere, Evers-Kiebooms, Welkenhusen, Denayer, & Claes, 2000). Education about genetics and genomics should avoid unnecessary technical jargon (Lea et al., 2011), and provide the opportunity to understand information that is salient to a particular individual (Syurina et al., 2011). Researchers have also highlighted the importance of developing strategies that are understandable and acceptable for diverse cultures and individuals of varying educational levels (Catz et al., 2005; Smerecnik et al., 2008). To this should be added the importance of considering individuals' genomic literacy in developing educational strategies.

In sum, research to date regarding the public's knowledge of genetics and genomics suggests that individuals generally do have an understanding of the multifactorial causation of common, complex conditions. However, while members of the general public may be familiar with genetics terms, they have knowledge gaps regarding basic facts about genes and DNA. A prior review of the literature on public understanding of genetics found that the public had a stronger understanding of the concepts of heredity and heritability than concepts related to molecular genetics (e.g., gene structure, function, cellular location) (Condit, 2010). Findings to date indicate that there may be differences in genetic knowledge across population subgroups, but there is a need for additional large studies examining genetic and genomic knowledge in representative population samples, with sufficient sample size to examine differences across subgroups. Such data could inform whether educational initiatives could target particular sociodemographic groups. In addition, the development of educational strategies for improving genetic and genomic knowledge that may be effective across population subgroups and for individuals with varying levels of genomic literacy is greatly needed.

LITERACY SKILLS AND PRINT AND ORAL COMMUNICATION

This section describes the limited literature to date regarding the effects of literacy skills on the print and oral communication of genetic and genomic information, including both studies that assessed genomic literacy skills and general health literacy skills. With respect to print communication, Thompson and colleagues (2004) conducted research to develop a print brochure to describe testing for the genes *BRCA1* and *BRCA2* to an audience with limited print literacy skills. Although participants generally liked the brochure, the findings showed continued difficulties with comprehension of genetics terms and the illustrations used in the brochure (Thompson et al., 2004). In a study conducted with 163 post-treatment breast cancer patients, Lillie and colleagues assessed the relationships between individuals' print literacy skills and recall of genomic information and decision-making preferences. Print literacy skills were assessed using a validated word recognition test of health literacy called the Rapid Estimate of Adult Literacy in Medicine (Davis et al., 1993). The study findings showed that individuals with lower literacy skills had lower recall of information about a genomic test that identified recurrence risk for breast cancer and had lower preference for active participation in decision making about the test (Lillie et al., 2007). In a subsequent study conducted with the population, Brewer and colleagues (2009) demonstrated that literacy skills affected the interpretation of visual recurrence risk information. In this experimental study, they found that women with lower literacy skills gave higher mean estimates of recurrence risk and more variable risk estimates. They also found most risk communication formats harder to understand than did women with higher health literacy. In addition, women with lower literacy skills were less sensitive to numeric recurrence risks when making hypothetical chemotherapy treatment decisions (Brewer et al., 2009). Kaphingst and colleagues (2015) found that, among 1,057 medically underserved patients, those with limited health literacy skills were more interested in receiving a genomic assessment than those with higher skills, but had lower intentions to change health habits in response to the genomic information (Kaphingst et al., 2015).

Only a few studies have assessed genetic or genomic literacy specifically and examined how this impacts print communication of genetic and genomic information. In a recent study, Hooker and colleagues (2014) assessed participants' genetic literacy using a measure they designed to assess familiarity with common genetic terms and concepts. They then randomized 257 patients with inflammatory bowel disease to receive a print vignette describing a risk assessment that was based on either genetic testing or standard blood testing. The study findings showed that, among participants randomized to the genetic testing vignette, those with lower genetic literacy were less likely to perceive the genetic test as useful and believed that they would have less control of the disease after testing (Hooker et al., 2014).

The question of how literacy skills affect oral communication of genetic and genomic information has received limited attention to date. To examine this issue, Erby and colleagues (2008) developed a word recognition measure of genetic literacy called the Rapid Estimate of Adult Literacy in Genetics (REAL-G). The authors showed that individuals with limited genetic literacy skills as assessed by the REAL-G had lower knowledge scores after viewing videotaped genetic counseling sessions, indicating less learning from this verbally presented genetic information (Erby, Roter, Larson, & Cho, 2008). In a subsequent study, the research team systematically examined the difficulty of information presented orally by a genetic counselor during genetic counseling sessions. This study showed that more difficult oral language was associated with less satisfaction among clients (Roter, Erby, Larson, & Ellington, 2007).

In summary, the existing literature in this area does suggest that literacy skills will be critical to consider in developing print and oral strategies for communicating genetic and genomic information. These prior studies have indicated that individuals with limited literacy skills may understand less of written and oral genetic and genomic information and may also be less willing to engage in discussions with healthcare providers about the information. It is possible, therefore, that having limited literacy skills may lessen the potential of genetic and genomic information to motivate behavior change and inform health decision making (McBride, Koehly, Sanderson, & Kaphingst, 2010), although the research base in this area is still quite limited. In particular, there is a pressing need for research investigating how genomic literacy skills affect individuals' understanding of information about genomic risk for common, complex disease outcomes and results generated by next-generation sequencing technologies. It is also critical to note that how genetic and genomic information is often currently presented may be too complex for most adults (Lachance, Erby, Ford, Allen, & Kaphingst, 2010; Wang, Gallo, Fleisher, & Miller, 2011), and research is greatly needed to develop more effective educational strategies to provide this information.

The last component of genomic literacy described here is numeracy, which relates to individuals' skills with mathematical concepts and their applications (Goldbeck, Ahlers-Smith, & Paschal, 2005; Lea et al., 2011). Numeracy is thought to encompass both individuals' capacities with the form of information (e.g., graphic, decimals, fractions) and how they process and interpret numeric information (Goldbeck et al., 2005). Greater numeracy skills facilitate mathematical operations, encourage greater depth of information processing, and improve interpretation of numeric information (Lea et al., 2011; Lipkus & Peters, 2009). Data from a nationally representative sample showed that only 13 percent of US adults had proficient numeracy skills, while over half had limited numeracy skills (National Center for Education Statistics, 2006). Numeracy has been increasingly recognized as critical to medical decision making and the communication of risk information, and is, therefore, likely to be important in understanding numeric genetic and genomic risk information and responses to this information (Lea et al., 2011). This section will describe prior research to date in this area.

Numeracy has been shown to influence how individuals interpret medical risk information (Peters, Hart, & Fraenkel, 2011) and risk information presented in graphic form (Brown et al., 2011). In an experimental study conducted with adult men, Rolison and colleagues (2012) found that few men correctly interpreted lifetime risk information when the estimates referred to an absolute risk of cancer; those with higher numeracy skills were more likely to correctly interpret the information (Rolison, Hanoch, & Miron-Shatz, 2012). In a study conducted with overweight patients at risk for type 2 diabetes, Vassy and colleagues (2012) observed that although patients reported that hypothetical high-risk genetic results would motivate them to adopt healthy lifestyle changes, responses to low-risk genetic results varied by numeracy, as well as literacy. The study findings showed that participants with lower numeracy skills had an anticipated increase in motivation for changing their lifestyle in response to low-risk genetic results (Vassy et al., 2012). Numeracy has also been shown to interact with literacy in affecting learning of orally communicated information during genetic counseling sessions. Portnoy and colleagues found that higher numeracy skills improved learning for those with higher literacy skills, but did not affect learning among those with lower literacy skills (Portnoy, Roter, & Erby, 2010).

Prior research, therefore, suggests that numeracy skills will affect how individuals interpret genetic and genomic risk information. While some guidance exists for the presentation of numeric risk information (Fagerlin, Zikmund-Fisher, & Ubel, 2011), the development of genetic and genomic risk communication strategies appropriate for individuals with limited numeracy and literacy skills is needed. In addition, research is needed to examine the interactions of conceptual knowledge, print and oral literacy, and numeracy in the context of genetics and genomics (Lea et al., 2011), and how these domains will together affect the communication of genetic and genomic information.

FUTURE DIRECTIONS FOR RESEARCH

This review of the existing literature into the effects of genomic literacy on communication of genetic and genomic information highlights a number of priority areas for future research (see Table 10.1). In order to use genetic and genomic information to improve healthcare and disease prevention efforts, research is critically needed to better understand how individuals understand and use such information (Green et al., 2011). The development of communication approaches for genomic information has been identified as a top priority for translational research in genomics (McBride et al., 2010a). Of particular importance here, it is not known what level of genomics-related knowledge and skills are needed to make informed decisions about genetic testing, understand genomic risk messages, or motivate behavior change (Kaphingst & McBride, 2010; McBride et al., 2010b) and how genomic literacy impacts individuals' responses to genomic information (Lea et al., 2011). The effect of genomic literacy on information processing and subsequent decision making will become increasingly important as genomic information reaches larger and more diverse populations (Lea et al., 2011), and all domains of genomic literacy will be important to investigate.

In the domain of conceptual knowledge, empirical research is needed to determine what knowledge is necessary to make decisions at both the individual and societal levels (Green et al., 2011). For example, what do individuals need and want to know about genetics and genomics in order to make informed decisions about genetic testing (Kaphingst et al., 2010; McBride, et al., 2010b)? Condit (2010) has emphasized the importance of considering theory-based components of knowledge in such research (Condit, 2010), and to facilitate this research, development of theory-based measures of

TABLE 10.1 *Research priorities in genomic literacy and its component domains*

Conceptual knowledge
  - What knowledge is needed to act upon genetic and genomic information?
  - What information about genetics and genomics do individuals want to know?
  - How should conceptual knowledge about genetics and genomics be measured?

Literacy skills
  - How do oral literacy skills impact responses to verbally presented genetic and genomic information?
  - How do literacy skills impact responses to print and oral information?
  - What educational approaches are effective for individuals with limited literacy skills?

Numeracy skills
  - How do limited numeracy skills impact responses to genetic and genomic risk information?
  - How does comprehension of risk information affect individuals' actions based on the information?
  - What formats and presentations of genetic and genomic risk information will facilitate comprehension of the information?

Genomic literacy
  - How do individuals understand and use genetic and genomic information?
  - How does genomic literacy impact individuals' responses to genetic and genomic information?
  - What level of genomic literacy is needed to act upon genetic and genomic information?

genomic knowledge are needed. In prior research, different genetic knowledge items have generally been used in different studies, making comparisons across studies more difficult, and few studies have focused on genomic knowledge (Kaphingst et al., 2012). There is a clear need to create and validate theory-based scales to assess knowledge of genomic concepts (Christensen et al., 2010; Kaphingst et al., 2012).

Major goals of this area of research should be to investigate what types of knowledge are needed for which health decisions and identify possible sources of each of these types of knowledge. It will be important to examine both the knowledge that adults need to make health decisions for themselves and the knowledge that adults and minor children need to make pediatric health decisions. While related research has examined issues such as enrollment of children into genetic susceptibility research studies (Friedman Ross et al., 2013; Geller, Tambor, Bernhardt, Fraser, & Wissow, 2003; Wilfond & Friedman Ross, 2009), views about pediatric genetic testing (Borry et al., 2009; Tercyak et al., 2011), and parents' attitudes toward returning findings from genetic research to children (Kleiderman et al., 2014; Sapp et al., 2014), our understanding of the specific types of knowledge needed to use genetic and genomic information in health decision making across the life course is still quite limited.

In the domain of literacy skills, there is a need to develop and evaluate genomic communication approaches that can be understood and used by individuals with varying levels of genomic literacy (Hooker et al., 2014; Lea et al., 2011). Patients want genomic information presented in ways that are understandable and do not use complex terminology and jargon (Hooker et al., 2014). Kaphingst and colleagues (2012) found that a feedback strategy for multiple genetic susceptibility risk feedback developed based on plain language and health literacy principles led to high levels of recall among 199 patients, and that patients were unlikely to interpret the genetic susceptibility results as deterministic for common, chronic health outcomes (Kaphingst et al., 2012). Orally presented genomic information may also be an important approach for those with limited print literacy skills (Kaphingst et al., 2011). However, little research has examined individuals' responses to orally presented genetic and genomic information. Communication research could also examine whether and how genomic literacy affects health behavior change and health decision making as a result of the provision of genomic information (Kaphingst et al., 2015). The development of effective strategies for communicating genetic and genomic information to individuals with varying genomic literacy levels will likely require stakeholders from many fields, including adult learning, education, public health, health communication, and genetic counseling (Hurle et al., 2013).

Numeracy issues also clearly need more research, such as studies examining individuals' responses to different formats and presentations of genomic risk information. Few such studies have been conducted (Dorval et al., 2013), and such research has generally not focused on individuals with limited numeracy skills (Lea et al., 2011). The development of genetic and genomic risk communication strategies appropriate for individuals with limited numeracy skills is greatly needed. The effects of comprehension of numeric genomic risk information on health behavior change and health decision making following the receipt of risk information has not been investigated and this is also a critical area for research (Lea et al., 2011). The complexity and uncertainty inherent in much genomic risk information will also likely pose challenges to individuals' comprehension of risk information (Dougherty et al., 2014; Lea et al., 2011; Vassy et al., 2012), and this is another priority area as the provision of genomic information increases.

Finally, a number of researchers have highlighted the importance of educational efforts to improve the public's genomic literacy (Christensen et al., 2010; Green et al., 2011; Manolio et al., 2013). Educational efforts at many levels will be critical for improving genomic literacy, including strengthening public outreach and educational efforts for families and adults (Green et al., 2011; Hurle et al., 2013). Research will be needed to identify the educational needs of different audiences (Manolio et al., 2013) and develop best educational practices (Hurle et al., 2013). Such efforts should focus on improving the knowledge and skills needed to make personal and family health decisions (Hurle et al., 2013) and participate in public discourse about genomics (Green et al., 2011). New approaches to genetics and genomics education for the public are needed (Dougherty et al., 2014), such as novel educational approaches combining formal and informal science learning through community-based participatory learning (Haga et al., 2013). Improving genomic literacy will likely require active engagement between genomics experts and public stakeholders (Hurle et al., 2013).

Genomic science education in K–12 has been identified as a critical means of increasing genomic literacy in our society (Hurle et al., 2013). However, a number of barriers have been identified to teaching genetics and genomics in K–12, including insufficient emphasis on genomics in national standards, lack of funding and curriculum materials, standardized testing, time constraints, and misconceptions about genetics and genomics (Dougherty, Pleasants, Solow, Wong, & Zhang, 2011; Hurle et al., 2013; Quinn, Schweingruber, & Keller, 2012). A working group assembled by the NHGRI identified a series of research opportunities in K–12 science education to improve genomic literacy, including (1) developing reliable and validated measures to assess students' understanding of genetics and genomics concepts; (2) identifying common misconceptions and roadblocks in genetic and genomic education; and (3) evaluating students' retention of key ideas via longitudinal studies (Hurle et al., 2013). The working group also emphasized the importance of developing methods of integrating learning about genetics and genomics across the life course so that individuals are prepared to make health-related decisions based on genetic and genomic information (Hurle et al., 2013).

CONCLUSION

This chapter has summarized the existing state of the literature on genomic literacy and its component domains of genetic and genomic conceptual knowledge, literacy skills, and numeracy skills. Prior literature has identified gaps in genetic knowledge, although larger quantitative studies are needed to better understand knowledge in different population subgroups. Research is also needed to identify areas of conceptual knowledge that individuals need in order to act at the individual and societal levels. The research conducted to date on how literacy skills affect individuals' understanding of print and oral genetic and genomic information is limited, and little research has examined how numeracy skills impact interpretation of genetic and genomic risk information. While the existing research suggests that literacy and numeracy skills will be critical to the communication of genetic and genomic information, numerous areas for future investigations have been identified. Future studies should examine how individuals with limited genomic literacy understand and use genomic information presented in verbal, print, and graphic channels. This type of research is essential to the development of effective strategies for communicating genetic and genomic information that are appropriate for individuals at different ages with a range of genomic literacy levels. Finally, future research is needed that examines how the domains of print literacy, oral literacy, numeracy, and conceptual knowledge interact in affecting individuals' responses to genetic and genomic information. Individuals with limited literacy skills may likely have limited numeracy skills and limited conceptual knowledge, and few studies have examined the effects of this type of interaction on decision making, accessing of genetic services, and health behavior change.

Advances in genomic technologies will likely mean that researchers and healthcare providers will increasingly be called upon to provide genomic information to patients and the general public. Concomitantly, these technological advances will mean increasingly complex issues in the communication of genetic and genomic information. Research in how to provide genetic and genomic information in ways that are understandable and usable by individuals with varying levels of genomic literacy will be critical to translating the advances in genomics to improvements in the public's health (Hay et al., 2007; Lea et al., 2011). Such research will require new partnerships between educational systems, healthcare systems, the government, community organizations and the public in order to meet translational goals (Lea et al., 2011).

REFERENCES

Ashida, S., Goodman, M., Pandaya, C., Koehly, L. M., Lachance, C., Stafford, J. D., et al. (2011). Age differences in genetic knowledge, health literacy, and causal beliefs for health conditions. *Public Health Genomics, 14*(4–5), 307–316.

Bates, B. R., Templeton, A., Achter, P. J., Harris, T. M., & Condit, C. M. (2003). What does "a gene for heart disease" mean? A focus group study of public understanding of genetic risk factors. *American Journal of Medical Genetics Part A, 119A,* 156–161.

Berkman, N., Sheridan, S., Donahue, K., Halpern, D., Viera, A., Crotty, K., et al. (2011). *Health literacy interventions and outcomes: An updated systematic review.* Evidence Report/Technology Assessment No 1999. Rockville, MD: Agency for Healthcare Research and Quality.

Biesecker, L. G., Mullikin, J. C., Facio, F. M., Turner, C., Cherukuri, P. F., Blakesley, R. W., et al. (2009). The ClinSeq Project: Pilot large-scale genome sequencing for research in genomic medicine. *Genome Research, 19,* 1665–1674.

Borry, P., Evers-Kiebooms, G., Cornel, M. C., Clarke, A., Dierickx, K., & Public and Professional Policy Committee of the European Society of Human Genetics. (2009). Genetic testing in asymptomatic minors: Background considerations toward ESHG recommendations. *European Journal of Human Genetics, 17,* 711–719.

Bowling, B. V., Acra, E. E., Wang, L., Myers, M. F., Dean, G. E., Markle, G. C., et al. (2008). Development and evaluation of a genetics literacy assessment instrument for undergraduates. *Genetics, 178,* 15–22.

Brewer, N. T., Tzeng, J. P., Lillie, S. E., Edwards, A. S., Peppercorn, J. M., & Rimer, B. K. (2009). Health literacy and cancer risk perception: Implications for genomic risk communication. *Medical Decision Making, 29,* 159–166.

Brown, S. M., Culver, J. O., Osann, K. E., MacDonald, D. J., Sand, S., Thornton, A. A., et al. (2011). Health literacy, numeracy, and interpretation of graphical breast cancer risk estimates. *Patient Education and Counseling, 83*(1), 92–98.

Callanan, N., Bloom, D., Sorenson, J. R., DeVellis, B., & Cheuvront, B. (1995). CF carrier testing: Experience of relatives. *Journal of Genetic Counseling, 4,* 83–95.

Carlsbeek, H., Morren, M., Bensing, J., & Rijken, M. (2007). Knowledge and attitudes towards genetic testing: A two year follow-up study in patients with asthma, diabetes mellitus and cardiovascular disease. *Journal of Genetic Counseling, 16*(4), 493–504.

Catz, D. S., Green, N. S., Tobin, J. N., Lloyd-Puryear, M. A., Kyler, P., Umemoto, A., et al. (2005). Attitudes about genetics in underserved, culturally diverse populations. *Community Genetics, 8,* 161–172.

Christensen, K. D., Jayaratne, T. E., Roberts, J. S., Kardia, S. L. R., & Petty, E. M. (2010). Understandings of basic genetics in the United States: Results from a national survey of Black and White men and women. *Public Health Genomics, 13,* 467–476.

Cohen, L., Fine, B., & Pergament, E. (1998). An assessment of ethnocultural beliefs regarding the cause of birth defects and genetic disorders. *Journal of Genetic Counseling, 50,* 15–29.

Condit, C. (2010). Public understandings of genetics and health. *Clinical Genetics, 77,* 1–9.

Davis, T. C., Long, S. W., Jackson, R. H., Mayeaux, E. J., George, R. B., Murphy, P. W., et al. (1993). Rapid Estimate of Adult Literacy in Medicine: A shortened screening instrument. *Family Medicine, 25,* 391–395.

Decruyenaere, M., Evers-Kiebooms, G., Welkenhusen, M., Denayer, L., & Claes, E. (2000). Cognitive representations of breast cancer, emotional distress and preventive health behaviour: A theoretical perspective. *Psycho-Oncology, 9*, 528–536.

Dorval, M., Bouchard, K., Chiquette, J., Glendon, G., Maugard, C. M., Dubuisson, W., et al. (2013). A focus group study on breast cancer risk presentation: One format does not fit all. *European Journal of Human Genetics, 21*, 719–724.

Dougherty, M., Pleasants, C., Solow, L., Wong, A., & Zhang, H. (2011). A comprehensive analysis of high school genetics standards: Are states keeping pace with modern genetics? *CBE Life Sci Educ, 10*, 318–327.

Dougherty, M. J., Lontok, K. S., Donigan, K., & McInerney, J. D. (2014). The critical challenge of educating the public about genetics. *Current Genetic Medicine Reports, 2*, 48-55..

Edwards, A., Gray, J., Clarke, A., Dundon, J., Elwyn, G., Gaff, C., et al. (2008). Interventions to improve risk communication in clinical genetics: Systematic review. *Patient Education and Counseling, 71*, 4–25.

Emery, J., Kumar, S., & Smith, H. (1998). Patient understanding of genetic principles and their expectations of genetic services within NHS: A qualitative study. *Community Genetics, 1*, 78–83.

Erby, L., Roter, D., Larson, S., & Cho, J. (2008). The Rapid Estimate of Adult Literacy in Genetics (REAL-G): A means to assess literacy deficits in the context of genetics. *American Journal of Medical Genetics, 146A*, 174–181.

Fagerlin, A., Zikmund-Fisher, B. J., & Ubel, P. A. (2011). Helping patients decide: Ten steps to better risk communication. *Journal of the National Cancer Institute, 103*, 1436–1443.

Friedman Ross, L., Saal, H. M., David, K. L., Anderson, R. R., & American Academy of Pediatrics & Genomics. (2013). Technical report: Ethical and policy issues in genetic testing and screening of children. *Genetics in Medicine, 15*(3), 234–245.

Furr, L. A., & Kelly, S. E. (1999). The genetic knowledge index: Developing a standard measure of genetic knowledge. *Genetic Testing, 3*(2), 193–199.

Geller, G., Tambor, E. S., Bernhardt, B. A., Fraser, G., & Wissow, L. S. (2003). Informed consent for enrolling minor in genetic susceptibility research: A qualitative study of at-risk children's and parents' views about children's role in decision-making. *Journal of Adolescent Health, 32*(4), 260–271.

Goldbeck, A. L., Ahlers-Smith, C. R., & Paschal, A. M. (2005). A definition and operational framework for health numeracy. *American Journal of Preventive Medicine, 29*(4), 375–376.

Green, E. D., Guyer, M. S., & National Human Genome Research Institute. (2011). Charting a course for genomic medicine from base pairs to bedside. *Nature, 470*, 204–213.

Green, M. J., Peterson, S. K., Baker, M. W., Harper, G. R., Friedman, L. C., Rubinstein, W. S., et al. (2004). Effect of a computer-based decision aid on knowledge, perceptions, and intentions about genetic testing for breast cancer susceptibility: A randomized controlled trial. *JAMA, 292*, 442–452.

Guttmacher, A. E., & Collins, F. S. (2002). Genomic medicine – A primer. *New England Journal of Medicine, 347*, 1512–1520.

Haga, S. B., Barry, W. T., Mills, R., Ginsberg, G. S., Svetkey, L. P., Sullivan, J., et al. (2013). Public knowledge of and attitudes toward genetics and genetic testing. *Genetic Testing and Molecular Biomarkers, 17*(4), 327–335.

Haga, S. B., Rosanbalm, K. D., Boles, L., Tindall, G. M., Livingston, T. M., & O'Daniel, J. M. (2013). Promoting public awareness and engagement in genome sciences. *Journal of Genetic Counseling, 22*, 508–516.

Hall, M. J., Forman, A. D., Montgomery, S. V., Rainey, K. L., & Daly, M. B. (2014). Understanding patient and provider perceptions and expectations of genomic medicine. *Journal of Surgical Oncology, 111*(1), 9–17.

Hay, J. L., Meischke, H. W., Bowen, D. J., Mayer, J., Shoveller, J., Press, N., et al. (2007). Anticipating dissemination of cancer genomics in public health: A theoretical approach to psychosocial and behavioral challenges. *Annals of Behavioral Medicine, 34*(3), 275–286.

Henneman, L., Timmermans, D. R. M., & van der Wal, G. (2004). Public experiences, knowledge and expectations about medical genetics and the use of genetic information. *Community Genetics, 7*, 33–43.

Hooker, G. W., Peay, H., Erby, L., Bayless, T., Biesecker, B. B., & Roter, D. L. (2014). Genetics literacy and patient perceptions of IBD testing utility and disease control: A randomized vignette study of genetic testing. *Inflammatory Bowel Diseases, 20*(5), 901–908.

Hughes, C., Gomez-Caminero, A., Benkendorf, J., Kerner, J., Isaacs, C., Barter, J., et al. (1997). Ethnic differences in knowledge and attitudes about BRCA1 testing in women at increased risk. *Patient Education and Counseling, 32*(1–2), 51–62.

Hurle, B., Citrin, T., Jenkins, J. F., Kaphingst, K. A., Lamb, N., Roseman, J., et al. (2013). What does it mean to be genomically literate? National Human Genome Research Institute meeting report. *Genetics in Medicine, 15*, 658–663.

Ishiyama, I., Nagai, A., Muto, K., Tamakoshi, A., Kokado, M., Mimura, K., et al. (2008). Relationship between public attitudes toward genomic studies related to medicine and their level of genomic literacy in Japan. *American Journal of Medical Genetics, 146A*, 1696–1706.

Jallinoja, P., & Aro, A. A. (1999). Knowledge about genes and heredity among Finns. *New Genetics and Society, 18*(1), 101–110.

Kaphingst, K., & McBride, C. (2010). Patient responses to genetic information: Studies of patients with hereditary cancer syndromes identify issues for use of genetic testing in nephrology practice. *Seminars in Nephrology, 30*(2), 203–214.

Kaphingst, K. A., Blanchard, M., Milam, L., Pokharel, M., Elrick, A., & Goodman, M. S. (2016). Relationships between health literacy and genomics-related knowledge, self-efficacy, perceived importance, and communication in a medically underserved population. *Journal of Health Communication, 21*(Suppl. 1), 58–68.

Kaphingst, K. A., Facio, F. M., Cheng, M.-R., Brooks, S., Eidem, H., Linn, A., et al. (2012). Effects of informed consent for individual genome sequencing on relevant knowledge. *Clinical Genetics, 82*(5), 408–415.

Kaphingst, K. A., Lachance, C. R., Gepp, A., D'Anna, L. H., & Rios-Ellis, B. (2011). Educating underserved Latino communities about family health history using lay health advisors. *Public Health Genomics, 14*(4–5), 211–221.

Kaphingst, K. A., McBride, C. M., Wade, C. H., Alford, S. H., Brody, L. C., & Baxevanis, A. D. (2010). Consumers' use of web-based information and their decisions about multiplex genetic susceptibility testing. *Journal of Medical Internet Research, 12*(3), e41.

Kaphingst, K. A., McBride, C. M., Wade, C. H., Baxevanis, A. D., Reid, R. J., Larson, E. B., et al. (2012). Patients' understanding of and responses to multiplex genetic susceptibility test results. *Genetics in Medicine, 14*, 681–687.

Kaphingst, K. A., Stafford, J. D., McGowan, L. D. A., Seo, J., Lachance, C. R., & Goodman, M. S. (2015). Effects of racial and ethnic group and health literacy on responses to genomic risk information in a medically underserved population. *Health Psychology, 34*(2), 101–110.

Kelly, K., Leventhal, H., Marvin, M., Toppmeyer, D., Baran, J., & Schwalb, M. (2004). Cancer genetics knowledge and beliefs and receipt of results in Ashkenazi Jewish individuals receiving counseling for BRCA1/2 mutations. *Cancer Control, 11*(4), 236–244.

Kessler, L., Collier, A., & Halbert, C. H. (2007). Knowledge about genetics among African Americans. *Journal of Genetic Counseling, 16*(2), 191–200.

Khoury, M. J. (2003). Genetics and genomics in practice: The continuum from genetic disease to genetic information in health and disease. *Genetics in Medicine, 5*(4), 261–268.

Khoury, M. J., Gwinn, M., Yoon, P. W., Dowling, N. F., Moore, C. A., & Bradley, L. A. (2007). The continuum of translation research in genomic medicine: How can we accelerate the appropriate integration of human genome discoveries into health care and disease prevention? *Genetics in Medicine, 9*, 665–674.

Kleiderman, E., Knoppers, B. M., Fernandez, C. V., Boycott, K. M., Ouellette, G., Wong-Rieger, D., et al. (2014). Returning incidental findings from genetic research to children: Views of parents of children affected by rare diseases. *Journal of Medical Ethics, 40*, 691–696.

Kutner, M., Greenberg, E., Jin, Y., Paulsen, C., & White, S. (2006). *The health literacy of America's adults: Results from the 2003 National Assessment of Adult Literacy.* Washington, DC: National Center for Education Statistics.

Lachance, C., Erby, L. H., Ford, B. M., Allen, V. C., & Kaphingst, K. A. (2010). Informational content, literacy demands, and usability of websites offering health-related genetic tests directly to consumers. *Genetics in Medicine, 12*(5), 304–312.

Lafayette, D., Abuelo, D., & Passero, M. T. U. (1999). Attitudes toward cystic fibrosis carrier and prenatal testing and utilization of carrier testing among relatives of individuals with cystic fibrosis. *Journal of Genetic Testing, 8*, 17–36.

Lanie, A. D., Jayaratne, T. E., Sheldon, J. P., Kardia, S. L. R., Anderson, E. S., Feldbaum, M., et al. (2004). Exploring the public understanding of basic genetic concepts. *Journal of Genetic Counseling, 13*(4), 305–320.

Lea, D. H., Kaphingst, K. A., Bowen, D., Lipkus, I., & Hadley, D. W. (2011). Communicating genetic information and genetic risk: An emerging role for health educators. *Public Health Genomics, 14*(4–5), 279–289.

Lerman, C., Biesecker, B., Berkendorf, J. L., Kerner, J., Gomez-Caminero, A., Hughes, C., et al. (1997). Controlled trial of pretest education approaches to enhance informed decision-making for BRCA1 gene testing. *Journal of the National Cancer Institute, 89*(2), 148–157.

Lillie, S. E., Brewer, N. T., O'Neill, S. C., Morrill, E. F., Dees, E. C., Carey, L. A., et al. (2007). Retention and use of breast cancer recurrence risk information from genomic tests: The role of health literacy. *Cancer Epidemiology, Biomarkers and Prevention, 16*(2), 249–255.

Lipkus, I. M., & Peters, E. (2009). Understanding the role of numeracy in health: Proposed theoretical framework and practical insights. *Health Education and Behavior, 36*(6), 1065–1081.

Manolio, T. A., Chisholm, R. L., Ozenberger, B., Roden, D. M., Williams, M. S., Wilson, R., et al. (2013). Implementing genomic medicine in the clinic: The future is here. *Genetics in Medicine, 15*(4), 258–267.

McBride, C. M., Bowen, D., Brody, L. C., Condit, C. M., Croyle, R. T., Gwinn, M., et al. (2010a). Future health applications of genomics: Priorities for communication, behavioral, and social sciences research. *American Journal of Preventive Medicine, 38*(5), 556–561.

McBride, C. M., Hensley-Alford, S., Reid, R. J., Larson, E. B., Baxevanis, A. D., & Brody, L. C. (2008). Putting science over supposition in the arena of personalized genomics. *Nature Genetics, 40*(8), 939–942.

McBride, C. M., Koehly, L. M., Sanderson, S. C., & Kaphingst, K. A. (2010b). The behavioral response to personalized genetic information: Will genetic risk profiles motivate individuals and families to choose more healthful behaviors? *Annual Review of Public Health, 31*, 89–103.

McInerney, J. (2002). Education in a genomic world. *Journal of Medical Philosophy, 27*, 369–390.

Meilleur, K. G., & Littleton-Kearney, M. T. (2009). Interventions to improve patient education regarding multifactorial genetic conditions: A systematic review. *American Journal of Medical Genetics, 149A*, 819–830.

Mesters, I., Ausems, A., & De Vries, H. (2005). General public's knowledge, interest and information needs related to genetic cancer: An exploratory study. *European Journal of Cancer Prevention, 14*, 69–75.

Molster, C., Charles, T., Samanek, A., & O'Leary, P. (2009). Australian study on public knowledge of human genetics and health. *Community Genetics, 12*(2), 84–91.

National Center for Education Statistics. (2006). *National Assessment of Adult Literacy (NAAL): A first look at the literacy of America's adults in the 21st century*. Washington, DC: National Center for Education Statistics.

Nielsen-Bohlman, L., Panzer, A. M., Kindig, D. A., eds. (2004). *Health literacy: A prescription to end confusion*. Washington, DC: National Academies Press.

Parrott, R., Silk, K., Krieger, J. R., Harris, T., & Condit, C. (2004). Behavioral health outcomes associated with religious faith and media exposure about human genetics. *Health Communication, 16*(1), 29–45.

Parrott, R. L., Silk, K. J., & Condit, C. (2003). Diversity in lay perceptions of the sources of human traits: Genes, environments, and personal behaviors. *Social Science and Medicine, 56*, 1099–1109.

Pearson, Y. E., & Liu-Thompkins, Y. (2012). Consuming direct-to-consumer genetic tests: The role of genetic literacy and knowledge calibration. *Journal of Public Policy and Marketing, 31*(1), 42–57.

Peters, E., Hart, P. S., & Fraenkel, L. (2011). Informing patients: The influence of numeracy, framing, and format of side effect information on risk perceptions. *Medical Decision Making, 31*(3), 432–436.

Portnoy, D. B., Roter, D., & Erby, L. H. (2010). The role of numeracy on client knowledge in BRCA genetic counseling. *Patient Education and Counseling, 81*(1), 131–136.

Quinn, H., Schweingruber, H., Keller, T., eds. (2012). *A framework for K-12 science education: Practices, crosscutting concepts, and core ideas.* Washington, DC: Committee on Conceptual Framework for the New K-12 Science Education Standards; Board on Science Education; Division of Behavioral and Social Sciences and Education; National Research Council.

Richards, M., & Ponder, M. (1996). Lay understanding of genetics: A test of a hypothesis. *Journal of Medical Genetics, 33*, 1032–1036.

Rolison, J. J., Hanoch, Y., & Miron-Shatz, T. (2012). What do men understand about lifetime risk following genetic testing? The effect of context and numeracy. *Health Psychology, 31*(4), 530–533.

Rose, A., Peters, N., Shea, J. A., & Armstrong, K. (2005). The association between knowledge and attitudes about genetic testing for cancer risk in the United States. *Journal of Health Communication, 10*, 309–321.

Roter, D. L., Erby, L. H., Larson, S., & Ellington, L. (2007). Assessing oral literacy demand in genetic counseling dialogue: Preliminary test of a conceptual framework. *Social Science and Medicine, 65*(7), 1442–1457.

Sapp, J. C., Dong, D., Ivey, L. E., Hooker, G., Biesecker, L. G., & Biesecker, B. B. (2014). Parental attitudes, values, and beliefs toward the return of results from exome sequencing in children. *Clinical Genetics, 85*(2), 120–126.

Smerecnik, C. M., Mesters, I., de Vries, N. K., & de Vries, H. (2008). Educating the general public about multifactorial genetic disease: Applying a theory-based framework to understand current public knowledge. *Genetics in Medicine, 10*(4), 251–258.

Syurina, E. V., Brankovic, I., Probst-Hensch, N., & Brand, A. (2011). Genome-based health literacy: A new challenge for public health genomics. *Public Health Genomics, 14*, 201–210.

Tercyak, K. P., Alford, S. H., Emmons, K. M., Lipkus, I. M., Wilfond, B., & McBride, C. M. (2011). Parents' attitudes toward pediatric genetic testing for common disease risk. *Pediatrics, 127*(5), e1288–1295.

Thompson, H. S., Wahl, E., Fatone, A., Brown, K., Kwate, N. O. A., & Valdimarsdottir, H. (2004). Enhancing the readability of materials describing genetic risk for breast cancer. *Cancer Control, 11*(4), 245–253.

Vassy, J. L., O'Brien, K. E., Waxler, J. L., Park, E. R., Delahanty, L. M., Florez, J. C., et al. (2012). Impact of literacy and numeracy on motivation for behavior change after diabetes genetic risk testing. *Medical Decision Making, 32,* 606–615.

Wang, C., Bowen, D. J., & Kardia, S. L. R. (2005). Research and practice opportunities at the intersection of health education, health behavior, and genomics. *Health Education & Behavior, 32,* 686–701.

Wang, C., Gallo, R. E., Fleisher, L., & Miller, S. M. (2011). Literacy assessment of family health history tools for public health prevention. *Public Health Genomics, 14*(4–5), 222–237.

Wilfond, B., & Friedman Ross, L. (2009). From genetics to genomics: Ethics, policy, and parental decision-making. *Journal of Pediatric Psychology, 34*(6), 639–647.

11

Legal Issues Associated with the
Introduction of Genetic Testing to the
Education System

DAVID PELOQUIN AND MARK BARNES

In the summer of 2010, the University of California, Berkeley ("Berkeley"), mailed saliva collection kits to each of its incoming freshmen and provided them with the opportunity to submit a DNA sample that could be tested for genes affecting the body's metabolization of milk, alcohol, and folic acid (Sanders, 2010). The goal of the program, entitled "Bring Your Genes to Cal," was to spur discussion among students regarding the use of genetic testing in the field of "personalized medicine." Anticipating opposition to the program, Berkeley selected genes for testing that it thought would be of interest to students without causing them to make serious health or lifestyle changes (Lewin, 2010). Despite these precautions, after becoming aware of the program, the California Department of Public Health indicated to Berkeley that it would not permit the return of individualized results to students. The Department contended that both federal and California law required that individualized results be generated in a laboratory certified under the Clinical Laboratory Improvement Amendments of 1988 ("CLIA"), a federal statute that sets standards for clinical laboratories performing medical tests and was enacted to ensure the reliability and accuracy of the results of such tests (Sanders, 2010). Berkeley had planned to process the test results in a research laboratory lacking CLIA certification. As a result of the California Department of Public Health's interpretation of CLIA, Berkeley modified the program and, rather than providing students with their individualized results, aggregated results for use in group discussion.

The problem that Berkeley faced in attempting to introduce genetic testing into the college curriculum demonstrates but one of the many legal issues that could arise in response to any attempt to introduce genetic testing into the education system. The present chapter is intended to provide an introduction to some of the relevant laws and regulations in this area. In order to organize our discussion of legal concepts in line with the themes of this book, we analyze the issues raised by two of the volume's major themes: (1) heritability for achievement, and (2) direct-to-consumer (DTC) testing. Given the breadth of laws that might apply in this area, we focus primarily on federal law. Despite our federal focus, it is important to keep in mind that the public education system in the United States is funded primarily at the state and local level, and thus any attempt to use genetic testing in the public school system on a wide scale would likely need to grapple with myriad, disparate state laws. Moreover, as the technology underpinning genetic testing continues to evolve, so does the state of the laws that could be implicated by the use of such testing in the education system.

HERITABILITY FOR ACHIEVEMENT

The concept of using genetic testing to measure the capacity for achievement of school children enrolled in public education – for example, to evaluate a child for enrollment in special education classes or gifted and talented courses – raises several legal issues that would need to be addressed before such tests could be introduced into the public education system on a systematic basis. We analyze concerns that arise under the US Constitution, those arising under state statutes governing genetic testing generally, and those that arise under special education law.

Constitutional Concerns

The use of genetic tests on a wide scale in public schools could potentially raise a number of constitutional issues. The first of these is the Fourth Amendment right to be free from "unreasonable" searches and seizures (U.S. Const., amd. 4). A genetic test, a procedure for which some amount of genetic material must be collected, whether in the form of saliva, hair, or blood, falls within the ambit of the Fourth Amendment if performed by a public school because it involves the invasion of the person by the government, even if for only a brief instant (Skinner v. Railway Labor Executives Association, 1989; New Jersey v. T.L.O., 1985).[1] Generally, such searches must be subject to notice and the consent of the person being searched, or be the result of legal process, such as a search warrant (Ferguson v. City of Charleston, 2001; Schmerber v. State of California, 1996).

---

[1] Skinner recognized that state-compelled collection of urine is a search within the meaning of the Fourth Amendment, whereas in New Jersey v. T.L.O., the Supreme Court held that public school officials are government officers subject to the requirements of the Fourth Amendment.

While the courts have not had occasion to address the Fourth Amendment implications of laws mandating genetic testing in the school system, the US Supreme Court has twice addressed Fourth Amendment concerns related to a more traditional use of medical testing in the schools: suspicionless drug testing of school athletes and other participants in non-athletic extracurricular activities (Vernonia School District 47J v. Acton, 1995; Board of Education v. Earls, 2002). The challenges in such cases have been brought by parents who refused to consent to the drug testing of their child, arguing that it constituted an unreasonable search. The Court rejected the Fourth Amendment challenges in each instance after balancing the nature of the privacy interest at issue, the character of the intrusion, and the nature and immediacy of the governmental concern. It first noted that public school students, as minors who are subject to the control of the school, have a "diminished expectation" of privacy (Vernonia, 1995, p. 658, n. 2). The Court also analogized to other types of invasions of privacy, including vaccinations and physical examinations to which students are generally required to submit in order to enroll in school. The Court determined that the giving of the urine sample involved no more invasion of privacy than is normally encountered when using a public restroom, and highlighted the fact that the results of the test would be distributed only to a "limited class of school personnel who have a need to know" (Vernonia, 1995, p. 658). When analyzing the governmental interest at play, the Court looked to evidence of the detrimental effect of illegal drug use on students' health, particularly that of student athletes who may suffer an increased risk of injuries on the field as a result of their illegal drug use, and noted the school's "special responsibility" to care for the children under its supervision (Vernonia, 1995, p. 830). Also important to the Court's finding that suspicionless drug testing did not violate the Fourth Amendment was its observation that students retained the option of avoiding the drug test altogether by abstaining from participation in extracurricular activities.

The Court's analysis of suspicionless drug testing in schools offers several lessons to guide the possible introduction of genetic testing into the school system. Most importantly, it seems likely that any attempt to integrate genetic testing into schools will need to occur on a "voluntary" rather than a "mandatory" basis. This is because when analyzing the drug testing programs, the Supreme Court put great emphasis on the fact that students could avoid an unwanted drug test by choosing not to participate in athletic or other extracurricular activities. If, however, as seems likely, genetic testing is used as a method of screening children for participation in "special education" programs, the consequences of opting out of the test would be much more severe than simply not participating in an extracurricular program. Sports and other extracurricular activities can be considered as "ancillary" to the primary purpose of the school system: education of students. For students who suffer from a learning disability, however, special education classes may comprise the vast majority of their educational curriculum. Accordingly, in order to avoid the genetic test, such students would essentially be forced to opt out of public education altogether, something a court would likely find unacceptable.

Furthermore, part of the concern leading to the court's acceptance of mandatory testing in the drug context was the court's belief that school children using illegal drugs could harm themselves or other students. This could be the case, for example, if the student's altered behavior stemming from the use of illicit substances causes him or her to become violent toward others, or if the student encourages other students to use illicit substances. These same types of safety concerns are not present in the use of genetic testing in order to place students into different educational tracks.

In addition to Fourth Amendment search concerns, any introduction of mandatory genetic testing would also be likely to raise concerns related to the First Amendment and the "free exercise" of religion, as well as to the Fourteenth Amendment "due process" right of parents to be able to control the education of their children (US Const., amd. 1, amd. 14). Mandatory newborn screening laws that require babies to undergo certain genetic tests provide an example of the First and Fourteenth Amendment considerations at play here. While almost every state requires screening of newborns for certain serious, treatable genetic conditions such as Phenylketonuria (PKU), nearly all state laws on mandatory newborn testing include an optout for parents who object to such screening; in some states the optout is restricted to parents who object to the screening on religious grounds, whereas in others it extends to any type of objection a parent may have.[2] Similarly, nearly every state requires public school children to obtain certain

---

[2] For example, Massachusetts regulations allow parents to opt out of mandatory newborn screening if they "object thereto on the grounds that such test(s) conflict(s) with their religious tenets and practices" (Code of Massachusetts Regulations, Title 105 § 270.006). In such cases the regulation provides that "the attending physician shall document such objections to the mandated newborn blood screening in the newborn's medical record." Minnesota law provides an example of a state where parents may opt-out of newborn blood screening on any grounds, not limited to those with a religious basis (Minnesota Statutes § 144.125, subdiv. 4). One of the few states that does not have any sort of parental opt-out from newborn genetic screening is Nebraska. Nebraska's newborn genetic screening statute, Nebraska Revised Statutes § 71–519, has been challenged multiple times by parents as an infringement of their First Amendment right to "free exercise" of religion and a Fourteenth Amendment "due process" right to raise their children as they see fit.

vaccinations prior to school enrollment, and like the newborn screening statutes, most of these statutes permit parents to object, either on religious grounds or on the basis of other "philosophical" beliefs.[3]

Based on the fact that laws related to mandatory genetic testing of newborns and mandatory vaccination of school children nearly all include optouts that permit objecting parents to avoid such tests, it seems advisable that any mandatory genetic testing of school children would also contain such an optout. The case for an optout from the use of genetic testing in the education system is arguably even stronger than that for an optout

These challenges have been uniformly rejected, with the Nebraska Supreme Court and a federal district court both holding that the Nebraska state legislature had a "rational basis" for enacting the newborn screening requirement since the early diagnosis permitted by the testing would allow for "prevention of death and disability in children" as well as easing the "potential social burdens created by children who are not identified and treated" (Douglas County v. Anaya, 2005; Spiering v. Heineman, 2006).

[3] For example, Massachusetts' law on vaccinations permits parents to exempt their children from the vaccination requirement due to a conflict between the vaccination requirement and "sincere religious beliefs" (Mass. Gen. Laws ch. 76, § 15), whereas Arkansas law provides an example of a growing trend toward permitting "philosophical" objections in addition to religious ones, after two federal district courts in that state held that permitting solely a religious exemption to mandatory vaccinations violated the "free exercise" clause of the First Amendment by permitting the state to determine which religious groups would gain state "recognition" (Ark. Code § 6-18-702(d)(4); Boone v. Boozman, 2002; McCarthy v. Boozman, 2002). In a sign of an increased focus on newborn screening at the federal level, the US Congress in 2014 passed legislation stating that federally funded research on newborn dried blood spots should be considered research on human subjects and, therefore, must meet the requirement of the federal human subject research regulations known as the "Common Rule", even if the bloodspots in question were no longer identifiable (Newborn Screening Saves Lives Reauthorization Act of 2014, H.R. 1281, § 12) (45 C.F.R. pt. 46). The legislation required that the informed consent of the subject (or, more likely, his or her parent) be obtained prior to using such blood spots for research by stating that an Institutional Review Board (IRB) may not waive the requirement of informed consent for research on newborn bloodspots. The legislation contained a provision stating that its requirements regarding research use of newborn bloodspots would no longer have effect following the effective date of revisions to the Common Rule. In January 2017, the revised version of the Common Rule issued and is expected to take effect in January 2018. The revised Common Rule contains a provision stating that research on newborn dried blood spots would be treated like research on all other types of biological specimens, *i.e.*, research using newborn dried blood spots collected during the course of routine clinical care would be considered "research" and thus require informed consent of the child's parent or guardian only if the specimen were identifiable. In addition, research involving such specimens would remain eligible for a waiver of the informed consent requirement by an IRB assuming that all criteria for grant of a waiver are satisfied. (82 Fed. Reg. 7,149, 7,179-70 (Jan. 19, 2017)). While not directly applicable to genetic tests used outside of the research context, the legislative and regulatory activity in this area indicates an ongoing societal debate about the extent to which parents should be able to exercise control over testing performed on their children or biological specimens obtained from their children.

from mandatory newborn screening or school vaccination, as the use of genetic tests to predict a student's aptitude for learning is an area that is far less developed than the use of genetic tests for the types of diseases, such as PKU, that are commonly included in mandatory newborn screenings, nor does it have the direct benefits to other students that mandatory vaccination provides. Accordingly, the government's interest in mandating genetic testing of school children is not as strong as the government's interest in mandating these other, more established types of testing.

Informed Consent Laws

A second legal issue arising with the use of genetic testing in the schools is what type of informed consent would be required for the testing. As noted earlier, constitutional considerations would require informed consent of parents for a child's participation; however, statutes governing informed consent for medical testing would also be implicated. There is no federal law that would require consent for genetic testing performed on school children. The set of federal regulations referred to as the "Common Rule" that govern human subjects research conducted or supported by most federal agencies would likely not apply, despite the fact that many public school systems receive funding from the federal government. This is because the Common Rule require consent for "research," which is defined as "a systematic investigation, including research development, testing and evaluation, designed to develop or contribute to generalizable knowledge" (45 C.F.R. § 46.102). If genetic tests are used to evaluate children for inclusion in special education or gifted and talented education classes, the Common Rule would not apply because the tests are not being administered for the purpose of contributing to "generalizable knowledge." Rather, the tests are being conducted in order to evaluate a specific child for entrance into a given educational program.

Nevertheless, while there may not be any federal laws requiring consent in this situation, several state legislatures have enacted laws that impose special consent requirements for genetic testing and/or restrict the disclosure of genetic test results. While a fifty-state survey of genetic testing laws is beyond the scope of this chapter, we provide two instances of state genetic testing laws that would be particularly relevant to any use of genetic testing in the schools. For example, in New York State, no person may perform a genetic test on a biological sample taken from another individual without the prior written informed consent of such individual or the signature of a person authorized to consent for such individual (New York Civil Rights

Law § 79-l(2)). The written consent must contain several elements, includ-ing "a general description of the test," "a statement of the purpose of the test," "a general description of each specific disease or condition tested for," "a statement that the individual may wish to obtain professional genetic counseling prior to signing the informed consent," and "the level of cer-tainty that a positive test result for that disease or condition serves as a predictor of such disease" (New York Civil Rights Law § 79-l(2)(b)).

Without a specific exemption in the law for educational uses of the test, a school wishing to perform genetic testing on students would almost certainly be subject to this statute, because the testing of a child for an intellectual disability that qualifies the child for special education classes would likely be considered testing for a "condition." Accordingly, the school system would need to provide a great deal of information to students who are subject to the genetic test, or more likely, to their par-ents as legal guardians. Several requirements of this statute may hinder schools in obtaining consent for genetic testing from students or their parents. For instance, the statement that individuals may wish to undergo professional genetic counseling may cause some parents to delay permit-ting their child to undergo the testing until after speaking with a coun-selor, and after speaking with the counselor they may decide no longer to pursue testing. The requirement that the consent indicate the level of certainty that the test is a predictor of a given disease or condition would likely require the school to inform students and their parents of the school's level of certainty that the test results indicate that a given student has the intellectual ability that would qualify him or her for participation in special education classes.[4]

While most states do not have statutes with pre-testing requirements as stringent as those of New York, many states do have consent require-ments governing the release of genetic test results that could impact the use of such tests in schools. For example, Illinois law requires that in gen-eral genetic test results only be released after execution of a "written legally effective authorization for release of the test results" by the subject of the test or the subject's legally authorized representative (Illinois Compiled Statutes ch. 410 § 513/30). Accordingly, under this statute the student, or more likely his or her parent, would need to consent to release of genetic test results to schools.

---

[4] New York's statute provides some flexibility regarding the requirement that a level of cer-tainty be established, indicating that if no level of certainty has been established for a given test, the requirement of providing a level of certainty can be disregarded.

Given the myriad state statutes regulating genetic testing, any move toward the use of genetic test results in schools would need to proceed on a state-by-state basis, taking into account the conflicting laws that exist regarding consent for genetic testing and release of results. If genetic testing is to proceed on a wide scale in public schools, it may be advisable for supporters of such testing to lobby state legislators to streamline the consent requirements such that genetic tests administered in schools would not be subject to the general state laws on genetic testing.

Education Law

In addition to federal constitutional considerations and informed consent considerations, special education in the United States is subject to certain federal laws that would certainly impact the use of genetic testing as a tool to evaluate students for inclusion in such programs, requiring that such tests be validated prior to being used on a wide scale. Chief among these laws is the Individuals with Disabilities Education Act ("IDEA") (20 U.S.C. § 1400 *et seq.*). Congress enacted IDEA's predecessor statute, the Education for all Handicapped Children Act of 1975, after finding that (i) children with disabilities often did not receive appropriate educational services or were excluded entirely from the public school system, and (ii) undiagnosed disabilities could prevent children from having a successful educational experience (20 U.S.C. § 1400(c)(2)). IDEA requires that students with disabilities be provided a "free appropriate public education" ("FAPE") that is "tailored to the[ir] unique needs ... by means of an 'individualized educational program (IEP)'" (20 U.S.C. § 1414(d)(1)(A)). IDEA further mandates that to the maximum extent appropriate, children with disabilities be educated with children without disabilities (20 U.S.C. § 1412(a)(5)).

The types of disabilities addressed by IDEA include "mental retardation, a hearing impairment (including deafness), a speech or language impairment, a visual impairment (including blindness), a serious emotional disturbance ... an orthopedic impairment, autism, traumatic brain injury, an other health impairment, a specific learning disability, deaf-blindness, or multiple disabilities" (34 C.F.R. § 300.8(a)(1)). IDEA additionally mandates specific procedures that must be used when determining whether a child has a disability that would qualify him or her for special education in accordance with IDEA. These regulations require that any evaluation of a child for a disability must "[u]se a variety of assessment tools and strategies to gather relevant functional, development, and academic information about the child" and admonishes that a public agency determining whether a child has a disability must "[n]ot use any single measure or assessment as the sole criterion for determining whether a child is a child with a disability" (34 C.F.R. § 300.304). Accordingly, if genetic testing is to be used in public schools as a means of determining whether a given child has a disability that qualifies him or her for special education, to meet the requirements of IDEA, the genetic test would need to be used in conjunction with other tests to determine disability.

Prior to being used in the school system, any genetic test would need to be validated to ensure that it accurately measures disability and does not lead to a disparate impact on children from certain racial or ethnic backgrounds. Indeed, there is a long history of litigation under IDEA and its predecessor statutes challenging the use of certain tests to categorize children as disabled. For instance, in a 1984 case the parents of several African American students in the San Francisco school district challenged that school district's use of a certain IQ test to measure whether students were disabled, arguing that use of the test had a disparate impact on African American students, causing them to be placed in special education classes at a rate far higher than their peers of other races. As a result of this lawsuit, a federal court of appeals invalidated the San Francisco school district's use of the IQ tests at issue because it found they had not been properly validated for the specific purpose for which they were used (Larry P. By Lucille P. v. Riles, 1984). More recently, courts analyzing the use of tests to place children in special education classes have recognized a "fundamental tension in special education law – that between ensuring that all disabled children have access to educational opportunity and ensuring that non-disabled children are not improperly identified as disabled. This tension is particularly salient for minority students, who historically have been over-identified as disabled and disproportionally placed in segregated educational settings, due in part to biased IQ tests" (E.M. ex rel. E.M. v. Pajaro Valley Unified Sch. Dist. Office of Admin. Hearings, 2011).

Disturbingly for any proposed use of genetic testing as a method of placing students in special education classes, school districts have in the past used genetic arguments to justify the disparate impact of IQ tests used to determine eligibility for special education under IDEA. For example, in the Larry P. case discussed earlier, one of the arguments the San Francisco school district used to justify the disparate impact of the test results was a "genetic argument" contending that African American students came from a weaker gene pool, and thus required special education classes.

The fact that, in the past, references to genetics have been used to support such racist arguments is likely to make courts skeptical about uses of genetic testing to classify students into different educational tracks, such as special education or gifted and talented courses. Schools should thus expect challenges to any use of genetic testing to qualify students for inclusion within special education classes, and thus must be prepared to counter such challenges with data demonstrating the efficacy of the tests at issue. Furthermore, because of the extensive procedural mechanisms it provides for parents to challenge the decisions of school districts with regard to special education placement, IDEA has been described as creating an expectation that "parents serve as private 'attorneys general' that enforce the law, protect their children's rights, and ensure access to services" (Pergament, 2013, p. 85). Any use of genetic tests in the schools will thus need to be done with an eye toward responding to claims raised by these private "attorneys general."

DTC GENETIC TESTING

In addition to the legal requirements stemming from laws governing the education system in the United States that could impact the proposals found in this volume's chapters on "heritability for achievement," the second of the major issues explored in this volume, DTC genetic testing, also raises a variety of legal issues, some of which have recently gained a great deal of publicity. Laws governing DTC testing would come into play if, for example, a school decided to use DTC genetic testing kits in its genetic testing of students or if a classroom teacher decided to use a DTC genetic test in his or her classroom as a way of teaching students about genetics.

DTC genetic testing in the United States is currently subject to two complementary bodies of federal law, both of which are administered by agencies located within the federal Department of Health and Human Services: (1) the Food and Drug Administration's ("FDA") power to regulate medical devices; and (2) the Centers for Medicare and Medicaid Services' ("CMS") authority under CLIA to regulate clinical laboratories. FDA's general authority in this area stems from its power under section 201(h) of the Food, Drug and Cosmetic Act ("FDCA") to regulate any "device," defined as "an instrument, apparatus, implement, machine, contrivance, implant, *in vitro reagent*, or other similar or related article, including any component, part, or accessory, which is ... (2) intended for use in the *diagnosis of disease or other conditions*, or in the cure, mitigation, treatment, or prevention of disease, in man or other animals" (21 U.S.C. § 321(h)) (emphasis added).

FDA Regulation

As part of its statutory authority to regulate "devices," FDA can require device manufacturers to undergo pre-market review or clearance procedures prior to selling a product through interstate commerce. Thus a key question with respect to genetic tests offered directly to the consumer is whether they are "devices" within the meaning of the FDCA, a question with which both FDA and industry lobbying groups have grappled for several years.

Part of the confusion in this area seems to be grounded in semantics, i.e., a genetic test offered in a laboratory setting does not strike one as a "device" within the common meaning of the term in the same way as tangible items such as heart stents or pacemakers. Adding to the confusion, FDA's historical practices regarding regulation of laboratory tests have cast doubt on its ability or desire to regulate such items. This is because for many years FDA has divided laboratory tests into two categories: in vitro diagnostic devices[5] developed by one company and then sold to laboratories via kits shipped in interstate commerce, and so-called "laboratory-developed tests" ("LDTs") which are generally defined as tests that are "manufactured, including being developed and validated, and offered, within a single laboratory" (FDA, 75 Fed. Reg. 34,463). FDA has long exercised jurisdiction over the first type of tests. FDA's approach to LDTs, by contrast, has been one of "enforcement discretion," meaning that while FDA has announced at various instances that it has authority to regulate LDTs, declaring that LDTs are merely a subset of in vitro diagnostic devices, it has exercised its discretion by declining to regulate actively such products (FDA Analyte Specific Reagents Final Rule, 1997; Shuren, 2010).

FDA's approach to regulating LDTs is critical to any discussion of the regulation of genetic testing because most genetic tests on the market today are LDTs. That is, they are often developed in house, for use by a single laboratory, rather than for sale to other laboratories in the form of a test kit. The LDT definition includes DTC tests for which a specimen collection kit is offered directly to the general public so long as the analysis of the specimen is performed in the same laboratory where the test was developed.[6] As a consequence of FDA's exercise of enforcement discretion with regard to LDTs, the courts have not had occasion to address the issue of whether FDA does indeed possess the ability to regulate LDTs, and thus there remains a great deal of debate concerning whether such authority in fact exists.[7]

---

[5] "In vitro diagnostic devices" include reagents, instruments, and systems intended for use in the collection, preparation, and examination of human specimens and that are further intended for the diagnosis of disease or other conditions, including a determination of the state of health, in order to cure, mitigate, treat or prevent disease or its sequelae (21 C.F.R. § 809.3.).

[6] By contrast, a genetic test would cease to be an LDT, and thus would fall clearly into FDA's jurisdiction, if a laboratory developed the test in house but then sold a kit containing all the materials necessary for performance of the test to *other laboratories*.

[7] For example, two leading advocacy organizations for clinical laboratories in the United States have opposing views on the question of whether FDA possesses authority to regulate

In recent years, simultaneous with the development of DTC genetic tests, FDA has suggested that its exercise of enforcement discretion with regard to LDTs is coming to an end. By way of example, in the summer of 2010, FDA published a notice of public meeting and request for comments regarding oversight of LDTs announcing that "the agency believes it is time to reconsider its policy of enforcement discretion over LDTs" (FDA Oversight of Laboratory Developed Tests, 2010). That same summer in comments before Congress, FDA's Director of the Center for Devices and Radiological Health, Jeffrey Shuren, discussed FDA's rationale for pursuing regulation of LDTs, noting that

> [t]he nature of the laboratory-developed tests has changed over the last 30 years, but most dramatically in the last few years. Today, LDTs are increasingly used to assess high-risk but relatively common diseases and conditions, often are used to provide critical information for patient treatment decisions, rely on novel (sometimes preliminary) scientific findings to support their usefulness … and are performed in commercial laboratory settings that are geographically separate from the patient's primary health care professional and health care setting. (Shuren, 2010)

Shuren also addressed DTC genetic testing in particular, noting that

> Marketing genetic tests directly to consumers can increase the risk of a test because a patient may make a decision that adversely affects their health, such as stopping or changing the dose of a medication or continuing an unhealthy lifestyle, without the intervention of a learned intermediary.

Importantly, Dr. Shuren was careful to highlight that not all DTC genetic testing would be subject to FDA's power to regulate "devices," noting that where the purpose of the testing was not to diagnose disease or other conditions in man or animals, it would not be a device. Accordingly, tests for ancestry would not be considered devices subject to FDA jurisdiction.

LDTs. The American Clinical Laboratory Association has historically argued that FDA lacks jurisdiction over LDTs because LDTs are not "devices"; rather, they are "proprietary procedures for performing a diagnostic test using reagents and laboratory equipment," and thus are properly regulated by CMS under CLIA rather than FDA under the FDCA (American Clinical Laboratory Association, 2013). The College of American Pathologists, on the other hand, has argued that FDA has jurisdiction to regulate LDTs and has urged FDA to issue rulemaking adopting a new framework for regulation of such tests (College of American Pathologists, 2010). The American Clinical Laboratory Association (ACLA) has since engaged former United States Solicitor General Paul Clement and Harvard Law School professor Lawrence Tribe to advise ACLA on potential legal challenges to FDA regulation in the LDT space. (ACLA News, 2014).

Also in the summer of 2010, FDA sent letters to several companies that were offering DTC genetic testing, most notably 23andMe, indicating that by distributing their test kits they were marketing "devices" without having received the necessary pre-market approvals from FDA, and inviting a response from the companies (FDA Letter to 23andMe, 2010).

Following this flurry of activity, FDA entered a period of relative inactivity on the question of LDTs that lasted almost three years, thus prompting speculation as to whether FDA was still planning to revise its policy of enforcement discretion with regard to LDTs. This period of inactivity came to an abrupt end in the summer of 2013 when then FDA commissioner Margaret Hamburg indicated in the context of remarks to the American Society of Clinical Oncology that the fact that LDTs do not undergo pre-market review "can be a problem" as there is no regulatory oversight of whether a given test is "safe and effective" or "clinically valid" (Hamburg, 2013). In late 2013, renewed visibility for LDT regulation arrived when FDA issued a warning letter to 23andMe indicating that the company's Personal Genome Service ("PGS"), a DTC service which allows users to provide a saliva sample from which health reports on 254 diseases and conditions can be generated, is a device within the meaning of the FDCA. Importantly, this letter, unlike FDA's 2010 letter, provided for the first time FDA's views on the classification of 23andMe's PGS, indicating that the PGS would be placed in class III within FDA's three-class system of categorizing medical devices, which is intended for devices that pose the greatest risk. Risks identified in the letter included the possibility that a patient would make serious lifestyle or medical decisions based solely on the PGS' identification of his or her risk factors for developing cancer or his or her response factor for certain drugs. The letter demanded that the company immediately discontinue marketing the PGS until it receives FDA marketing authorization for the device (FDA 23andMe Warning Letter, 2013). The 23andMe warning letter was important from the standpoint of LDT regulation because it signaled FDA's willingness to pursue regulatory enforcement of DTC tests that fall within the definition of an LDT.

More recently, in the summer of 2014, FDA issued draft guidance regarding LDTs stating that FDA plans to cease its exercise of enforcement discretion over most LDTs and instead will begin to regulate such tests as it does other medical devices through placing such tests into three categories: class I, II, or III based on the device's intended use, technological characteristics, and the risk to patients if the device were to fail, with class I devices being the lowest-risk devices and class III devices being the highest-risk devices. The draft guidance stated definitively that FDA does not plan to exercise enforcement discretion over LDTs that are offered on a DTC basis (FDA Framework for Regulatory Oversight of Laboratory Developed Tests, 2014).

In early 2016, the House Appropriations Committee included language in the Congressional Report for an FDA appropriation bill directing FDA to "suspend further efforts to finalize the LDT guidance and continue working with Congress to pass legislation that addresses a new pathway for regulation of LDTs in a transparent manner" (House of Representatives Report, 2016, p. 70). Then, in November 2016, shortly after the election of President Donald Trump, FDA announced that it would not finalize the draft guidance. Instead, in January 2017, FDA issued a "discussion paper" on LDTs in the hopes of spurring further dialogue on the issue. The discussion paper outlined a risk-based system of phased-in oversight over four years that would impose premarket review requirements on higher-risk LDTs, while "grandfathering" some existing LDTs and exempting from the review process certain categories of LDTs, such as low risk LDTs, LDTs for rare diseases, and LDTs intended solely for forensic use. FDA was careful to note that the "discussion paper" does not represent the formal position of the FDA and is not enforceable at this time (FDA, Discussion Paper on Laboratory Developed Tests (LDTs) January 13, 2017).

The above discussion of the regulatory environment surrounding LDTs illustrates some of the hurdles that would have to be overcome with regard to any use of genetic testing in the education system. Given FDA's increased interest in regulating DTC LDTs, any test used in the education system that is offered directly to students would likely face scrutiny and be required to undergo pre-marketing approval prior to any use that would affect the educational curriculum offered to a particular student. For instance, a genetic test assessing a student's learning disability would almost certainly be seen by FDA as falling within the definition of "device," as it is intended to diagnose a "condition," i.e., a specific learning disability. Given that diagnosis with a learning disability could have a profound impact on a student's future educational opportunities while also impacting medical treatment decisions, it is very likely that FDA would classify any such device as "high risk," thus placing it in class III of the FDA's device classification scheme. The same would be true for tests of other conditions that could qualify a child for special education under IDEA, for example, "emotional disturbance" or "developmental delay." Accordingly, the test would likely be subject to FDA's pre-market approval process ("PMA"), the most stringent type of device marketing application required by FDA.

The need for a test to obtain PMA could pose particular challenges for those offering genetic tests. Receiving PMA generally requires submission of "sufficient valid scientific evidence to assure that the device is safe and effective for its intended use(s)" (FDA Premarket Approval). FDA published guidance in 2007 specifically discussing PMA applications for genetic tests. The guidance notes that PMAs must provide valid scientific evidence establishing reasonable assurance of the safety and effectiveness of the device, generally including the submission of clinical data. If the test is for new genetic markers, the test must be shown as safe and effective for its intended use, whereas if the test involves established genetic markers, appropriate information in the published medical literature may be used to support the test.

FDA guidance contains a specific discussion of the uses of "screening" tests, which noted that where the disorder screened for has a low prevalence in a given population, it may be difficult to amass sufficient evidence of the test's effectiveness, as there may be more "false positive" results than "true positive" results in the population studied (FDA Guidance for Industry, 2007). The particular challenges inherent in demonstrating the effectiveness of genetic tests had earlier been recognized by the Secretary's Advisory Committee on Genetics, Health & Society ("SACGHS"), an expert committee chartered to provide advice to the US Secretary of Health and Human Services regarding genetic testing, which observed that "[p]rospective data of a test's clinical validity, however, are often unavailable or incomplete for years after a test is developed, especially for predictive or presymptomatic tests. As such, numerous challenges remain for the demonstration of clinical validity, such as the collection of postmarket data and the sharing of information among laboratories" (Secretary's Advisory Committee on Genetics, Health & Society, 2008, p.4).

While the foregoing discussion highlights some of the challenges involved in creating a genetic test made available to the public at large, certain genetic tests that could perhaps play a role in the education system have been able to satisfy FDA's PMA requirement. For instance, in January 2014 FDA announced that it would allow marketing of a "first-of-its-kind" test to diagnose developmental delays and intellectual disabilities in children. The test in question, the Affymetrix CytoScan Dx Assay, was not developed as an LDT, but rather as a testing kit to be sold to other laboratories, thus falling squarely within FDA's historical exercise of jurisdiction over in vitro diagnostic tests that are distributed as test kits. FDA's announcement regarding the test noted that it "should not be used for stand-alone diagnostic purposes," but rather test results should be used only "in conjunction with other clinical and diagnostic findings, consistent with professional standards of practice" (FDA News Release, 2014). Physicians commenting on use of the tests have noted that the test should be used when a physician believes that something about the child's development appears unusual and points to a potential genetic lesion. In the words of Dr. Annemarie Stroustup, an assistant professor of pediatrics at Mount Sinai Hospital, the Affymetrix CytoScan Dx Assay "is not a screening test to be done on all newborns to predict how they are going to do in school when they are 5" (Perrone, 2014).

FDA's approach to the Afymetrix CytoScan Dx Assay test is significant because it underscores the notion, discussed earlier in the section on heritability for achievement, that if genetic tests are to be used in the schools as a tool for predicting future academic performance or assessing a child's need for special education classes, they should be combined with other, more traditional measures of assessing academic performance or learning disability. Indeed, the case of the Afymetrix CytoScan Dx Assay test suggests that FDA may provide approval for genetic tests used to evaluate future academic performance only if the instructions for use of such tests explicitly indicate that the test should be used in combination with other measures of assessing academic performance.

CLIA Regulations

In addition to FDA regulations, the other regulatory area that would likely impact the use of DTC genetic testing in the schools is CLIA, the statute that upset Berkeley's plans to share individualized genetic testing results with its incoming freshmen. CLIA is administered by CMS and regulates clinical laboratories that perform testing on patient specimens with the goal of ensuring accurate and reliable test results.[8] LDTs are one area in which FDA and CLIA jurisdiction overlaps.

In contrast to FDA regulations, which focus on the safety and *clinical validity* of a test, and which involve review prior to the marketing of a given test, CLIA focuses on the *analytical validity* of the test in the context of a given laboratory (SACGHS, 2008). Put slightly differently, in analyzing clinical validity FDA focuses on the accuracy with which a given test "identifies, measures, or predicts the presence or absence of a clinical condition or predisposition in a patient," whereas by focusing on analytical validity,

---

[8] Two states, New York and Washington, are exempt from CLIA because CMS has determined that they have state regulations governing laboratories that are at least as stringent as those of CLIA (SACGHS, 2008, p.3).

CLIA addresses the narrower question of whether "a specific test finds what it is supposed to find (i.e., the analyte it is intended to detect)," but it does not seek to answer the question of whether the presence of a given analyte accurately indicates a given medical condition (CMS, 2013). Accordingly, before reporting test results to patients, laboratories performing LDTs must establish certain "performance specifications," including (1) accuracy, (2) precision, (3) analytical sensitivity, (4) analytical specificity to include interfering substances, (5) reportable range of test results for the test system, (6) reference intervals (normal values), (7) and any other performance characteristic required for test performance (42 C.F.R. § 493.1253(b)(2)).

In recent years, CMS has shown interest in increasing the stringency of performance specifications for laboratories performing genetic tests. It had initially proposed adding a new genetic testing specialty to CLIA that would contain specific standards for laboratories performing genetic tests, though it has now abandoned that approach, and it remains unclear if the agency will develop specific standards for genetic tests (Sarata & Johnson, 2014). Nevertheless, as Berkeley discovered through its attempt to introduce genetic testing to the undergraduate curriculum, if results are to be returned to students, schools offering such tests should ensure that the tests are performed by a CLIA-certified laboratory.

CONCLUSION

Given the breadth of the subject matter, the foregoing discussion covers only the major federal laws that would likely impact any attempt to introduce genetic testing into the education system. Additional laws, not addressed herein, such as the education laws of individual states, would also be implicated. What the foregoing discussion has demonstrated, however, is that to comply with current law, any attempt to introduce genetic testing to the education system would require that the validity of the tests be demonstrated prior to their introduction, that genetic tests are used in a holistic manner with more traditional measures of intellectual capability, that parental and/or student consent is obtained prior to administration of any tests, and that the tests themselves comply with regulatory requirements imposed by FDA and CMS.

REFERENCES

American Clinical Laboratory Association. (2013, June 4). Citizens petition to Food & Drug Administration.

American Clinical Laboratory Association. (2014). ACLA retains Attorneys Paul D. Clement and Lawrence H. Tribe to represent ACLA in opposing the FDA's proposal to treat laboratory developed tests (LDTs) as medical devices. Retrieved from www.acla.com/acla-retains-attorneys-paul-d-clement-and-laurence-h-tribe-to-represent-acla-in-opposing-the-fdas-proposal-to-treat-laboratory-developed-tests-ldts-as-medical-devices/.

Arkansas Code § 6-18-702(d)(4).

Blood Screening of Newborns for Treatable Diseases and Disorders, Code of Massachusetts Regulations Title 105 § 270.006.

Board of Education v. Earls, 536 U.S. 822 (2002).

Boone v. Boozman, 217 F. Supp. 2d 938 (E.D. Ark. 2002).

Centers for Medicare and Medicaid Services. (2013, October 22). LDT and CLIA FAQs. Retrieved from www.cms.gov/Regulations-and-Guidance/Legislation/CLIA/Downloads/LDT-and-CLIA_FAQs.pdf.

Clinical Laboratory Improvement Amendments of 1988, 42 U.S.C. § 263a *et seq.*

College of American Pathologists. (2010, August 15). Comments to the Food and Drug Administration on oversight of laboratory developed tests.

Common Rule Definitions, 45 C.F.R. § 46.102.

Confidentiality of Records of Genetic Tests, New York Civil Rights Law § 79-l.

Douglas County v. Anaya, 694 N.W.2d 601, 608 (Neb. 2005).

E.M. *ex rel.* E.M. v. Pajaro Valley Unified School District Office of Administrative Hearings, 652 F.3d 999, 1004 (9th Cir. 2011).

Ferguson v. City of Charleston, 532 U.S. 67, 76 n.9 (2001).

Federal Food, Drug and Cosmetic Act, 21 U.S.C. § 301 *et seq.*

Food and Drug Administration. (1997, November 21). Analyte Specific Reagents – Final Rule. 62 Fed. Reg. 62,243, 62,249.

Food and Drug Administration (2007, June 19). Guidance for industry and FDA staff:Pharmacogenetictestsandgenetictestsforheritablemarkers.Retrievedfrom www.fda.gov/downloads/MedicalDevices/DeviceRegulationandGuidance/GuidanceDocuments/ucm071075.pdf.

Food and Drug Administration. (2010, June 10). Letter to 23andMe. Retrieved from www.fda.gov/downloads/MedicalDevices/ResourcesforYou/Industry/UCM215240.pdf.

Food and Drug Administration. (2014, January 17). FDA allows marketing for first-of-its-kind post-natal test to help diagnose developmental delays and intellectual disabilities in children. Retrieved from www.fda.gov/newsevents/newsroom/pressannouncements/ucm382179.htm.

Food and Drug Administration. (2014, October 3). Draft guidance for industry, Food and Drug Administration staff, and clinical laboratories. Framework for regulatory oversight of laboratory developed tests (LDTs). Retrieved from www.fda.gov/downloads/MedicalDevices/DeviceRegulationandGuidance/GuidanceDocuments/UCM416685.pdf.

Food and Drug Administration. (2010, June 17). Oversight of Laboratory Developed Tests – Public Meeting & Request for Comments. 75 Fed. Reg. 34,463.

Food and Drug Administration. Premarket approval. Retrieved www.fda.gov/Medicaldevices/Deviceregulationandguidance/Howtomarketyourdevice/Premarketsubmissions/Premarketapprovalpma/Default.Htm.

Food and Drug Administration. (2013, November 22). Warning letter to 23andMe. Retrieved from www.fda.gov/iceci/enforcementactions/warningletters/2013/ucm376296.htm.

Food and Drug Administration. (2017, January 13). Discussion Paper on Laboratory Developed Tests (LDTs). Retrieved from www.fda.gov/downloads/MedicalDevices/ProductsandMedicalProcedures/InVitroDiagnostics/LaboratoryDevelopedTests/UCM536965.pdf.

Food and Drug Administration Safety and Innovation Act, Pub. L. No. 112–144, § 1143.

Genetic Information Privacy Act, Illinois Compiled Statutes ch. 410 § 513.

Hamburg, M. (2013, June 2). Commissioner's address, Annual Meeting of the American Society of Clinical Oncology. Retrieved from www.fda.gov/NewsEvents/Speeches/ucm354888.htm.

Individuals with Disabilities Education Act, 20 U.S.C. § 1400 *et seq.*

Individuals with Disabilities Education Act Regulations, 34 C.F.R. pt. 300 *et seq.*

Larry P. by Lucille P. v. Riles, 793 F.2d 969, 976 (9th Cir. 1984).

Lewin, T. (2010, May 18). College bound, DNA swab in hand. *New York Times*, p. A14. Retrieved from www.nytimes.com/2010/05/19/education/19dna.html?_r=1&.

Massachusetts General Laws ch.76 § 15.

McCarthy v. Boozman, 212 F. Supp. 2d 945 (W.D. Ark. 2002).

Minnesota Statutes § 144.125, subdivision 4.

Nebraska Revised Statutes § 71–519.

Newborn Screening Saves Lives Reauthorization Act of 2014, H.R. 1281.

New Jersey v. T.L.O., 469 U.S. 325, 336–37 (1985).

Pergament, D. (2013, Spring). What does choice really mean? Prenatal testing, disability, and special education without illusions. *Health Matrix*, 23, 55–117.

Perrone, M. (2014, January 17). FDA approves genetic test that screens infants for predictors of mental disabilities. *The Associated Press*. Retrieved from www.ctvnews.ca/health/fda-approves-genetic-test-that-screens-infants-for-predictors-of-mental-disabilities-1.1644306#.Uw9UiFoshj8.email.

Sanders, R. (2010, August 12). U.C. Berkeley alters DNA testing program. Retrieved from http://newscenter.berkeley.edu/2010/08/12/dna_change/.

(2010, September 10). Tempest in a spit cup. Retrieved from http://newscenter.berkeley.edu/2010/09/10/tempest_over_dna_testing/.

Sarata, A., & Johnson, J. (2014, March 27). Regulation of clinical tests: In vitro diagnostic (IVD) devices, laboratory developed tests (LDTs), and genetic tests. Retrieved from www.fas.org/sgp/crs/misc/R43438.pdf.

Shuren, J. (2010, July 22). Statement before the Subcommittee on Oversight and Investigations, U.S. House of Representatives. Retrieved from www.fda.gov/newsevents/testimony/ucm219925.htm.

Schmerber v. State of California, 384 U.S. 757, 767–68 (1996).

Secretary's Advisory Committee on Genetics, Health & Society. (2008). U.S. system of oversight of genetic testing: A response to the charge of the Secretary of Health and Human Services. Retrieved from http://osp.od.nih.gov/sites/default/files/SACGHS_oversight_report.pdf.

Skinner v. Railway Labor Executives Assn., 489 U.S. 602, 618 (1989).

Slaughter, L. et al. (2013, August 9). Letter to the Honorable Sylvia Mathews Burwell, Director, Office of Management and Budget. Retrieved from www.pharma-medtechbi.com/~/media/Supporting%20Documents/The%20Gray%20Sheet/39/33/OMB%20Letter%20080913.pdf.

Spiering v. Heineman, 448 F. Supp. 2d 1129 (D. Neb. 2006).

US Constitution, Amendments 1, 4, 14.

U.S. Department of Health & Human Services. (2017, January 19). Federal Policy for the Protection of Human Subjects (the "Common Rule") Final Rule. 82 Fed. Reg. 7,149.

US House of Representatives, Report No. 114 – xxx, Agriculture, Rural Development, Food and Drug Administration, and Related Agencies Appropriations Bill, 2017 (2016, April 19).

Vernonia School District 47J v. Acton, 515 U.S. 646 (1995).

12

Ethical Risks and Remedies in Social-Behavioral Research Involving Genetic Testing

CELIA B. FISHER

The Human Genome Project and rapid technological advances in genetic testing have begun to enrich the contributions of educational and social-behavioral research to understanding the unique and interacting roles of genetic and environmental factors in academic achievement, education-related traits, and responsivity to educational and development-promoting interventions (Haworth & Plomin, 2012). For example, in just the past few years, studies have identified specific gene variants associated with academic achievement and intelligence test scores, self-perceived ability, general anxiety, and math cognition and school engagement (Calvin et al., 2012; Davies, Marioni, Liewald, Hill, & Hagenaars, 2016; Greven, Harlaar, Kovas, Chamorro-Premuzic, & Plomin, 2009; Kraphol et al., 2014; Laet, Colpin, Leeuwen, Noortgate, & Claes, 2016; Rimfeld, Kovas, Dale, & Plomin, 2015; Wang et al., 2014). Research on gene–environment interactions within the family environment has also shown that both quality of family relations and parenting style may interact with underlying genetic factors to influence children's school performance and related mental health conditions. Studies have indicated a moderating effect of the *HTR2A* gene on the association between maternal education and children's grade point averages (Keltikangas-Järvinen et al., 2010). Research also indicates that genetic factors associated with how children experience chaotic home life moderates family influences on school achievement (Hanscombe, Haworth, Davis, Jaffee, & Plomin, 2011) and that interactions between maltreatment and genotype are predictive of ADHD (Kim-Cohen et al., 2006; Wilmot et al., 2016).

With knowledge gleaned from gene–environment interaction studies, prevention scientists have begun incorporating identified genetic markers into trial studies in order to better understand the underlying mechanisms influencing differential response success in targeted interventions. Research has indicated that children with genetic susceptibility to conditions such as substance and alcohol abuse, sexual risk behaviors, and externalizing behaviors not only have the poorest outcomes in uncontrolled environments but also respond most strongly to prevention programs (Brody, Chen, Beach, Philibert, & Kogan, 2009; Bakermans-Kranenburg, van IJzendoorn, Mesman, Alink, & Juffer, 2008; Schlomer et al., 2015). The promise of susceptibility theory and emerging technologies for studying candidate gene × environment (cG×C) interactions is beginning to be applied to better our understanding of individual variations in children's responses to specific school-based interventions. These studies are designed to address academically relevant behavioral and mental health problems, especially within urban centers where academic risks associated with behavioral problems are particularly prevalent (Institute of Education Sciences, 2012; Kellam, 2008). For example, Musci et al. (2013) found longitudinal effects of two elementary school-based interventions to be moderated by the brain-derived neurotropic factor (*BDNF*) gene linked with aggression and impulsive behaviors. In light of such findings, the complex susceptibilities resulting from gene–environment interactions must be viewed not merely as vulnerabilities to disorder development but as potentially valuable sensitivities that may increase prevention success (Belsky & Pluess, 2009).

INFORMATIONAL RISK IN GENOMIC RESEARCH

Differential sensitivity intervention research holds out the promise for understanding variation in response to school-based interventions and the possibility of individualized, targeted prevention strategies. However, recent research suggests that hundreds of genes may be responsible for the heritability of school-relevant behavioral problems, that different gene variants can influence similar psychologically and educationally relevant phenotypes, and that for complex psychological traits genetic effects may account for only a small proportion of individual variance (Plomin & Davis, 2009). Moreover, most prevention research will include "susceptivity" variants that are neither necessary nor sufficient to predict individual responses to intervention (Beskow & Burke, 2009).

The high level of inter-individual genetic variability and within-person processes, the relatively small sample size of available study populations at risk for specific school-related health disorders in particular contexts, and the variability of program components implemented in natural settings limit the efficiency and statistical power of intervention evaluation studies and even those results that appear significant are not often replicated

(Henderson, 2008). Thus, while gene–environment interactions can help to elucidate factors placing certain individuals at increased risk for school-related problems, many prevention studies involving genetic information will lack clinical or personal relevance to individual study participants. Similarly, research on genetic factors influencing school achievement can lead to the over or underestimation of intervention responsivity for specific, already marginalized groups, which in turn can influence policy cost–benefit estimates for funding of interventions in underserved communities. Thus the goods and harms of personal or public disclosure is a critical issue for ethical deliberation.

The field's mixed success in implicating specific candidate genes with the development of psychological and behavior disorders (Salvator & Dick, 2015; Schlomer, Cleveland, Vandenbergh, Fosco, & Feinberg, 2015), and the probabilistic nature of genetic influences within the context of gene–environment interactions point to ethical challenges as largely "informational" (Beskow et al., 2009). This chapter will illuminate these ethical challenges as well as steps that can be taken to ensure the responsible conduct of research involving genetic testing. The first two sections address the need to incorporate principles of genetic literacy into informed consent practices and the unique ethical issues that arise for guardian permission and child assent procedures in cross-sectional and longitudinal studies and research involving data depositories and secondary analysis. Next are sections on the tension between ensuring adequate privacy protections and the risks and benefits of disclosing research derived personal genetic information to individual participants and their family members. The section that follows focuses on ethical challenges of disseminating the results of susceptibility and intervention responsivity studies, with particular attention to the potential impact on marginalized populations. The chapter concludes with a discussion of the value of participant and community perspectives in enhancing scientific validity and the responsible conduct of research.

Guardian Permission and Genetic Literacy

Under federal regulations (Department of Health and Human Services, 2009), the autonomy and privacy rights of minors are protected by the requirement that informed guardian permission is obtained before a child can participate in research. In addition to understanding the purpose and nature of research and participant rights and protections, guardian permission for research involving use of a child's genetic information requires attention to the guardian's level of genetic literacy. Fisher and McCarthy

(2013) define "genetic literacy" as the degree to which prospective participants or their guardians are familiar with and can apply information about
the use of genetic data in research to make informed, rational, and voluntary participation decisions. Genetic literacy is of particular importance in
school-based research incorporating genetic data because new technologies, rapid scientific advances, and evolving federal law can outpace the
average person's knowledge, expectations, and ability to process informed
consent information.

ENHANCING GENETIC LITERACY

Enhancing genetic literacy during the guardian permission process should
include explanation about: (1) the level of severity and how genetic risk
or intervention responsivity is studied; (2) the probabilistic nature of
genetic and environmental influences on the specific school-based abilities and behaviors under investigation; (3) the purpose and nature of
random assignment to intervention and control conditions, including an
explanation of the nature, rationale for and risks and benefits of random
assignment (within or across schools); (4) how genetic information will be
collected from the child; (5) whether analysis of specimens will be limited
to specific candidate genes or genome-wide associations that may give rise
to incidental findings; (6) where and for how long genetic material will be
stored and when destroyed; (7) limitations on the right to withdraw their
child's data once it has been analyzed or deidentified; (8) whether personal
genetic information is to be disclosed to the guardian and the rationale for
the disclosure decision; and (9) the availability of genetic counseling (see
Annas, 1995; Fisher & McCarthy, 2013; Wolf, Bouley & McCulluch, 2010, for
additional discussion of these options). Guardians asked to provide permission for pedigree or genetic linkage studies should also be informed about
the family privacy implications of this research, including the fact that the
child's genetic profile may reveal private genetic information about them
and that during the course of the study it may be determined that some
family members are not in fact genetic relatives.

Child Assent, Longitudinal Studies, and Data Repositories

In addition to guardian permission, the developing autonomy and privacy rights of children are further protected through regulations requiring
that provisions are made for soliciting the assent of the child, when he or
she is capable (DHHS 2009). Under federal regulations child assent may

be waived if the IRB determines that the capability of some or all of the children is so limited that they cannot reasonably be consulted or that the intervention or procedure involved in the research holds out a prospect of direct benefit that is important to the health or well-being of the children and is available only in the context of the research. Child assent should never be sought if the child's dissent would be overridden by guardian permission. In such cases children with sufficient cognitive maturity should be informed about relevant procedures and their preferences considered (Masty & Fisher, 2008).

When obtaining assent, investigators should consider ways to create a goodness-of-fit between children's level of maturation and consent enhancing educational approaches that lead to age-appropriate understanding of research procedures and protections and genetic literacy (Fisher, 2003, 2006a). In determining appropriate assent procedures, investigators should consider: (1) child participants' current cognitive capacity to understand and emotional readiness to make participation decisions about the issues posed by the specific research problems and design, and (2) the potential for information regarding genetic risk probabilities to create identity confusion, distress or self-fulfilling prophecies (Fisher & McCarthy, 2013).

LONGITUDINAL STUDIES

Longitudinal studies are a critical tool for examining potential genetic factors that may drive differential impact of school-based interventions (Musci et al., 2013). Such studies involve multiple testing periods over months or years that can include age-appropriate shifts in data collection methods from parent and teacher reports, to behavioral observations, to self-reporting procedures as participants mature (Fisher, 2006a) on academic performance, interpersonal interactions, and other school-related environmental contexts. In some studies, genetic information is collected retrospectively following an intervention (e.g., Brody et al., 2009). Given the lapses between data collection periods, the changing nature of data collection, and children's developing genetic literacy and consent capacities, the rights of child participants in longitudinal studies are best protected by viewing guardian permission and child assent as ongoing processes repeated at appropriate periods (Fisher, 2006a). Reconsent procedures should be conducted when: (1) it is reasonable to assume that participants or their guardians would not remember essential features of the study to ensure their continued informed, rational and voluntary consent; (2) new methods for collecting or analyzing genetic material are introduced into the research design; and (3) the length of time between testing periods is sufficient to assume cognitive and emotional maturation of child participants requiring a shift to more age-appropriate consent procedures. Since reassent procedures may lead some participants to withdraw at different phases of longitudinal research, Chen et al. (2003) suggest that researchers plan for this type of attrition in the research design phase and factor such possibility into initial sample size estimates.

Longitudinal research necessarily requires identifiers for tracking participant behaviors and linking genetic and social-behavioral data sets over time. While many investigators will have genetic material stored offsite in biobanks or other appropriate storage facilities, the identification of each participant's genetic profile will by necessity be included in their own records for data analysis. Investigators should ensure that the highest level of data security is applied to protect the confidentiality of the data and that a Certificate of Confidentiality has been obtained that protects investigators from being forced or compelled by law enforcement or subpoena to disclose personally identifiable genetic information that could damage the child's future insurability, employability, or social standing (http://grants .nih.gov/grants/policy/coc/index.htm). In addition, reconsent procedures should reiterate confidentiality protections and risks and describe any new challenges to privacy that may have arisen and how these challenges have been addressed. Potential confidentiality risks can be minimized once data collection has been completed, as the use of deidentifying procedures can effectively disassociate participants with identifiable personal information. However, deidentification alone cannot fully eliminate privacy concerns, as results regarding participants identified at risk for school behavioral problems within specific demographic or geographic populations may increase the likelihood of individual or group identification (Fisher, 2006a; Fisher & McCarthy, 2013). Thus investigators should carefully consider how to continue to protect participant identities in published materials and dissemination of results to schools and the public.

STORAGE OF GENETIC DATA IN REPOSITORIES

Long-term data storage and data analytic technologies have and will continue to undergo rapid advances. These advances contribute to new ways of analyzing previously unidentified genetic markers and linking such markers to social-behavioral data years after a data collection has been completed. At the same time, rapidly emerging technologies mean that original deidentification data security protections may become obsolete. When identifiable genetic material will be stored in databases and available for continued or secondary analysis, the Society for Research in Child Development (SRCD) Committee on the Common Rule, recommended that investigators should develop procedures to ensure that researchers who have access to data in the future will be bound by the best practices in confidentiality protections at the time data was collected *and* new protections as they emerge (Fisher et al., 2013). The SRCD committee also recommended that once a study is concluded the guardian permission (and child assent when applicable) obtained during the final stages of data collection should be considered a default permission for continuation of the use of data after the child has turned 18 years of age as long as (a) adequate security protections continue to be maintained and updated; (b) the level of informational risk has not increased with changes in societal attitudes or public policies; and (c) the original or other investigators will not be seeking to link the archival data set with new data collection (Fisher et al., 2013).

In consideration of the above, appropriate guardian permission and child assent procedures should explain: (1) how confidentiality of genetic material stored in databases will be protected both during the study and in the future; (2) how the use of data stored in biobanks may in the future differ from the study's original focus on specific gene candidates or specific mental health risks; (3) whether children, when they become legal adults, will be notified as to where their biological materials are stored and conditions in which they will or will not have a right to reconsent or withdraw permission for further use of their data at that time; and (4) as discussed further in the section below, whether steps will be taken to disclose to children at age eighteen personal genetic information relevant to their current or future mental health.

Sharing Pediatric Research-Derived Genetic Information with Parents

The increasing ability of research studies to include next-generation sequencing (see Kornilov, this volume) is raising complex questions regarding which genomic results to return to research participants. In medicine the rapid adoption of genomic testing technologies is creating a powerful shift toward the promotion of studies that can be applied to genetically personalized medical treatments. This in turn has created considerable debate in the medical community on the blurring of research and practice ethics in the decision to provide research participants with individual genetic information related to the aims of the study or to incidental findings not directly related to study objectives (Henderson, Juengst, King, Kuczynski, & Michie,

2012). While educational and social-behavioral research has not yet seen the same pressures as personalized medicine, many of the recent publications on children's genetic susceptibility to community and school-based interventions suggest the application of findings may someday result in risk prevention and development-promoting programs tailored to genetic group characteristics.

Educational and social-behavioral genomic research involving children present unique ethical challenges for determining when it is ethically appropriate to share the child's genetic information with parents. Restrictions on a minor's legal and cognitive ability to consent to research participation and the requirement for guardian permission will often lead to situations in which the child does not play a role in decisions regarding whether investigators will share results of genetic testing with parents. When guardians are provided their child's research-derived personal genetic information, family members will be privy to one of the most private elements of individuality – one's genetic makeup, of which the child is unaware and may remain unaware as he or she gets older (Fisher, 2006a; Fisher & McCarthy, 2013).

Investigators have been rightly cautious about sharing individual research-generated genetic or social-behavioral results with parents for reasons based on ethical principles of scientific integrity and the duty to do no harm (Fisher, 2013; Fisher & McCarthy, 2013; Fisher, Higgins-D'Alessandro, Rau, Kuther & Belanger, 1996). At present research on genetic susceptibility to school-based interventions are exploratory in nature and even when results demonstrate significant associations the relevance to an individual child's academic achievement or psychological well-being is unknown or uncertain (Janvier & Farlow, 2014). In this context sharing genetic information with no known practical relevance is not ethically justified and may cause harm by leading parents to become overprotective or overly pessimistic about the child's future, resulting in a restriction of activities and opportunities the child might have otherwise been afforded (Fisher, 2002; Wilfond & Friedman Ross, 2009).

Informing parents about a child's genetic vulnerability to school problems or behavioral or mental health disorders discovered during research participation can be ethically appropriate when there is adequate empirical data demonstrating that individual gene profiles are linked to critical aspects of academic performance or school-related mental health disorders *and* that interventions exist that can reduce vulnerability or enhance development (Grandjean & Sorsa, 1996; Jarvick et al., 2014; Ravitsky & Wilfond, 2006). The growth of next-generation sequencing and long-term

biobanking and re-analysis also raises disclosure questions regarding incidental findings – evidence of genetic risk that was not the primary focus of
the research investigation.

THE RIGHT-NOT-TO-KNOW

There is consensus in the medical community that adult research participants have a right to decline the receipt of genomic results, even when
such knowledge might be critical to their health care (Jarvik et al., 2014).
However, there is no similar consensus on whether child research participants have such a similar right, especially in situations in which their assent
to participate is not required. The public often perceives genetic markers to
be more deterministic than other risk factors in predicting human potential (Austin & Honer, 2005; Kendler, 2005). Thus a unique ethical concern
in sharing a child's personal genetic information with their family members is the way such findings can influence the child's own identity or the
identity assigned by others (Juengst, 2004). Further, disclosure of genetic
risk, however probabilistic, may inflict participants and family members
with the burden of information that could result in the restriction of the
child's future educational or employment opportunities or economic costs
of health insurance.

Some have argued that guardian authority to control the genetic risk
information about themselves that their child receives deprives children of
the opportunity afforded adult participants to refuse to be informed about
their genetic risk status, and violates the child's right to refuse invasive data
collection procedures, or their right to withhold information from others
that may be detrimental to their self-interests (Fisher, 2006a; Grandjean and
Sorsa, 1996). Feinberg (1992) has called children's moral claim to accept or
reject knowledge that will affect their lives a "right to an open future" and
that no adult (whether a parent or investigator) should make decisions that
irrevocably limit that right (Fisher & McCarthy, 2013; Wilfond & Friedman
Ross, 2009). The right-not-to-know is particularly relevant when genetic
information analyzed by the investigator is incidental to or secondary to the
condition under study. In such cases, neither the guardian nor the child has
had the opportunity to provide a preference for disclosure. When making a
disclosure decision, researchers should consider the potentially irreversible
risk to the child's self-concept, and social, economic, and psychological future
wellbeing. When research includes the possibility of uncovering incidental
findings about a genetic or mental health condition that may have potentially important health and treatment implications, guardian permission

should include prospectively soliciting such a preference via checkboxes on consent documents (Berkman & Hull, 2014; Fisher, Higgins-D'Allesandro, Rau, Kuther, & Belanger, 1996; Levy et al., 2010). In soliciting parent preferences, investigators need to pay careful attention to wording. In a study on parent perspectives on the return of research results, Ziniel et al. (2014) found that while almost all parents expressed a desire to receive genetic research results in response to a general question, almost half chose to receive only a subset of research results when presented with specific types of disclosure options.

DETERMINING WHEN TO DISCLOSE GENETIC INFORMATION TO GUARDIANS

Fisher and McCarthy (2013) have made several recommendations for assisting investigators studying genetic influences on children's behaviors in decisions regarding ethical justification for sharing a child's personal genetic information with guardians. First, investigators need to determine whether there is sufficient scientific data from prior research or their own investigation to indicate a reliable genetic association with and high predictive validity for school achievement, related psychological or behavioral disorders, or responsivity to intervention. If this first criterion is met, investigators need to consider whether the genetic condition indicates a serious disorder that can be treated or prevented and weigh the probability of goods derived from guardian action against the probability of social, educational, or economic stigmatization for the child or family members. In research involving initially asymptomatic child participants, researchers need to determine whether there is a clinically validated threshold of serious genetic risk that will call for alerting guardians or older youth if such evidence emerges (see also Jarvick et al., 2014).

If the potential benefits of disclosing information to guardians outweigh the risks of harm, investigators need to consider several other ethical procedures to follow. First, they must determine if minor participants have the cognitive, experiential and emotional capacity to understand and respond adaptively to information regarding their individual genetic risk; or if not whether it is ethically appropriate to develop a policy shared with parents and children during informed consent that provides the child access to the genetic information when he or she reaches legal majority. Second, when explaining personal risk they need to communicate in language guardians and participants can understand the probabilistic nature of gene influences and what is known about the relative contribution of genes and environment in terms of what is communicated. Finally, the disclosure needs to include appropriate genetic counseling and referral services available to assist guardians' and child participants' ability to make informed future decisions regarding the child's genetic risk.

Epigenetic changes that influence and are influenced by social and environmental conditions early in human life can have effects throughout the lifecourse (Heijmans et al., 2008). Increasingly, school-based research utilizes longitudinal designs to identify the developmental trajectories of mental health disorders and the single or joint effects of heredity and experience including substance abuse disorders (Beach, Brody, Lei, & Philibert, 2010; Benner, Kretsch, Harden, & Crosnoe, 2014), aggressive behavioral disorders (Frazzetto et al., 2007), and attentional disorders (Kim-Cohen et al., 2006). By its nature, such research often involves asymptomatic children who will or will not develop the school-related problems under investigation or who will develop a problem previously not known to be associated with a genetic variant. In some cases assessment instruments or interviewing techniques may yield information on developmental risk or harmful behaviors that have previously been undetected. The decision to take no action when confronted with data on a child's individual risk may be ethically justified when research measures were not designed for individual diagnostic validity. However, when research measures include clinically validated thresholds for mental health disorders, investigators need to determine: (a) the probability and magnitude of psychological, social, legal, or economic harm to the child if confidentiality is protected or if information is disclosed; (b) the availability of empirically evaluated treatments or services to support a disclosure decision; and (c) whether an adolescent participant's maturity and available community supports suggest a referral for services rather than disclosure would provide the greatest benefit (Fisher, 2002, 2003; Fisher & Goodman, 2009; Fisher et al., 1996).

Social Justice and Genetic Research

Along with the contributions that researchers have made to understanding the intersecting roles of genetic and contextual factors in explaining children's academic achievement and mental health, the current shift to genetics as a major explanatory framework for responsivity to prevention runs the risk of reinforcing negative social stereotypes about individual and family characteristics that may inadvertently serve to maintain entrenched social inequities based on class and race/ethnicity (Fisher, Busch, Jopp, & Brown, 2012; Fisher & McCarthy, 2013).

DEFINITIONS OF RACE AND ETHNICITY

The scientific value and validity of genetic research findings aimed at examining and preventing risk across various racial/ethnic/cultural groups is threatened by the absence of clear definitions of what these terms mean, how they are continuously shaped and redefined by social and political forces, and in disregard for the historical coupling of these terms with political beliefs regarding the inherent superiority of a particular group (Duster, 2005; Fisher et al., 2002; Goldenberg et al., 2013). There is growing consensus that broadly worded racial/ethnic classifications to describe research populations, is an inaccurate and insufficient measure of genetic variation and an inadequate stand-in for unmeasured social, cultural, and economic contextual factors that play a powerful role in mental health disparities (Collins, Green, Gutterman, & Guyer, 2003; Ossorio & Duster, 2005; Smedley et al., 2005). It is also important to recognize that participants' self-reported racial/ethnic identity incorporates the same complexity of biological, cultural, psychological and behavioral constructs as those created by society (Beskow & Burke, 2009; Bonham, Warshauer-Baker, & Collins, 2005). The use of broad "panethnic" labels to categorize participants can also mask the influence on mental health of racial, ethnic, mixed race, or bicultural self-identification (Fisher, 2014a; Fisher & Wallace, 2000; Fisher et al., 2002; Trimble & Fisher, 2006).

Failing to adequately recruit or describe members of different racial/ethnic groups can be particularly problematic in genetic association or candidate gene x environment studies on academic achievement and related childhood behavior that typically test for a correlation between a trait and genetic variation to identify candidate genes that directly contribute to or moderate or are moderated by the effect of environmental factors on that trait (Goldenberg et al., 2013; Liu et al., 2013). Associations that emerge as statistically significant may be spurious when participant populations lack homogeneity in genetic background due to population substructure (nonrandom mating between subgroups in a population) or admixture, e.g., when "ethnic/racial" populations in the United States have mixed genetic ancestry, e.g., European and African (Ziv & Burchard, 2003). The inclusion in a research study of only genetically and ancestrally homogenized populations can result in limited applicability to prevention science if the trait under consideration is more, less, or as relevant to admixed populations.

To ensure scientific validity and avoid ascribing race as a proxy for gene variation in academic and mental health disparities, investigators need to carefully consider and explicitly describe the theoretical, empirical, and social frameworks driving the definitions of race, ethnicity, or culture used to select participant populations and to frame their research findings within the context of continuously changing scientific and societal conceptions of these definitions (Fisher, 2014b; Fisher et al., 2002; Fisher & McCarthy, 2013; Ali-Khan, Krakowski, Tahir, & Daar, 2011). This includes reports in publications about why specific racial/ethnic populations were selected, how racial/ethnic descriptors were chosen (e.g., self-reports of ethnic identity, on racial/ethnic/immigration history of family members), and the geographic region in which data was collected. These latter data are important owing to rapidly changing shifts in demography and migration (Takezawa et al., 2014).

PUBLIC ATTITUDES TOWARD GENETIC DETERMINISM

Public dissemination of research results often oversimplify or overstate the role of genetic factors, raising the possibility that ideas about human identity in popular culture are being overwhelmed by "genetic essentialism" (Henderson, 2008). Investigators disseminating research results should be alert to the public tendency toward stereotyping the role of genetic factors (Collins et al., 2003). The emerging gene-context risk and intervention paradigm also influences professional and societal definitions of normal and abnormal, and of health and disease. When a genetic influence on risk is reported, a new "genetized" disease may be created (Lippman, 1991) that leads to placing individuals in diagnostic categories based on risk, irrespective of the likelihood of developing this disorder or the absence of symptoms. Investigators conducting school-based interventions that require screening students for mental health or behavioral problems that qualify them for inclusion in a study on genetic risk, must also be wary of the potentially stigmatizing effect participation selection may have on the child's future school experiences. Teachers who are aware of the purpose of the study and the child's selection may see the student's participation as "scientific" confirmation of their subjective evaluations of the child's lack of academic potential, or believe it provides a genetic explanation for the child's disruptive classroom behaviors. In addition to stating the limitations in population generalizability of findings in a scholarly publication, Takezawa et al.

(2014) recommend that investigators prepare an easily understandable summary of research findings to enhance the accuracy of media reporting, and whenever possible point out mistakes or misinterpretations of findings.

CONCLUSION

This is an exciting time for increasing our understanding of how genes and environment interact to influence children's school performance and mental health. Technologies are moving rapidly toward increasing our ability to identify molecular pathways that underlie the effects of known risk factors, discover potent environmental influences on the emergence of gene–environment academic achievement and risk, and develop preventive interventions focusing on children identified at genetic risk. These advances lead to uncharted ethical territory involving informational risks associated with the potential for how children understand themselves, how they are treated by family and teachers, and the design of "personalized" social-behavioral interventions.

Investigators conducting genetic-based prevention research need to avoid the very human tendency toward genetic determinism in the design, interpretation and dissemination of research (Beskow & Burke, 2009) and take active steps to discourage "genetic reductionism" in the interpretation of their work by others. Ethical caution will be particularly critical for dissemination of research that runs the risk of attributing prevention responsivity or lack thereof to genetic dispositions in marginalized groups and to policies that ignore systemic institutional contributions (Fisher et al., 2012). Investigators working in schools on studies that target specific students for intervention need to also be aware of the potential stigmatization of selected participants and take steps to ensure that school personnel and family members understand the probabilistic nature of genetic and environmental influences on behavior and when the magnitude of a gene–disorder association is not sufficiently large to justify assumptions about a child's future potential.

The Importance of Guardian and Participant Perspectives

Protecting and enhancing the rights and welfare of children and youth participating in academic and mental health research involving genetic testing does not occur in a vacuum, but within the context of public, federal, and institutional priorities and policies. When identifying appropriate protections for participant rights and welfare, investigators most often draw upon their own moral compasses, the advice of colleagues, and recommendations from institutional review boards (IRBs); however, the views of the scientific establishment may differ from the moral values and ethical concerns of participants and their families (Fisher, 1999). Gathering perspective opinions from guardians, child participants, and their cultural, economic or health communities can help investigators identify factors within scientifically worthwhile studies that may inadvertently cause participant or family stress, produce individual or group stigma, or lead to negative consequences due to unintended breaches of confidentiality or privacy (Anderson et al., 2012; DuBois et al., 2011, Fisher & Masty, 2006; Fisher et al., 2002, 2012; O'Sullivan & Fisher, 1997) Engaging participants in dialog about the responsible conduct of research presents an opportunity to correct biases and misperceptions that arise when research ethics decision making is restricted to the perspectives of investigators, IRB members and regulators. To correct an institutionally biased imbalance in moral perspectives, Fisher and her colleagues (Fisher, 1999, 2002, 2006b, 2014b; Fisher & Goodman, 2009; Fisher & Ragsdale, 2006) recommend using a co-learning model of community–participant dialog in which investigators share with prospective participants their expertise about the scientific method, extant empirical knowledge and federal and professional standards for the responsible conduct of research, and participants share their insider expertise on the social validity of the study, how they have or will react to planned procedures, the subjective risk–benefit balance of the research, and the moral and cultural frameworks informing their perspectives. This sharing of expertise leads to the construction of recruitment, data collection, and dissemination procedures that can avoid exacerbating stigmatizing school or community attitudes toward participants and their family members or accidentally creating public awareness of participants' maladaptive behaviors. Engaging guardians, school professionals, and, when appropriate, prospective participants in the initial design of informed consent and confidentiality procedures can also help to ensure guardian permission and child assent practices that enhance genetic literacy, understanding of research procedures and their research rights, and fit the content and format to participants' cultural traditions, and can assist in identifying unforeseen participant research risks and, where possible, fitting potential research benefits to participant needs.

Enhancing the Responsible Conduct of Research

This chapter has discussed ethical challenges and sought to identify recommendations that can enhance the responsible conduct of educational and social-behavioral genomic research conducted with children and youth. Investigators contributing to our understanding of youth development can protect the rights and welfare of participants by: (a) creating opportunities for enhancing guardian and participant genetic literacy and informed consent; (b) remaining up-to-date on continuing advances in data analytic technologies that can jeopardize deidentification and procedures that can be used to protect against such threats; and (c) taking steps to develop guidelines and monitor data that may suggest a severity of genetic risk or mental health disorder that requires disclosure. As the racial, ethnic, and cultural geography continues to rapidly change, investigators can also address the societal implications of public interpretation of research results and their application to public policy through attention to population substructure and admixture in their research questions, recruitment strategies, analyses, and dissemination. Federal funding for genetic research is often driven by economic and political concerns (e.g., urban crime, school misconduct) that may have little to do with the concerns and social circumstances of targeted groups and the social value of prevention research on genetic risk for school behaviors and mental health disorders may be viewed very differently by researchers and the individuals or groups they wish to study (Fisher, 1999; Fisher & Wallace, 2000; Ziniel et al., 2014). By engaging communities in dialog on the goods and harms of research to the real-world contexts in which participants live, researchers can enhance scientific validity and participant protections in ways that reflect the values and merits the trust of participants and their families.

REFERENCES

Ali-Khan, S. E., Krakowski, T., Tahir, R., & Daar, A. S. (2011). The use of race, ethnicity and ancestry in human genetic research. *The HUGO Journal, 5*, 47–63.

Anderson, E. E., Solomon, S., Heitman, E., DuBois, J. M., Fisher, C. B., Kost, R. G., et al. (2012). Research ethics education for community-engaged research: A review and research agenda. *Journal of Empirical Research on Human Research Ethics, 7*, 3–19.

Annas, G. J., Glantz, L. H., & Roche, P. A. (1995). Drafting the Genetic Privacy Act: Science, policy, and practical considerations. *Journal of Law, Medicine & Ethics, 23*, 360–366.

Austin, J., & Honer, W. (2005), The potential impact of genetic counseling for mental illness. *Clinical Genetics, 67*, 134–142. doi:10.1111/j.1399-0004.2004.00330.x

Bakermans-Kranenburg, M. J., Van IJzendoorn, M. H., Mesman, J., Alink, L. R. A., & Juffer, F. (2008). Effects of an attachment-based intervention on daily cortisol moderated by DRD4: A randomized control trial on 1-3-year-olds screened for externalizing behavior. *Development & Psychopathology, 20*, 805–820.

Beach, S. H., Brody, G. H., Lei, M., & Philibert, R. A. (2010). Differential susceptibility to parenting among African American youths: Testing the DRD4 hypothesis. *Journal of Family Psychology, 24*(5), 513–521.

Belsky, J., & Pluess, M. (2009). Beyond diathesis stress: Differential susceptibility to environmental influences. *Psychological Bulletin, 135*(6), 885–908.

Benner, A. D., Kretsch, N., Harden, K. P., & Crosnoe, R. (2014). Academic achievement as a moderator of genetic influences on alcohol use in adolescence. *Developmental Psychology, 50*(4), 1170–1178. doi:10.1037/a0035227

Beskow, L. M., & Burke, W. (2009). Ethial issues in genetic epidemiology. In S. S. Coughlin, T. L. Beachamp, & D. L. Weed, *Ethics and epidemiology* (2nd ed., pp. 182–193). New York: Oxford University Press.

Berkman, B. E., & Hull, S. C. (2014). The "right not to know" in the genomic era: Time to break from tradition? *American Journal of Bioethics, 14*, 28–31.

Bonham, V. L., Warshauer-Baker, E., & Collins, F. S. (2005). Race and ethnicity in the genome era: The complexity of the constructs. *American Psychologist, 60*(1), 9–15.

Brody, G. H., Chen, Y., Beach, S. R. H., Philibert, R. A., & Kogan, S. M. (2009). Participation in a family-centered prevention program decreases genetic risk for adolescents' risky behaviors. *Pediatrics, 124*(3), 911–917.

Calvin, C. M., Deary, I. J., Webbink, D., Smith, P., Fernandes, C., Lee, S. H., et al. (2012). Multivariate genetic analyses of cognition and academic achievement from two population samples of 174,000 and 166,000 school children. *Behavior Genetics, 42*(5), 699–710. doi:10.1007/s10519-012-9549-7

Chen, D. T., Miller, F. G., & Rosentstein, D. L. (2003). Ethical aspects of research into the etiology of autism. *Mental Retardation and Developmental Disability Research Reviews, 9*, 48–53.

Collins, F. S., Green, E. D., Guttmacher, A. E., & Guyer, M. S. (2003). A vision for the future of genomics research. *Nature, 422*, 835–884.

Davies, G., Marioni, R. E., Liewald, D. C., Hill, W. D., & Hagenaars, S. P. (2016). Genome-wide association study of cognitive functions and educational attainment in UK biobank. *Molecular Psychiatry, 21*, 758–767.

Department of Health and Human Services (DHHS). (2009). *Title 45 Public welfare, Part 46, Code of Federal Regulations: Protection of human subjects.* Washington, DC: Government Printing Office. Retrieved from www.hhs.gov/ohrp/humansubjects/guidance/45cfr46.html

DuBois, J., Baily-Curch, B., Bustillos, D., Campbell, J., Cottler, L., Fisher, C. B., et al. (2011). Ethical issues in mental health research: The case for community engagement. *Current Opinion in Psychiatry, 24*(3), 208–214.

Duster, T. (2005). Race and reification in science. *Science, 307*(5712), 1050–1051.

Feinberg, J. (1992). The child's right to an open future. In J. Feinberg (Ed.), *Freedom & fulfillment. Philosophical essays* (pp. 76–97). Princeton, NJ: Princeton University Press.

Fisher, C. B. (1999). Relational ethics and research with vulnerable populations. *Reports on research involving persons with mental disorders that may affect decision-making capacity.* Papers commissioned by the National Bioethics Advisory Commission, 2, 29–49. Retrieved October 26, 2009, from www.bioethics.gov/reports/past_commissions/nbac_mental2.pdf.

(2002). Participant consultation: Ethical insights into parental permission and confidentiality procedures for policy relevant research with youth. In R. M. Lerner, F. Jacobs, & D. Wertlieb (Eds.), *Handbook of applied developmental science. 4* (pp. 371–396). Thousand Oaks, CA: Sage.

(2003). Adolescent and parent perspectives on ethical issues in youth drug use and suicide survey research. *Ethics & Behavior, 13*(4), 303–332.

(2006a). Privacy and ethics in pediatric environmental health research – Part I: Genetic and prenatal testing. *Environmental Health Perspectives, 114*, 1617–1621.

(2006b). Privacy and ethics in pediatric environmental health research – Part II: Protecting families and communities. *Environmental Health Perspectives, 114*, 1622–1625.

(2013). Confidentiality and disclosure in non-intervention adolescent risk research. *Applied Developmental Science. 17*(2), 88–93.

(2014a). Multicultural ethics in professional psychology practice, consulting, and training. In F. T. L. Leong (Ed.), *APA handbook of multicultural psychology, 2* (pp. 35–57). Washington, DC: American Psychological Association.

(2014b). HIV prevention research ethics: An introduction to the special issue. *Journal of Empirical Research on Human Research Ethics, 9*(1), 1–5.

Fisher, C. B., Brunnquell, D. J., Hughes, D. L., Liben, L. S., Maholmes, V., Plattner, S., et al. (2013). Preserving and enhancing the responsible conduct of research involving children and youth: A response to proposed changes in federal regulations. *Social Policy Report, 27*(1), 1, 3–15.

Fisher, C. B., Busch-Rossnagel, N. B., Jopp, D. S., & Brown, J. L. (2012). Applied developmental science, social justice and socio-political well-being. *Applied Developmental Science, 16*(1), 54–64.

Fisher, C. B., & Goodman, S. J. (2009). Goodness-of-fit ethics for non-intervention research involving dangerous and illegal behaviors. In D. Buchanan, C. B. Fisher, & L. Gable (Eds.), *Research with high-risk populations: Balancing science, ethics, and law* (pp. 25–46). Washington, DC: APA Books.

Fisher, C. B., Higgins-D'Allesandro, A., Rau, J. M. B., Kuther, T., & Belanger, S. (1996). Referring and reporting research participants at risk: Views from urban adolescents. *Child Development, 67*, 2086–2099.

Fisher, C. B., Hoagwood, K., Boyce, C., Duster, T., Frank, D. A., Grisso, T., et al. (2002). Research ethics for mental health science involving ethnic minority children and youth. *American Psychology, 57*, 1024–1040.

Fisher, C. B., & McCarthy, E. L. (2013). Ethics in prevention science involving genetic testing [Special issue]. *Prevention Science, 14*(3), 310–318.

Fisher, C. B., & Ragsdale, K. (2006). A goodness-of-fit ethics for multicultural research. In J. Trimble & C. B. Fisher (Eds.), *The handbook of ethical research with ethnocultural populations and communities* (pp. 3–26). Thousand Oaks, CA: Sage.

Fisher, C. B., & Masty, J. K. (2006). Community perspectives on the ethics of adolescent risk research. In B. Leadbeater, T. Reicken, C. Benoit, M. Jansson, & A. Marshall (Eds.), *Research ethics in community-based and participatory action research with youth* (pp. 22–41). Toronto: University of Toronto Press.

Fisher, C. B., & Wallace, S. A. (2000). Through the community looking glass: Re-evaluating the ethical and policy implications of research on adolescent risk and sociopathology. *Ethics and Behavior, 10*(2), 99–118.

Frazzetto, G., Di Lorenzo, G., Carola, V., Proietti, L., Sokolowska, E., Siracusano, A., et al. (2007) Early trauma and increased risk for physical aggression: The moderating role of MAOA. *PLoS One, 2*(5), e486.

Goldenberg, A. J., Hartmann, C. D., Morello, L., Brooks, S., Colón-Zimmermann, K., & Marshall, P. A. (2013). Gene-environment interactions and health inequalities: Views of underserved communities. *Journal of Community Genetics, 4*(4), 425–434. doi:10.1007/s12687-013-0143-3

Grandjean, P., & Sorsa, M. (1996). Ethical aspects of genetic predisposition to environmentally-related disease. *Science of the Total Environment, 184*, 37–43.

Greven, C. U., Harlaar, N., Kovas, Y., Chamorro-Premuzic, T., & Plomin, R. (2009). More than just IQ: School achievement is predicted by self-perceived abilities – but for genetic rather than environmental reasons. *Psychological Science, 20*(6), 753–763.

Hanscombe, K. B., Haworth, C. M. A, Davis, O. S. P., Jaffee, S. R., & Plomin, R. (2011). Chaotic homes and school achievement: A twin study. *Journal of Child Psychology and Psychiatry and Allied Disciplines, 52*(11), 1212–1220. doi:10.1111/j.1469-7610.2011.02421.x

Haworth, C. M. A., & Plomin, R. (2012). Genetics and education: Toward a genetically sensitive classroom. In K. Harris, S. Graham, & T. Urdan (Eds.), *APA educational psychology handbook: Volume 1 – Theories, constructs and critical issues* (pp. 529–599). Washington, DC: American Psychological Association. doi:10.1037/13273-018

Heijmans, B. T., Tobi, E. W., Stein, A. D., Putter, H., Blauw, G. J., Susser, E. S., et al. (2008). Persistent epigenetic differences associated with prenatal exposure to famine in humans. *Proceedings of the National Academy of Sciences, 105*(44), 17046–17049.

Henderson, G. E. (2008). Introducing social and ethical perspectives on gene-environment. *Sociological Methods & Research, 37*, 251–276.

Henderson, G. E., Juengst, E. T., King, N. M. P., Kuczynski, K., & Michie, M. (2012). What research ethics should learn from genomics and society research: Lessons from the ELSI Congress of 2011. *Journal of Law, Medicine and Ethics, 40*(4), 1008–1024. doi:10.1111/j.1748-720X.2012.00728.x

Institute of Education Sciences. (2012). Fast facts. Retrieved from http://nces.ed.gov/fastfacts/display.asp?id=16.

Janvier, A., & Farlow, B. (2014). Arrogance-based medicine: Guidelines regarding genetic testing in children. *American Journal of Bioethics, 14*(3), 15–16. doi:10.1080/15265161.2013.879951

Jarvik, G. P., Amendola, L. M., Berg, J. S., Brothers, K., Clayton, E. W., Chung, W., et al. (2014). Return of genomic results to research participants: The floor, the ceiling, and the choices in between. *American Journal of Human Genetics, 94*(6), 818–826. doi:10.1016/j.ajhg.2014.04.009

Juengst, E. T. (2004). FACE facts: Why human genetics will always provoke bioethics. *Journal of Law, Medicine & Ethics, 32*(2), 267–275.

Kellam, S. G., Brown, C. H., Poduska, J. M., Ialongo, N., Wang, W., Toyinbo, P., & Wilcox, H. C. (2008). Effects of a universal classroom behavior management program in first and second grades on young adult behavioral, psychiatric, and social outcomes. *Drug and Alcohol Dependence, 95,* S5–S28. doi:10.1016/j.drugalcdep.2008.01.004.

Keltikangas-Jarvinen, L., Jokela, M., Hintsanen, M., Salo, J., Hintsa, T., Alatupa, S., et al. (2010). Does genetic background moderate the association between parental education and school achievement? *Genes, Brain and Behavior, 9*(3), 318–324. doi:10.1111/j.1601-183X.2009.00561.x

Kendler, K. S. (2005). "A gene for…": The nature of gene action in psychiatric disorders. *American Journal of Psychiatry, 162,* 1243–1252.

Kim-Cohen, J., Caspi, A., Taylor, A., Williams, B., Newcombe, R., Craig, I. W., et al. (2006). MAOA, maltreatment, and gene-environment interaction predicting children's mental health: New evidence and a meta-analysis. *Molecular Psychiatry, 11,* 903–913.

Krapohl, E. Rimfeld, K., Shakeshaft, N. G., Trzaskowski, M., McMillan, A., Pingault, J. B., et al. (2014). The high heritability of educational achievement reflects many genetically influenced traits, not just intelligence. *Proceedings of the National Academy of Sciences, 111*(42), 15273–15278.

Laet, S., Colpin, H., Leeuwen, K., Noortgate, W. Claes, S. Janssens, A., et al. (2016). Transactional links between teacher-student relationships and adolescent rule-breaking behavior and behavioral school engagement: Moderating role of a dopaminergic genetic profile score. *Journal of Youth & Adolescence, 45,* 1226–1244.

Levy, D., Splansky, G. L., Strand, N. K., Atwood, L. D., Benjamin, E. J., Blease, S., et al. (2010). Consent for genetic research in the Framingham Heart Study. *American Journal of Medical Genetics. Part A, 152A*(5), 1250–1256. doi:10.1002/ajmg.a.33377

Lippman, A. (1991). Prenatal genetic testing and screening: Constructing needs and reinforcing inequities. *American Journal of Law and Medicine, 17,* 15–50.

Liu, J., Lewinger, J. P., Gilliland, F. D., Gauderman, W. J., & Conti, D. V. (2013). Confounding and heterogeneity in genetic association studies with admixed populations. *American Journal of Epidemiology, 177*(4), 351–360. doi:10.1093/aje/kws234

Masty, J., & Fisher, C. B. (2008). A goodness of fit approach to parent permission and child assent pediatric intervention research. *Ethics & Behavior, 13,* 139–160.

Musci, R. J., Bradshaw, C. P., Maher, B., Uhl, G. R., Kellam, S. G., & Ialongo, N. S. (2013). Reducing aggression and impulsivity through school-based prevention programs: A gene by intervention interaction. *Prevention Science, 15*(6), 831–840. doi:10.1007/s11121-013-0441-3

Ossorio, P. N., & Duster, T. (2005). Race and genetics: Controversies in biomedical, behavioral, and forensic sciences. *American Psychologist, 60,* 115–128.

O'Sullivan, C., & Fisher, C. B. (1997). The effect of confidentiality and reporting procedures on parent-child agreement to participate in adolescent risk research. *Applied Developmental Science, 1,* 185–197.

Plomin, R., & Davis, O. S. (2009). The future of genetics in psychology and psychiatry: Microarrays, genome-wode association, and non-coding. *Journal of Child Psychology and Psychiatry, 50*, 63–71.

Ravitsky V. & Wilfond B.S. (2006). Disclosing individual genetic results to research participants. *American Journal of Bioethics, 6*, 8–17.

Rimfeld, K., Kovas, Y., Dale, P. H., & Plomin, R. (2015). Pleioropy across academic subjects at the end of compulsory education. *Science Reports, 5*, 1–12.

Salvatore, J. E., & Dick, D. M. (2015). Gene-environment interplay: Where we are, where we are going. *Journal of Marriage and Family, 77*(2), 344–350. doi:10.1111/jomf.12164

Schlomer, G. L., Cleveland, H. H., Vandenbergh, D. J., Fosco, G. M., & Feinberg, M. E. (2015). Looking forward in candidate gene research: Concerns and suggestions. *Journal of Marriage and Family, 77*(2), 351–354. doi:10.1111/jomf.12165

Smedley, A. & Smedley, B. D. (2005). Race as biology is fiction, racism as a social problem is real: Anthropological and historical perspectives on the social construction of race. *American Psychologist, 60*, 16–26.

Takezawa, Y., Kato, K., Oota, H., Caulfield, T., Fujimoto, A., Honda, S., et al. (2014). Human genetic research, race, ethnicity and the labeling of populations: Recommendations based on an interdisciplinary workshop in Japan. *BMC Medical Ethics, 15*(33), 1–5. doi:10.1186/1472-6939-15-33

Trimble, J., & Fisher, C. B. (2006). Our shared journey: Lessons from the past to protect the future. In J. Trimble & C. B. Fisher (Eds.), *The handbook of ethical research with ethnocultural populations and communities* (pp. xv–xxix). Thousand Oaks, CA: Sage.

Wang, Z., Hart, S. A., Kovas, Y., Lukowski, S., Soden, B., Thompson, L. A., et al. (2014). Who is afraid of math? Two sources of genetic variance for mathematical anxiety. *Journal of Child Psychology & Psychiatry, 55*(9), 1056–1064.

Wilfond, B., & Friedman Ross, L. F. (2009). From genetics to genomics: Ethics, policy, and parental decision-making. *Journal of Pediatric Psychology, 34*(6), 639–647.

Wilmot, B., Fry, R., Smeester, L., Musser, E. D., Mill, J., & Nigg, J. T. (2016). Methylomic analysis of salivary DNA in childhood ADHD identifies altered DNA methylation in VIPR2. *Journal of Child Psychology & Psychiatry, 57*, 152–160.

Wolf, L. E., Bouley, T. A., & McCulloch, C. E. (2010). Genetic research with stored biological materials: Ethics and practice. *IRB: Ethics & Human Research, 32*(2), 7–18.

Ziniel, S., Savage, S. K., Huntington, N., Amatruda, J., Green, R. C., Weitzman, E. R., et al. (2014). Parents' preferences for returning results in pediatric genomic research. *Public Health Genomics, 17*(2), 105–114. doi:10.1159/000358539

Ziv, E., & Burchard, E. G. (2003). Human population structure and genetic association studies. *Pharmacogenomics, 4*(4), 431–441. doi:10.1517/phgs.4.4.431.22758

13

Development of the Personal
Genomics Industry

JORGE L. CONTRERAS AND VIKRANT G. DESHMUKH

INTRODUCTION

Today, numerous commercial services offer genetic testing, genotyping, and genome sequencing services both to medical providers and directly to the public. Twenty-five years ago, such offerings would have been unthinkable, both in terms of cost and medical practice. This chapter describes the development of the personal genomics industry and its evolving business models and goals.

A recent study has found that, between the beginning of the Human Genome Project (HGP) in 1990 and 2004, 470 different private firms in twenty-five countries began to offer products and services based on genomic technology or data (Wiechers, Perin, & Cook-Deegan, 2013). These commercial offerings included the sale of genome-sequencing equipment and reagents, the development of drugs and vaccines using genomic data (pharmacogenomics), testing for disease susceptibility, and a host of data-driven applications (Wiechers et al., 2013). Researchers at the Battelle Memorial Institute have estimated (Battelle, 2013) that by 2012, more than 47,000 individuals in the United States alone were employed by the genomics industry, which they divide into six primary sectors (bioinformatics, testing, reagents, instrumentation, R&D, and pharmacogenomics). Thus, although the commercial genomics industry has existed for only twenty-five years, it is large and complex, with widely varying product offerings, business models, and strategies.

In this chapter, we focus only on those segments of the commercial genomics industry that offer products and services to consumers, either directly or through intermediaries such as physicians, genetic counselors, or testing laboratories, a sector that we collectively refer to as "personal genomics" (Khoury et al., 2009). Our focus will further be limited to those products and services that provide genetic or genomic *information* to consumers, as opposed to drugs, vaccines, or treatment regimens that may have been discovered using genomic information, or the administration of which may be influenced by a recipient's genomic characteristics. But even limited thus, the field is complex and multifaceted.

GENETICS AND GENOMICS

Each strand of human DNA consists of approximately 3.2 billion paired nucleotide bases, the sum of which is referred to as the human "genome." Some fraction of these bases is organized into contiguous sub-units called "genes," ranging in size from as few as a hundred to more than two million base pairs. It is currently estimated that human DNA contains approximately 20,000 genes. Genes are responsible for the inheritance of traits from one generation to the next and encode the many proteins responsible for biochemical functions within the cell. Each human genome is approximately 99.9 percent identical, and very small differences account for much of the variability in human physical and physiological traits (Feuk, 2006; NHGRI, 2014a), along with epigenetic variation (Issa, 2002).

While hypotheses regarding the existence of biochemical mechanisms for the heredity of human traits have existed since the nineteenth century, it was not until Watson and Crick's landmark discovery of the structure of DNA in 1953 that modern genetics was born. Throughout the 1960s and 1970s, genetic studies became increasingly sophisticated, until by the mid-1970s, technology had evolved to a point at which researchers could begin to identify individual genes responsible for diseases such as cystic fibrosis and Huntington disease (MacDonald et al., 1993; Rommens, 1989). Even so, each of these discoveries took years of painstaking work and a measure of good luck to achieve. In 1983 a revolutionary new process for determining the order of bases within a DNA molecule emerged. This process, called the polymerase chain reaction (PCR), enabled researchers to unravel the genetic code of humans and many other organisms more efficiently (Mullis, 1987), and earned its inventor, Kary Mullis at Cetus Corporation, the Nobel Prize.

The advent of PCR technology soon gave rise to an ambitious plan to sequence not only genes identified with specific diseases but the entire human genome (Watson & Jordan, 1989). The decision by the US government to form an international consortium to undertake this monumental project in the late 1980s signaled the birth of the field now known as genomics, the study not of individual genes but of the entire genome.

THE HGP AND DATA-DRIVEN BUSINESS MODELS

For most of the twentieth century, the bulk of genetic research was carried out at academic institutions and government laboratories. The scale, sophistication and speculative nature of such research generally made it unattractive to commercial enterprises well into the first decade of the twenty-first century. The HGP was officially launched in 1990 as a joint project of the US National Institutes of Health (NIH) and the US Department of Energy, with support from the Wellcome Trust in the United Kingdom and funding agencies in the United Kingdom, France, Germany, and Japan.

The initial stages of the HGP were devoted to refining the instrumentation needed to sequence the human genome and undertaking pilot sequencing projects on simpler organisms such as the *E. coli* bacterium (Durham, 1997). By 1998, after the expenditure of nearly $2 billion, the HGP prepared to begin work on the human genome. Then, in May of that year, J. Craig Venter, a former NIH scientist, famously proclaimed that he, with substantial commercial backing, would utilize state-of-the-art equipment, together with much of the HGP's publicly released data, to sequence the entire human genome in only three years, a full four years before the HGP was scheduled to complete its work (Wade, 2000). Venter's announcement sent a shock wave through the genomics community and led to a widely publicized "arms race" between his new company, Celera Genomics, and the HGP (Roberts, 2001).

Ultimately, a truce was brokered by the journal *Science* (Jasny, 2013) and, in June 2000, Francis Collins, the leader of the HGP, and Venter were invited to the White House to announce jointly that a "first draft" of the human genome had been completed (Wade, 2000). In his remarks, President Clinton emphasized the role of commercial enterprises in the new field of genomics, declaring that "[w]e must discover the function of these genes and their protein products, and then we must rapidly convert that knowledge into treatments that can lengthen and enrich lives. I want to emphasize that biotechnology companies are absolutely essential in this endeavor" (Clinton, 2000).

Unlike the public HGP, Celera's goal in sequencing the human genome was not to release genomic data to the public, but to profit from licensing this data to pharmaceutical and biotechnology companies. Like many companies, Celera sought to facilitate the emerging pharmacogenomics industry, which, it was hoped, would develop new and more effective drugs and vaccines guided by genomic information. Thus, while the public HGP regularly uploaded its DNA sequence data to the public GenBank database maintained by the US National Library of Medicine, Celera made its data available solely on its commercial web site. The company allowed scientists from non-profit and academic institutions to access the data without charge, but required researchers who wished to use the data for commercial purposes to enter into a license agreement (Marshall, 2000).

Celera's approach outraged much of the scientific community and led to a highly publicized debate over private ownership of human genome data (Marshall, 2000). When Celera and the HGP announced the completion of their first drafts of the human genome in 2000, Celera committed that it would make its data broadly available, though it still required payment by commercial users (Marshall, 2001b). Celera's subscription-based data business was ultimately unsuccessful, and in 2005, the company exited the business and released its genomic data to GenBank (Kaiser, 2005). It is likely that Celera's data-driven business failed, in large part, due to the competing public efforts that released large quantities of similar, if not identical, data to the public.

In the 1990s Celera was just one of several firms that attempted to capitalize on potentially profitable uses of genomic sequence data. Even before the completion of the HGP, firms including Incyte Pharmaceuticals in Palo Alto, California, and Human Genome Sciences in Rockville, Maryland, were actively pursuing a business strategy of patenting, and seeking to license, short gene sequences known as expressed sequence tags (ESTs) and other genetic data (Marshall, 2001a). By the time the first EST patent was issued to Incyte in 1998, that company alone had filed patent applications claiming more than 1.2 million DNA sequence fragments (Murry, 1999). These early efforts were eventually thwarted by a combination of factors including judicial and administrative decisions limiting the patentability of ESTs (Demaine, 2002), as well as earlier efforts to place large quantities of similar EST data into the public domain. The most notable of these earlier efforts was the "Merck Gene Index," a project led by pharmaceutical giant Merck in collaboration with Lawrence Livermore National Laboratory and Washington University (Contreras, 2011). By 1998, the Merck Gene Index had released over 800,000 ESTs through GenBank, substantially limiting the ability of companies to license the same or similar data to the pharmaceutical industry in a profitable manner.

A similar effort known as the SNP Consortium was conducted in conjunction with the HGP by a group of pharmaceutical and information technology companies, with additional support from the Wellcome Trust. The SNP Consortium sponsored research to identify and map genetic markers referred to as "single nucleotide polymorphisms" (SNPs), which it then released to the public domain (Contreras, 2011; Holden, 2002). SNPs are common genetic variations that occur throughout a person's DNA, some of which are important in the study of human health (Genetics Home Reference, 2014a). The SNP Consortium ultimately mapped 1.4 million SNPs, all of which were free from patents and made publicly accessible without charge.

It is likely that public data release efforts by the HGP and associated private sector projects such as the Merck Gene Index and the SNP Consortium limited the market for general purpose genomic databases, though, as we discuss below, there may still be substantial value in mutation databases associated with particular diseases.

THE GENETIC TESTING SECTOR

The business models discussed in the preceding section were based on the private generation or collection of large quantities of genomic data, with the goal that this data then be licensed on a commercial basis to pharmaceutical and biotechnology companies engaged in pharmacogenomics research. A different business model developed from more narrowly focused efforts to identify genetic mutations associated with particular diseases. As of this writing, more than 37,000 different genetic tests are available from 1,600 laboratories and clinics in the United States for nearly 4,000 genetic disorders (GeneTests, 2014). The vast majority of these genetic tests are available to patients only through a physician or clinical setting.

The first disease-specific genetic test was developed for cystic fibrosis, a debilitating condition that affects approximately 30,000 children and adults in the United States, and 70,000 worldwide (Cystic Fibrosis, 2014). A mutation in the *CFTR* gene that is strongly correlated with cystic fibrosis was discovered in 1989 by teams at the University of Michigan, Johns Hopkins University, and the Hospital for Sick Children in Montreal. The discovery was patented, but each of the institutions holding patent rights elected to license its rights on a non-exclusive basis (meaning that the patent holder permitted multiple laboratories to perform testing, rather than only a single laboratory) (Chandrasekharan, Heaney, James, Conover, & Cook-Deegan, 2010). As a result, in 2009 sixty-three different labs in the United States performed testing for *CFTR* mutations at relatively affordable prices (ibid.).

A different market structure developed with respect to tests in which controlling patents were licensed on an exclusive basis. For example, Athena Diagnostics of Marlborough, Massachusetts has exclusive rights to exploit patents covering genetic diagnostic tests for mutations of several genes associated with hearing loss and Alzheimer disease. In the case of the *APOE* gene, whose particular variants are associated with late-onset Alzheimer disease, Athena holds an exclusive license from Duke University and has actively sought to prevent other laboratories from offering testing based on this mutation (Skeehan, Heaney, & Cook-Deegan, 2010). One exception arose in 2008, when Athena licensed a small company called Smart Genetics to offer *APOE* testing directly to consumers via a mail-in kit in conjunction with telephone consultation. But Smart Genetics discontinued its APOE testing program after only a few months, apparently following the intervention of Duke University (Skeehan et al., 2010).

The best known example of commercial genetic testing is probably that of Myriad Genetics. Strong correlations between mutations in the *BRCA1/2* genes and breast/ovarian cancer in certain populations were identified in the early 1990s by research groups at the University of Utah and the US National Institutes of Environmental Health Sciences and two corporations: Myriad and Oncormed, Inc.. Myriad eventually obtained control over the patents covering the most significant *BRCA* mutations and elected not to license third parties to perform testing. As a result, by 2000 Myriad was the only US laboratory performing full *BRCA* testing, for which it charged approximately $3,000. Controversy arose due to the cost of Myriad's test and the fact that many US healthcare payors, including the federal Medicare system, declined to cover *BRCA* testing costs in many cases. Myriad's assertion of exclusive rights in the *BRCA* testing market led a coalition of patients, advocacy groups, physicians, and clinics to mount a legal challenge against Myriad's patents in 2009. The case led to a landmark 2013 decision by the US Supreme Court invalidating all of the challenged patent claims and establishing that DNA sequences occurring in the human body are not eligible for patent protection in the United States (Ass'n for Molecular Pathology v. Myriad Genetics, Inc., 2013).

Shortly after the *Myriad* ruling, a number of additional firms, including laboratory giants Ambry Genetics and Laboratory Corporation of America, entered the market for *BRCA* testing by offering tests at significantly lower price points than Myriad (Ambry, 2013; LabCorp, 2013). At this writing, Myriad is engaged in litigation seeking to enforce a new set of patents against these and other firms.

The degree to which *Myriad* will impact patents held by Athena and other diagnostic testing companies, and its overall effect on the genetic testing industry in the United States, is still unclear. Several commentators believe that single-gene patents of the type held by Myriad and others would, in any event, be of little use to exclude competitors from performing whole genome sequencing and other tests that involve more than the isolation of single genes (Rai & Cook-Deegan, 2013).

Outside of the United States, patents covering genetic testing appear to be on more solid footing. The Australian Supreme Court, for example, recently upheld a number of Myriad's patents covering BRCA mutations (Cancer Voices Australia v Myriad Inc [2014] FCAFC, 2014). The crux of the dispute was whether isolated DNA including complementary DNA (cDNA) was patent eligible. In holding both varieties of DNA to be patentable, the Australian Court arrived at a different conclusion than the US Supreme Court, which had held that isolated DNA was not patent eligible because it was a product of nature. In Europe, Myriad's *BRCA1* and *BRCA2* patents were challenged and substantially upheld by the European Patent Office. However, during successive proceedings, the claims were reduced in scope from the entire *BRCA1* gene sequence to individual mutations, cloning vectors and host cells, and methods for detecting mutations in *BRCA2* (Matthijs, Huys, Van Overwalle, & Stoppa-Lyonnet, 2013). The resulting claim scope in Europe is substantially narrower than in either the United States or Australia.

DIRECT-TO-CONSUMER GENOMICS

Following the completion of the HGP, rapid advances in gene sequencing technology coupled with a precipitous drop in the price of sequencing equipment, led to the emergence of a new market: direct-to-consumer (DTC) genomic testing. DTC genomics vendors offer services directly to the public, typically without the involvement of a healthcare provider. These services typically involve the consumer's submission of a DNA sample (usually a saliva swab) to a designated facility, and the vendor's performance of one of three types of analysis: genotyping, exome sequencing, or whole genome sequencing (see Kornilov, this volume).

The process of scanning a genome for known genetic markers or SNPs is called *genotyping*. Genotyping is typically performed by comparing a sample of human DNA to known markers on a commercially available panel or "chip." Vendors such as Illumina currently offer genotyping panels that contain over one million known markers. The presence or absence of the tested markers can support inferences about the subject's risk for certain diseases, ancestry, and physiological characteristics.

While genotyping can offer a substantial amount of information to consumers, even the largest million-marker chips contain only a tiny fraction (i.e., 0.03 percent) of the 3.2 billion base pairs comprising the full human genome. Sequencing the entire genome of an individual (whole genome sequencing) is a much larger task. The HGP spent approximately $3.8 billion over a decade to sequence the genomes of multiple individuals to develop a common genomic profile for human species. The cost of genome sequencing has dropped precipitously in the years since the completion of the HGP. For years, the "holy grail" of genome sequencing has been the *$1,000 genome*: the ability to sequence an entire human genome for a cost of only $1,000 (Davies, 2010). The National Human Genome Research Institute (NHGRI) tracks costs associated with whole genome sequencing at centers that it funds, and most recently estimated that this cost had dropped to between the $4,000 and $5,000 (Wetterstrand, 2014). And in 2014, Illumina announced that it "broke the sound barrier" of human genomics by enabling the $1,000 genome with a new sequencing platform (Illumina, 2014). Nevertheless, it is likely that the $1,000 genome still remains a few years away for the average consumer.

Meanwhile, companies like Gene by Gene offer a middle-road between genotyping and whole genome sequencing in the form of *exome* sequencing, which involves sequencing only those protein-coding fragments of an individual's genome. These 19,000 or so protein-coding regions represent about 1 percent of the entire human genome (Ng et al., 2009), but can support medically valuable incidental/secondary findings (Green et al., 2013). Thus, exome sequencing avoids the high cost of whole genome sequencing, while offering more information than genotyping.

Ancestry Information

Genetic ancestry testing (GAT) or genetic genealogy relies on an examination of variations in DNA to infer where a person's ancestors may have originated (Genetics Home Reference, 2014b). GAT is based on the idea that closely related individuals, families, or populations share more genetic variations with one another. GAT can complement traditional genealogical research by providing missing pieces of ancestry information, but cannot itself provide detailed information about specific ancestors.

One of the first organizations to offer GAT to the public was the Genographic Project, a collaboration between the National Geographic Society and IBM. The Genographic Project began to distribute genetic testing kits in 2005 for a $99 charge (Davies, 2010). The Project gave its participants information regarding their likely ancestry, along with historical perspectives on the migration paths followed by ancient peoples (Wells, 2005). The kits are still offered for $99, and at the time of this writing over

688,000 individuals from more than 140 countries have participated in this project.

A number of additional commercial GAT tests are offered for approximately $99 by companies including 23andMe, Ancestry.com, and Family Tree DNA (23andMe, 2014a; Ancestry, 2014; Family Tree DNA, 2014).

Nutritional Genomics

The field of nutritional genomics pertains to interactions between genes and the environment, particularly nutrients, chemicals, and other matter introduced into the body as part of dietary food consumption (Dudley, 2013). For example, Familial Hypercholesterolemia (FH), a condition characterized by severely elevated LDL cholesterol and increased risk of coronary artery disease, can be traced to mutations in the *APOB*, *LDLR*, and *PCSK9* genes. Individuals who carry these mutations are managed medically by a combination of diet and lifestyle changes, along with pharmacotherapy (Youngblom, 2014).

Some companies have tried to capitalize on public interest in nutritional genomics through consumer testing products. For example, Evidence Based Nutrition (EBN) based in Chula Vista, California, offers customers a "DNA Nutrition Action Plan" that "makes nutritional and lifestyle recommendations based on an understanding of the individual's unique genetic profile" (Spicer, 2008). EBN primarily sells nutritional supplements.

Sciona, Inc., headquartered in Aurora, Colorado, offers the MyCellf test, which it claims to be "designed to provide dietary and lifestyle recommendations gleaned from individual genetic data" (Davies, 2010). Author Kevin Davies took the MyCellf test and reported that Sciona recommended that he get more exercise, cut back on alcohol and caffeine, and increase his daily intake of vegetables (ibid.). Predictably, claims such as these from nutrigenetic testing firms have been criticized for being too generic at best, and at times misleading (Kutz, 2006).

DTC and Health Information

While using genomic information to provide information about ancestry and nutrition has commercial potential, these applications pale in comparison to the detection of health-related information hidden within the genomes of consumers. Home-based genetic paternity testing, as well as genetic testing for specific health conditions, have been available for years. Only with the increasing accessibility of genotyping and genomic sequencing, however, have consumers had the opportunity to obtain large quantities of data regarding their genomic make-up and its potential health implications.

In 1998, deCODE Genetics, a Reykjavík, Iceland-based biopharmaceutical company, lobbied for and won exclusive rights to revamp Iceland's national health record system, which included records containing patient diagnoses, treatments, results, complications, etc., as well as biospecimens, genotypes and genealogical records for the entire nation (Chadwick, 1999). deCODE discovered several genes associated with diseases in humans, and offered lab tests for various disease genotypes, in collaboration with Hoffman-LaRoche, Merck, and others, although a majority of these tests were not marketed to consumers.

In 2007, deCODE launched a web-based DTC genomics service called deCODEme, which was the first commercial offering of its kind. For less than $1,000, deCODE would analyze approximately 600,000 sites in a customer's DNA having a known influence on both common physical traits such as baldness, eye color, and tongue-rolling, as well as risk for certain diseases including diabetes and several cancers (Davies, 2010). deCODE conceptualized its offering as an educational service rather than a medical diagnostic, explaining "[w]e are not providing people with a genetic test. We are only allowing them to compare their genomes to the genomes of those who in the literature have been described as having a risk of a disease. We encourage people not to make medical decisions on the basis of results of this, but we point people to the possibility of taking results of this to their doctors" (Davies, 2010).

In 2006, David Agus, an oncologist, and Dietrich Stephan, a neuroscientist, founded Navigenics, Inc. Navigenics offered DTC tests for eighteen common illnesses including diabetes, heart disease, obesity, and certain cancers (Davies, 2010; Hall, 2007). One of the distinguishing features of Navigenics' approach was its heavy emphasis on genetic counseling. Recognizing the complexity and sensitive nature of individual genomic information, Navigenics required a "telephone consult" with one of its genetic counselors before giving a customer full access to his or her DTC testing results. Customers could also download their raw genomic data after signing a waiver.

The best known purveyor of DTC genomic testing today is probably 23andMe, a California-based company founded by Linda Avey, Paul Cusenza, and Anne Wojcicki in 2006 (23andMe, 2014a). In November 2007, 23andMe joined deCODE and Navigenics in offering the public DTC testing that included ancestry and health information (Hanahan,

2007). Wojciki, the spouse of Google co-founder Sergey Brin, explained, "23andMe is designed to provide our customers with scientifically accurate, high-quality information about their own genetic code in a format that is easy to understand and use." 23andMe's bold approach and knack for publicity quickly grabbed the public imagination, and *Time Magazine* named it the 2008 "Invention of the Year" (TIME Magazine, 2008). 23andMe has also attracted investment from other pharmaceutical and biotechnology companies since its founding, and recently announced a $60 million deal with biotech giant Genentech. A significant aspect of this investment will give Genentech access to the 23andMe database containing genotypic records of over 800,000 customers (Herper, 2015). In February 2015, the US Food and Drug Administration (FDA) allowed 23andMe to market the first DTC genetic carrier test for Bloom syndrome (US Food and Drug Administration, 2015), a rare genetic disorder characterized by short stature and increased predisposition to cancer.

Even independently of 23andMe, Google's interest in the field of genomics is not to be underestimated. In February 2014, the Internet giant launched Google Genomics, a web-based application for importing, storing, searching, analyzing, and sharing individual genomic data (Gruber, 2014). Harvard professor George Church predicts that Google's data mining software will open "huge new markets in wellness and precision medicine" (ibid.).

Alongside these purveyors of broad spectrum DTC genomic information are numerous smaller firms that offer genotyping for specific non-disease traits. Among these is Richmond, Virginia based American International Biotechnology, which offers a $200 test kit that "provides athletes and parents of young sports competitors a wealth of information about their athletic strengths" (AIBioTech, 2011). Some in the scientific community have questioned the validity of the conclusions that can be drawn from such information, and have criticized the marketing approach taken by AIBioTech and others (Collier, 2012).

Comparing DTC Offerings

With at least three major DTC personal genomics options available to consumers by the late 2000s, it became important to clarify how these services differed from one another. Whereas 23andMe took a self-service approach in providing the consumer with large quantities of easy-to-use information, Navigenics and deCODE preferred that consumers interact with genetic counselors to interpret their results. And while 23andMe offered more information on complex traits, deCODE offered greater contextual information for data relating to ancestry (Davies, 2010).

The similarities and differences among DTC genomics offerings also became the object of scientific study. One such study found over 99.6 percent concordance in the SNP genotypes provided by deCODE, 23andMe and Navigenics (most likely because they all used the same commercial genotyping technology), but noted large variations in the analysis of risks reported by these companies (Imai, Kricka, & Fortina, 2011). For example, one study found that the relative risks for rheumatoid arthritis that the companies reported ranged from 0.9 to 1.85, i.e., from having a protective effect to having a deleterious effect (ibid.). Similar findings were observed in other studies, and the differences were attributed, among other things, to the methods used in characterizing the underlying populations (Kalf et al., 2014).

While academic studies have provided a scientific comparison of DTC services, members of the media have also reported their experiences with DTC testing (Dickinson, 2008). Journalist Boonsri Dickinson compared her DTC test results from all three companies and received surprisingly divergent results. Having both Asian and European ancestry, she discovered that the risk information provided by the test vendors varied widely depending on whether she identified herself as belonging to one ethnic group or the other. Such critiques have cast increasing doubt on the usefulness and informative value of many DTC genomic services.

Whole Genome Sequencing for All?

Around the time that deCODE, 23andMe, and Navigenics were marketing DTC genomic tests based on known markers, others, like Harvard professor George M. Church, ventured in a different direction. Church was one of the first individuals to have his entire genome sequenced and joined other notable figures such as James Watson in releasing his genomic data to the public. Church founded Knome Inc. in 2007 to offer whole genome sequencing services to the general public, "on the recognition that the rapidly falling price of whole genome sequencing would create substantial market need for whole genome interpretation technologies and services" (Knome, 2014). Today, that vision has largely been realized; indeed, when Knome first offered whole genome sequencing, it cost a hefty $350,000, whereas more recent cost estimates are in the $6,000 range (Eisenberg, 2013).

Other notable companies in the whole genome sequencing arena included Helicos Biosciences and Complete Genomics. Helicos was founded in 2003 by Stephen Quake from the California Institute of Technology, and Stanley Lapidus and Noubar Afeyan of Flagship Ventures. Helicos specialized in a novel technique known as Single Molecule Sequencing, which allowed the sequencing of a single DNA molecule without the need for PCR, and offered an advantage over second-generation sequencing techniques at the time (Thompson & Milos, 2011).

Complete Genomics was founded by Clifford Reid and Radoje (Rade) Drmanac in 2006. In 2009 the company began to offer whole genome sequencing services not to consumers, but to pharmaceutical and biotechnology companies and academic medical centers at a cost of approximately $5,000 per genome (Lauerman, 2009). While Knome and Complete Genomics continue to offer whole genome sequencing services as of this writing, Helicos filed for bankruptcy in 2012.

The potential of DTC whole genome sequencing is significant. What cost the HGP $3.8 billion to produce over ten years could soon be available to every man, woman, and child for less than $1,000 through a mail-in kit. The greatest challenge of DTC whole genome sequencing is how to interpret the vast quantity of genomic data that will be delivered to consumers. Modern science has only scratched the surface of understanding the myriad functions of the 3.2 billion base pairs that constitute the human genome. It will likely be many years before scientific understanding catches up with the technical ability to generate whole genome sequence data.

Market Shake-Out for DTC Vendors

The financial crisis of 2008 put an enormous strain on many industries, and several DTC genomics providers went out of business. In November 2009, two years after initially offering its DTC service, deCODE filed for bankruptcy. deCODE got a fresh start in 2012, when it was acquired by biotech giant Amgen (Amgen, 2012). It is not clear, however, whether Amgen will restart deCODE's DTC genomics service, or simply use the Icelandic company's genetic resources and database to support its existing drug discovery and development businesses. In 2012, Navigenics also ceased to offer DTC genomic services after being acquired by Life Technologies, a subsidiary of equipment manufacturer Thermo Fisher Scientific. Whereas both deCODE and Navigenics struggled, 23andMe has flourished, attracting over $100 million in investment from the likes of Google, Johnson and Johnson, and Genentech, and major private investors such as Google co-founder Sergey Brin and Digital Sky Technologies co-founder Yuri Milner (CrunchBase, 2014).

Governmental Scrutiny of DTC Testing

In 2004, the American College of Medical Genetics issued a statement that "genetic testing should be provided to the public only through the services of an appropriately qualified health care professional," warning that "the self-ordering of genetic tests by patients over the telephone or Internet [could result in harms including] inappropriate test utilization, misinterpretation of test results, lack of necessary follow-up, and other adverse consequences" (ACMG, 2004). This perspective has shaped the policy debate over DTC genomic services in the United States, and has led to significant intervention by state legislatures and federal agencies over the DTC genomics industry.

In the United States, state governments are generally responsible for regulating consumer health and safety, as well as the practice of medicine and licensure of healthcare providers within their borders. Depending on the state, laws and regulations may either expressly permit genomic DTC testing, ban such services outright, impose partial regulations, or remain silent on the issue (Berman, 2007; Dick, 2012). Illustrating this range of regulatory regimes, Virginia explicitly allows direct reporting of DTC genomic results to individuals, California and New York require physician authorization, and Colorado and Utah classify DTC genomic testing outside the practice of medicine and thus beyond state licensure regulation (Dick, 2012).

In 2008, both California and New York began to require medical licensing for DTC genomics vendors and sent cease and desist letters prohibiting further sales to consumers without medical oversight (Langreth, 2008; Magnus, Cho, & Cook-Deegan, 2009; Pollack, 2008). The targeted DTC companies objected, arguing that patients had a right to receive their genetic information, that genetic testing is not diagnostic or medical in nature, and that patients deserved direct access to testing without a physician intermediary (Magnus et al., 2009). Nevertheless, most DTC companies eventually complied with state demands and stopped accepting DNA samples from New York and California (Pollack, 2008). 23andMe, however, took a creative approach, still shipping kits to consumers in New York, but requiring them to "affirm under penalty of law that the sample for the saliva kit has not been collected in or mailed from the state of New York" (23andMe, 2014b). On the other hand, in 2010, Navigenics became one of the first DTC genomics companies to obtain a New York Clinical Laboratory Permit (Sweeney, 2010).

The US federal government has also shown an interest in the activities of DTC genomics vendors. In 2006 the US Government Accountability

Office (GAO) launched an investigation of the practices of several DTC companies, and concluded in a subsequent investigation that they provided "medically unproven disease predictions" (Kutz, 2010). GAO's investigation involved the purchase of DTC tests by a number of fictitious consumers. These test subjects received disease and risk predictions that varied across four DTC companies for the same DNA samples, and contradicted known family history and other traits in the DNA donors.

DTC genetic testing is subject to the jurisdiction of at least three separate federal regulatory agencies: the Federal Trade Commission (FTC), the Centers for Medicare and Medicaid Services (CMS), and the US FDA (NHGRI, 2014b).

The FTC regulates consumer protection and polices false and misleading advertising under the Federal Trade Commission Act. There have been numerous critiques of consumer DTC genomics advertising, pointing to its potential distortion of DTC testings' risks and benefits, its limited educational value and its potential for disseminating misinformation to the public (Gollust, Hull, & Wilfond, 2002). As a result, the FTC has scrutinized DTC genomics advertising and has issued a consumer advisory on its website about DTC genetic tests, alongside other cautionary articles about "miracle health claims" and "cancer treatment scams" (Federal Trade Commission, 2014).

CMS regulates clinical laboratory testing (excluding research) throughout the United States under the Clinical Laboratory Improvement Amendments (CLIA) of 1988, which covers the educational requirements, quality control, and proficiency testing of individual labs. In order to offer medical tests to the public and return results to consumers, laboratories performing DTC genomic testing must have an appropriate CLIA certificate (CMS, 2014).

Separately from CMS' regulation of clinical laboratories, the FDA regulates medical devices marketed in the United States under the Medical Devices Amendments to the Federal Food, Drug, and Cosmetic Act. FDA's broad oversight of medical devices has been argued to extend to laboratory-developed tests and test kits, as well as related software (Javitt & Carner, 2014).

In 2009, the FDA, which had become increasingly concerned about the distribution of genomic diagnostic information by unregulated DTC vendors, began to seek more information regarding DTC practices and tests. The major DTC genomics vendors largely sidestepped the FDA's inquiries, and matters came to a head in the summer of 2010, when the FDA issued warning letters to several vendors including 23andMe, Navigenics, and deCODE (US Food and Drug Administration, 2010). According to the FDA, the kits and services offered by these vendors fell under the Medical Devices Amendments, and these firms had neither filed for pre-market approval, nor notified the agency of their intent to commercially distribute the respective devices. On this basis, the FDA ordered these companies to discontinue marketing DTC genomic services until they received marketing authorization from the FDA, and to provide documentation about specific corrective actions they had undertaken to address these issues (ibid.).

As a result of the FDA's actions, in December 2013, 23andMe announced that it was discontinuing the "23andMe Health" personal genomics service, which had provided consumers with health-related interpretive information along with the results of its genotyping analysis, pending an FDA regulatory review (Afarian, 2013). 23andMe switched instead to providing its customers with raw SNP data (i.e., the actual genotypic test results indicating the presence of particular SNPs at particular locations along the genome), along with ancestry information. Other DTC companies that offered similar tests followed suit, limiting their offerings to raw SNP data and leaving consumers to perform their own interpretations.

Despite the current unavailability of interpretive or health information from DTC vendors, consumers who obtain SNP and other raw genomic data about themselves are not without options. Free software tools such as *Promethease* (Promethease, 2014) can generate personal genomics reports using raw sequence data based on publicly available scientific literature in the public SNPedia (Cariaso & Lennon, 2012; SNPedia, 2014). *Promethease* can import data directly from the reports offered by vendors such as 23andMe, and can also decipher raw data files that most DTC vendors provide to consumers. The actual process involves matching raw SNP data with known SNPs in SNPedia, and using that knowledge base to generate meaningful interpretation. Other websites such as interpretome.com (Karczewski, 2012), livewello.com (Livewello, 2015), and geneticgenie.org (Genetic Genie, 2015) also offer "interpretation-only" services for DTC tests similar to *Promethease*. Due to the rapid pace of discovery and the relative newness of these services, there is no gold standard. Nevertheless, one recent study found *Promethease* results to be among the most detailed, although not as user-friendly as the services originally offered by 23andMe (Regalado, 2014). Ultimately, the accuracy of these tools is only as good as the underlying data sources. SNPedia, for instance, has at least four levels of data curation, three of which are directly performed on the database itself, whereas an additional fourth level of review is dedicated to the content used by *Promethease*, and helps screen errors in original research in the original data sources (Cariaso & Lennon, 2012).

Personal Genomics Outside the United States

Though, with the exception of deCODE in Iceland, the personal genomics initiatives of US firms have received most of the attention from the Western press, such efforts are not limited to the US market. Regulators around the world have weighed the benefits and risks of genetic testing, including DTC, for several years, and legal developments in Europe bear striking similarities to those in the United States. In 2008, following the approval of protocols related to genetic testing by the Committee of Ministers of the Council of Europe, some speculated that Europe might move toward banning DTC testing altogether (Borry, 2008). In 2009, Germany enacted legislation that effectively banned DTC genomic services by requiring that genetic tests be carried out exclusively by a physician following informed patient consent (The Associated Press, 2009). France, Portugal, and Switzerland have similar legislation restricting DTC genomics, while the Netherlands only has limited restrictions on genetic tests for detecting "incurable diseases" (Borry et al., 2012). The United Kingdom and Belgium currently have no specific restrictions on DTC genomic testing (ibid.).

In the EU, DTC testing is currently covered by European Directive 98/79, which proposes objectives, but is not directly binding on member states, many of which have national legislation that covers these tests (Kalokairinou, 2014). However, a new proposed regulation, if adopted by the EU, would be binding on member states. The regulation proposes a risk classification system for medical devices including DTC tests. It further provides that agencies must evaluate claims before such tests could be marketed (ibid.). Finally, the regulation would also provide that only medical professionals can order these tests. Such a change in regulatory posture could effectively ban DTC in Europe (ibid.).

Like the United States, China has seen the emergence of a significant personal genomics industry, offering consumers both genetic testing for common health risks (especially neonatal screens), as well as more comprehensive genomic data (Jia, 2014). It has been reported that many Chinese hospitals routinely offer genomic sequencing services to healthy patients during check-ups (ibid.). The increasing popularity of personal genomic services in China has led Chinese regulators to take notice, and it is likely that greater scrutiny and regulation of the personal genomics market in China will soon arrive.

Despite these moves toward greater regulation of DTC testing, there have also been calls for scaling back the initial precautionary approach taken by regulators in light of newer studies, which show that many of the public health concerns over DTC have not materialized (Vayena, 2013). In a recent study, most participants in multiplex genetic susceptibility testing recalled their results correctly, did not interpret results in an overly deterministic way, and appreciated that genetics and behavior both contribute to disease risk (Kaphingst, 2012). In another study, subjects who underwent DTC testing exhibited no measurable short-term changes in psychological health, diet and exercise behavior, or use of screening tests (Bloss, 2011).

While commercial genomics itself knows no boundaries, and consumers from different parts of the world could potentially send their DNA samples to vendors based in other countries, until regulatory frameworks around the world catch up with the technology, the industry may need to self-regulate in the consumers' interest (Gurwitz & Bregman-Eschet, 2009). Whereas many laboratory-developed tests would arguably fall within the definition of a "device" that is subject to FDA regulation, there is a lack of data to support pre-market clearance, and the FDA cannot control off-label uses of these tests (McGuire, 2010). Moreover, many DTC tests provide predictive and long-term information, whereas post-market surveillance may provide a better long-term strategy as long all tests are analytically valid, and all clinical claims are accurate and substantiated (ibid.).

In addition, typical DTC contracts include numerous legal disclaimers and limitations of liability (Phillips, 2015). Most consumers lack the training to interpret these terms, or do not invest the time to do so. While regulation may be necessary to protect consumer interests, in the short-term DTC companies can also improve the readability of their contracts and privacy policies to enhance consumer understanding and the consent process (ibid.). The lessons learned in early adopter countries like the United States and several European nations are pertinent to consumers and companies around the world due to their impact on the availability and demand for DTC and other forms of genetic testing.

CONCLUSION

Over the past twenty-five years, the increasing speed and decreasing cost of genotyping and genome sequencing have led to the emergence of new business models based on providing genetic information directly to the public. Genetic tests for individual diseases are widely available today in many developed countries at moderate cost, and are covered by many national and private insurance plans. DTC genomic testing offered by 23andMe and similar providers has increased in popularity, but the medical value of these services remains to be proven. Such services also face increasing governmental scrutiny and regulation, either as healthcare providers or diagnostic device vendors, making their future uncertain. Whole genome sequencing, which will soon be broadly affordable, offers consumers a wealth of information, but much of the information that is delivered is not likely to be understood for years. As such, the value of whole genome sequencing, as opposed to genotyping for known disease risks, remains questionable.

The global market for personal genomic products and services is likely to evolve rapidly over the next five to ten years. DTC companies have responded to changes in the regulatory environment by emphasizing services like ancestry information to maintain their primary revenue streams. Other companies have begun to offer interpretation-only services for personal genomics, which have not yet received the same level of scrutiny as similar services previously offered by DTC companies. Finally, in addition to the primary market for these products, collaborations between DTC and major biotechnology and pharmaceutical companies could create secondary markets and augment their primary business models along the lines of 23andMe. It is not clear which of today's players, if any, will remain active in the future, or whether new players will enter the market from other sectors such as computing, information technology and online services, and whether early US players will remain dominant in the face of market entry by competitors in China and elsewhere. What is certain, however, is that the amount of genomic information available to the general public will continue to increase rapidly, and novel legal, ethical, and economic solutions will need to keep pace with this remarkable technological growth.

REFERENCES

23andMe, Inc. (2014a). Genetic kit for ancestry. Retrieved September 21, 2014, from www.23andme.com/

(2014b). Availablility in the state of New York. Retrieved October 17, 2014, from https://customercare.23andme.com/hc/en-us/articles/202907950-Availablility-in-the-state-of-New-York

American College of Medical Genetics Directors. (2004). ACMG statement on direct-to-consumer genetic testing. *Genetics in Medicine*, 6(1), 60.

Afarian, C. (2013). 23andMe, Inc. provides update on FDA regulatory review [Press release]. Retrieved October 9, 2014, from http://mediacenter.23andme.com/blog/2013/12/05/23andme-inc-provides-update-on-fda-regulatory-review/

AIBiotech. (2011). Maximize Performance, Identify Undiagnosed Risk Factors – Test Enables Individuals to Customize Workout Programs Based on Genetic Results [Press Release]. Retrieved June 12, 2017, from www.prnewswire.com/news-releases/new-genetic-test-helps-athletes-maximize-performance-identify-undiagnosed-risk-factors-120795914.html

Ambry Genetics. (2013). Ambry Genetics launches BRCA 1 & 2: Single genes and NGS panel offerings [Press release]. Retrieved October 9, 2014, from www.ambrygen.com/press-releases/ambry-genetics-launches-brca-1-2-single-genes-and-ngs-panel-offerings

Amgen. (2012). Amgen to acquire deCODE Genetics, a global leader in human genetics [Press release]. Retrieved October 9, 2014, from www.amgen.com/media/media_pr_detail.jsp?releaseID=1765710

Ancestry. (2014). DNA tests for ethnicity & genealogical DNA testing at Ancestry DNA. Retrieved October 7, 2014, from https://dna.ancestry.com/

Ass'n for Molecular Pathology v. Myriad Genetics, Inc., 133 S.Ct. 2107 (2013).

Battelle, T. P. P. (2013). *The impact of genomics on the U.S. economy.* Prepared by Battelle Technology Partnership Practice for United for Medical Research.

Berman Institute of Bioethics, Genetics and Pulic Policy Center, & Johns Hopkins University. (2007). Survey of direct-to-consumer testing statutes and regulations. Retrieved October 14, 2014, from www.dnapolicy.org/resources/DTCStateLawChart.pdf

Bloss, C. S., Schork, N. J., & Topol, E. J. (2011). Effect of direct-to-consumer genomewide profiling to assess disease risk. *New England Journal of Medicine, 364*(6), 524–534. doi:10.1056/NEJMoa1011893

Borry, P. (2008). Europe to ban direct-to-consumer genetic tests? *Nature Biotechnology, 26*(7), 736–737.

Borry, P., van Hellemondt, R. E., Sprumont, D., Jales, C. F. D., Rial-Sebbag, E., Spranger, T. M., et al. (2012). Legislation on direct-to-consumer genetic testing in seven European countries. *European Journal of Human Genetics, 20*(7), 715–721.

Cancer Voices Australia v Myriad Inc [2014] FCAFC, 115 (2014).

Cariaso, M., & Lennon, G. (2012). SNPedia: A wiki supporting personal genome annotation, interpretation and analysis. *Nucleic Acids Research, 40*, D1308–1312. doi:10.1093/nar/gkr798

Centers for Medicare & Medicaid Services. (2014). Direct Access Testing (DAT) and the Clinical Laboratory Improvement Amendments (CLIA) Regulations. Retrieved October 10, 2014, from www.cms.gov/Regulations-and-Guidance/Legislation/CLIA/Downloads/directaccesstesting.pdf

Chadwick, R. (1999). The Icelandic database – Do modern times need modern sagas? *BMJ, 319*(7207), 441–444.

Chandrasekharan, S., Heaney, C., James, T., Conover, C., & Cook-Deegan, R. (2010). Impact of gene patents and licensing practices on access to genetic testing for cystic fibrosis. *Genet Med, 12*(1), S194–211.

Clinton, B. (2000). Text of remarks on the completion of the first survey of the entire Human Genome Project. Retrieved September 15, 2014, from http://clinton5.nara.gov/WH/New/html/genome-20000626.html

Collier, R. (2012). Genetic tests for athletic ability: Science or snake oil? *Canadian Medical Association Journal, 184*(1), E43–44. doi:10.1503/cmaj.109-4063

Contreras, J. L. (2011). Bermuda's Legacy: Policy, patents and the design of the Genome Commons. *Minnesota Journal of Law, Science & Technology, 12*, 61.

CrunchBase. (2014). 23andMe – Investors. Retrieved October 7, 2014, from www.crunchbase.com/organization/23andme/investors

Cystic Fibrosis Foundation. (2014). About CF: What is cystic fibrosis? Retrieved October 12, 2014, from www.cff.org/aboutcf/

Davies, K. (2010). *The $1,000 genome: The revolution in DNA sequencing and the new era of personalized medicine.* New York: Simon and Schuster.

deCODE Genetics, Inc. (2007). deCODE launches decODEme™ [Press release]. Retrieved October 9, 2014, from www.decode.com/decode-launches-decodeme/

Demaine, L. J. F., & Xavier, A. (2002). Reinventing the double helix: A novel and nonobvious reconceptualization of the biotechnology patent. *Stanford Law Review, 55,* 303.

Dick, H. C. (2012). Risk and responsibility: State regulation and enforcement of the direct-to-consumer genetic testing industry. *St. Louis University Journal of Health Law & Policy, 6*(1).

Dickinson, B. (2008). How much can you learn from a home DNA test? *Discover Magazine.* Retrieved September 23, 2014, from http://discovermagazine.com/2008/sep/20-how-much-can-you-learn-from-a-home-dna-test

Dudley, J. T. K., & Konrad, J. (2013). Personal genomics and the environment. *Exploring personal genomics* (Kindle ed., pp. Kindle location 4178). Oxford, UK: Oxford University Press.

Durham, S. (1997). *E. coli* genome reported [Press release]. Retrieved March 8, 2015, from www.nih.gov/news/pr/sept97/nhgra-04.htm

Eisenberg, A. (2013). Genomic analysis, the office edition. *The New York Times.* Retrieved October 7, 2014, from www.nytimes.com/2013/02/03/business/knomes-new-machine-to-aid-labs-in-genomic-analysis.html?_r=0

Family Tree DNA. (2014). Family tree DNA. Retrieved October 14, 2014, from www.familytreedna.com/

Federal Trade Commission. (2014). Direct-to-consumer genetic tests. *Health & Fitness: Treatment & Cures.* Retrieved October 17, 2014, from www.consumer.ftc.gov/articles/0166-direct-consumer-genetic-tests

Feuk, L., Carson, A. R., & Scherer, S. W. (2006). Structural variation in the human genome. *Nature Reviews Genetics, 7*(2), 85–97.

GeneTests. (2014). GeneTests. Retrieved September 15, 2014, from www.genetests.org/

Genetic Genie. (2015). Methylation and detox analysis from 23andMe results. Retrieved March 22, 2015, from http://geneticgenie.org/

Genetics Home Reference. (2014a). What are single nucleotide polymorphisms (SNPs) – Genetics home reference. Retrieved October 17, 2014, from http://ghr.nlm.nih.gov/handbook/genomicresearch/snp

(2014b). *Genetics home reference handbook: Help me understand genetics.* Lister Hill National Center for Biomedical Communications, US National Library of Medicine. Retrieved October 17, 2014, from http://ghr.nlm.nih.gov/handbook.pdf.

Gollust, S. E., Hull, S., & Wilfond, B. S. (2002). Limitations of direct-to-consumer advertising for clinical genetic testing. *JAMA, 288*(14), 1762–1767. doi:10.1001/jama.288.14.1762

Green, R. C., Berg, J. S., Grody, W. W., Kalia, S. S., Korf, B. R., Martin, C. L., et al. (2013). ACMG recommendations for reporting of incidental findings in clinical exome and genome sequencing. *Genetics in Medicine*, 15(7), 565–574. doi:10.1038/gim.2013.73

Gruber, K. (2014). Google for genomes. *Nature Biotechnology*, 32(6), 508. doi:10.1038/nbto614-508a

Gurwitz, D., & Bregman-Eschet, Y. (2009). Personal genomics services: Whose genomes. *European Journal of Human Genetics*, 17(7), 883–889.

Hall, J. (2007). Navigenics launches with preeminent team of advisers, partners and investors [Press release]. Retrieved October 9, 2014, from http://investor.affymetrix.com/phoenix.zhtml?c=116408&p=irol-newsArticle&ID=1073452&highlight=

Hanahan, E., & Kranhold, P. (2007). 23andMe launches web-based service empowering individuals to access and understand their own genetic information [Press release]. Retrieved October 9, 2014, from http://mediacenter.23andme.com/press-releases/23andme-launches-web-based-service-empowering-individuals-to-access-and-understand-their-own-genetic-information/

Herper, M. (2015). Surprise! With $60 million Genentech deal, 23andMe has a business plan – Forbes. Retrieved March 23, 2015, from www.forbes.com/sites/matthewherper/2015/01/06/surprise-with-60-million-genentech-deal-23andme-has-a-business-plan/

Holden, A., L. (2002). The SNP Consortium: Summary of a private consortium effort to develop an applied map of the human genome. *BioTechniques*, 32, 22–26.

Illumina. (2014). Illumina introduces the HiSeq X™ Ten Sequencing System [Press release]. Retrieved October 9, 2014, from http://investor.illumina.com/phoenix.zhtml?c=121127&p=irol-newsArticle&ID=1890696&highlight=

Imai, K., Kricka, L. J., & Fortina, P. (2011). Concordance study of 3 direct-to-consumer genetic-testing services. *Clinical Chemistry*, 57(3), 518–521. doi:10.1373/clinchem.2010.158220

Issa, J. P. (2002). Epigenetic variation and human disease. *Journal of Nutrition*, 132(8), 2388S–2392S.

Jasny, B. (2013). Realities of data sharing using the genome wars as case study – An historical perspective and commentary. *EPJ Data Science*, 2(1), 1.

Javitt, G. H., & Carner, K. S. (2014). Regulation of next generation sequencing. *Journal of Law, Medicine & Ethics*, 42(s1), 9–21. doi:10.1111/jlme.12159

Jia, H. (2014). China's watchdog clamps down on genome sequencing services. *Nat Biotech*, 32(6), 511. doi:10.1038/nbto614-511

Kaiser, J. (2005). Celera to end subscriptions and give data to public GenBank. *Science*, 308(5723), 775. doi:10.1126/science.308.5723.775a

Kalf, R. R. J., Mihaescu, R., Kundu, S., de Knijff, P., Green, R. C., & Janssens, A. C. J. W. (2014). Variations in predicted risks in personal genome testing for common complex diseases. *Genet Med*, 16(1), 85–91. doi:10.1038/gim.2013.80

Kalokairinou, L., Howard, H. C., & Borry, P. (2014). Changes on the horizon for consumer genomics in the EU. *Science*, 346(6207), 296–298.

Kaphingst, K. A., McBride, C. M., Wade, C., Alford, S. H., Reid, R., Larson, E., et al. (2012). Patients' understanding of and responses to multiplex genetic susceptibility test results. *Genetics in Medicine, 14*(7), 681–687.

Karczewski, K. J., Tirrell, R. P., Cordero, P., Tatonetti, N. P., Dudley, J. T., Salari, K., … Kim, S. K. (2012). Interpretome: A Freely Available, Modular, And Secure Personal Genome Interpretation Engine. *Pacific Symposium on Biocomputing,* 339–350.

Khoury, M. J., McBride, C. M., Schully, S. D., Ioannidis, J. P. A., Feero, W. G., Janssens, A. C. J. W., et al. (2009). The scientific foundation for personal genomics: Recommendations from a National Institutes of Health-Centers for Disease Control and Prevention multidisciplinary workshop. *Genetics in Medicine, 11*(8), 559–567.

Knome, Inc. (2014). About – Knome. Retrieved October 23, 2014, from www .knome.com/about/

Konrad, J. K., Robert, P. T., Pablo, C., Nicholas, P. T., Joel, T. D., Keyan, S., et al. (2012). K. I. M. Interpretome: A freely available, modular, and secure personal genome interpretation engine. *Biocomputing,* 339–350.

Kutz, G. (2010). Direct-to-consumer genetic tests: Misleading test results are further complicated by deceptive marketing and other questionable practices (GAO-10-847T). US Government Accountability Office. Retrieved October 14, 2014, from www.gao.gov/assets/130/125079.pdf

Kutz, G. D. (2006). Tests purchased from four web sites mislead consumers (GAO-06-977T). Retrieved October 14, 2014, from www.gao.gov/products/ GAO-06-977T

LabCorp, Inc. (2013). LabCorp to offer BRCAssure^{SM} breast cancer mutation tests [Press release]. Retrieved October 14, 2014, from http://phx.corporate-ir.net/ phoenix.zhtml?c=84636&p=irol-newsArticle&ID=1881008

Langreth, R. H. M. (2008). States crack down on online gene tests. Retrieved October 8, 2014, from www.forbes.com/2008/04/17/genes-regulation-testing-biz-cx_mh_bl_0418genes.html

Lauerman, J. (2009). Complete genomics drives down cost of genome sequence to $5,000. Retrieved October 9, 2014, from www.bloomberg.com/apps/news?pid =newsarchive&sid=aEUlnq6ltPpQ

Livewello. (2015). Homepage. Retrieved March 22, 2015, from https://livewello .com/

MacDonald, M. E., Ambrose, C. M., Duyao, M. P., Myers, R. H., Lin, C., Srinidhi, L., et al. (1993). A novel gene containing a trinucleotide repeat that is expanded and unstable on Huntington's disease chromosomes. *Cell, 72*(6), 971–983. doi:10.1016/0092-8674(93)90585-E

Magnus, D., Cho, M., & Cook-Deegan, R. (2009). Direct-to-consumer genetic tests: Beyond medical regulation? *Genome Medicine, 1*(2), 17.

Marshall, E. (2000). Storm erupts over terms for publishing Celera's sequence. *Science, 290*(5499), 2042–2043. doi:10.1126/science.290.5499.2042

(2001a). Bermuda rules: Community spirit, with teeth. *Science, 291*(5507), 1192. doi:10.1126/science.291.5507.1192

(2001b). Sharing the glory, not the credit. *Science, 291*(5507), 1189–1193. doi:10.1126/ science.291.5507.1189

Matthijs, G., Huys, I., Van Overwalle, G., & Stoppa-Lyonnet, D. (2013). The European BRCA patent oppositions and appeals: Coloring inside the lines. *Nature Biotechnology, 31*(8), 704–710. doi:10.1038/nbt.2644

McGuire, A. L., Evans, B. J., Caulfield, T., & Burke, W. (2010). Regulating direct-to-consumer personal genome testing. *Science, 330*(6001), 181–182.

Mullis, K. B., Erlich, H. A., Arnheim, N., Horn, G. T., Saiki, R. K., & Scharf, S. J. (1987). One of the first polymerase chain reaction (PCR) patents. US4683195.

Murry, J. (1999). Owning genes: Disputes involving DNA sequence patents. *Chicago Kent Law Review, 75*, 231.

National Human Genome Research Institute. (2014a). FAQ about genetic and genomic science. Retrieved March 22, 2015, from www.genome.gov/19016904#al-2

(2014b). Regulation of genetic tests. Retrieved October 13, 2014, from www.genome.gov/10002335

Ng, S. B., Turner, E. H., Robertson, P. D., Flygare, S. D., Bigham, A. W., Lee, C., et al. (2009). Targeted capture and massively parallel sequencing of 12 human exomes. *Nature, 461*(7261), 272–276.

Phillips, A. M. (2015). Think before you click: Ordering a genetic test online. *SciTech Lawyer, 11*(2).

Pollack, A. (2008). Gene testing questioned by regulators. *The New York Times*. Retrieved October 8, 2014, from www.nytimes.com/2008/06/26/business/26gene.html

Promethease. (2014). Homepage. Retrieved October 7, 2014, from https://promethease.com

Rai, A. K., & Cook-Deegan, R. (2013). Moving beyond "isolated" gene patents. *Science, 341*(6142), 137–138. doi:10.1126/science.1242217

Regalado, A. (2014). How a wiki is keeping direct-to-consumer genetics alive. The year in review: Health care. *MIT Technology Review*. Retrieved March 22, 2015, from www.technologyreview.com/featuredstory/531461/how-a-wiki-is-keeping-direct-to-consumer-genetics-alive/

Roberts, L. (2001). Controversial from the start. *Science, 291*(5507), 1182–1188. doi:10.1126/science.291.5507.1182a

Rommens, J. M., Iannuzzi, M. C., Kerem, B., Drumm, M. L., Melmer, G., Dean, M., et al. (1989). Identification of the cystic fibrosis gene: Chromosome walking and jumping. *Science, 245*(4922), 1059–1065. doi:10.1126/science. 2772657

Skeehan, K., Heaney, C., & Cook-Deegan, R. (2010). Impact of gene patents and licensing practices on access to genetic testing for Alzheimer disease. *Genetics in Medicine, 12*(1s), S71–82.

SNPedia. (2014). Homepage. Retrieved October 7, 2014, from www.snpedia.com/index.php

Spicer, D. (2008). Evidence Based Nutrition, Inc., first to offer Sciona MyCellf™ DNA personalized genetics analysis kit to chiropractors [Press release]. Retrieved October 9, 2014, from www.reuters.com/article/2008/02/19/idUS155080+19-Feb-2008+BW20080219

Sweeney, B. (2010). Navigenics receives state of New York clinical laboratory permit [Press release]. Retrieved October 14, 2014, from www.businesswire.com/news/home/20100112005741/en/Navigenics-Receives-State-York-Clinical-Laboratory-Permit

The Associated Press. (2009). Germany limits genetic testing. Retrieved October 9, 2014, from www.utsandiego.com/news/2009/apr/24/eu-germany-genetic-testing-042409/

Thompson, J., & Milos, P. (2011). The properties and applications of single-molecule DNA sequencing. *Genome Biology*, *12*(2), 217.

TIME Magazine. (2008). The retail DNA test – Best inventions of 2008. Retrieved September 21, 2014, from http://content.time.com/time/specials/packages/article/0,28804,1852747_1854493_1854113,00.html

US Food and Drug Administration. (2010). In vitro diagnostics – FDA. Retrieved September 24, 2014, from www.fda.gov/MedicalDevices/ProductsandMedical Procedures/InVitroDiagnostics/default.htm

(2015). FDA permits marketing of first direct-to-consumer genetic carrier test for Bloom syndrome [Press release]. Retrieved March 9, 2015, from www.fda.gov/NewsEvents/Newsroom/PressAnnouncements/UCM435003

Vayena, E., & Prainsack, B. (2013). The challenge of personal genomics in Germany. *Nature Biotechnology*, *31*(1), 16–17.

Watson, J. D., & Jordan, E. (1989). The human genome program at the National Institutes of Health. *Genomics*, *5*(3), 654–656. doi:10.1016/0888-7543(89) 90040-2

Wade, N. (2000). Genetic code of human life is cracked by scientists. *The New York Times*. Retrieved October 6, 2014, from http://partners.nytimes.com/library/national/science/062700sci-genome.html

Wells, S. (2005). The Genographic Project. Retrieved September 21, 2014, from https://genographic.nationalgeographic.com/about/

Wetterstrand, K. A. (2014). DNA sequencing costs: Data from the NHGRI Genome Sequencing Program (GSP). Retrieved October 7, 2014, from www.genome .gov/sequencingcosts/

Wiechers, I., Perin, N., & Cook-Deegan, R. (2013). The emergence of commercial genomics: Analysis of the rise of a biotechnology subsector during the Human Genome Project, 1990 to 2004. *Genome Medicine*, *5*(9), 83.

Youngblom, E. K. J. (2014). Familial hypercholesterolemia. In E. Youngblom, M. Pariani, & J. W. Knowles (Eds.), *GeneReviews*. Seattle, WA: University of Washington. Retrieved October 9, 2014, from www.ncbi.nlm.nih.gov/books/NBK174884/

14

Ethical Issues in Using Genomics to Influence Educational Practice

SUSAN BOUREGY AND KRISTA BOUREGY

INTRODUCTION

Student achievement continues to be a politically and socially charged issue that brings up a host of angst for many people for many reasons. Parents fret over preschool placements in the hope that it will facilitate their child's segue into top universities and a prosperous future. Teachers and school administrators worry about meeting mandated performance standards to ensure federal funding. Legislators want to ensure an educated workforce to enhance their respective constituency's place in the global economy. Ultimately all parties involved want to facilitate student performance, and schools have responded with program enhancements, curriculum revisions, and ubiquitous standardized testing. Growth in the field of genomics opens yet another avenue to identify students who may benefit from alterations in standard educational programs. Genomics and big data initiatives hold the promise of identifying the genetic underpinnings of learning, opening the door to a personalized education similar to that of personalized precision medicine being explored today (Collins & Varmus, 2015; Feero, 2014). Rapid advances of direct-to-consumer genetic/genomic testing (DTCGT) and advances in genomic research will move genetics not only from bench to bedside but from bench to desktop as well, operationalized to individual students and their classrooms. Joining the wealth of student performance data building in local school districts with the growth in individual genomic data, future education environments may begin to involve incorporating individualized genomic profiles of students' expected academic performance.

In this chapter, we explore the basis for this expectation and its implications for the classroom and ethical educational practice. In doing so, we consider whether there is a point at which the costs and ethical risks of measuring and enhancing student achievement through incorporating genetic testing information, whether from DTCGT or a systematic screening across all students, outweigh the gains in student performance. In particular, we consider whether genomics truly holds the potential to provide information that is significantly more meaningful than what we can currently glean from actual learning behavior, and whether those improvements merit the associated risks of accepting genetics in the educational milieu.

BACKGROUND ON GENETIC BASES OF LEARNING AND MEMORY

Many of us had our first, and perhaps only, exposure to basic genetic principles in high school biology classes that taught Mendelian genetics and punnett squares. At its most basic level, genetics was presented as involving a limited number of variants of a given trait which sorted themselves out among offspring as either dominant or recessive. Offspring of pure breeding parents exhibited the full-dominant phenotype, and only when these progeny mated did the recessive trait show itself again. Teaching Mendelian genetics affords an understanding of basic genetic principles of inheritance and is useful in improving genomic literacy as the concepts can be grasped easily, even if not fully encompassing the complexities of genetics (Meyer, Bomfim, & El-Hani, 2013).

Mendel was fortuitous in finding traits in his pea plants that segregated neatly from parent to offspring in a largely binary fashion. Unlike Mendel's peas, however, behavioral traits rarely, if ever, segregate so cleanly with single genes expressing recognizable phenotypes. Rather, behavioral traits generally arise from the interaction of multiple genes expressed to varying degrees and acting together in the context of varied environments. As an example, consider studies on the genetic basis of learning in the fruit fly, *Drosophila melanogaster*. *Drosophila* were instrumental in identifying basic principles of heredity (Morgan, 1988), and their rapid life cycle quickly led to the use of *Drosophila* to identify mutations impacting development and gene expression. Once identified, a *Drosophila* mutant phenotype could be traced back to the underlying genetic mutation using breeding and molecular techniques, allowing for the identification of the individual genes involved in the trait of interest.

Drosophila is well known as a genetic model for developmental biology, but *Drosophila* also displays a number of complex behaviors that are amenable to genetic dissection (e.g., Benzer, 1967; Manoli et al., 2005; Nagoshi et al., 2010). Most importantly, *Drosophila* can learn and remember (McGuire, Deshazer, & Davis, 2005) in a variety of paradigms, including olfactory and visual learning. Olfactory memory includes classical conditioning whereby an odor is paired with electric shock after which flies are allowed to choose between the shock-associated odor and a control odor (Tully & Quinn, 1985). This learning paradigm produces a strong learned response which can be sustained for at least twenty-four hours.

Using the mutagenic capacity of flies to identify the genetic basis of fly behaviors was pioneered by Seymour Benzer, who advocated the ability to identify individual genes impacting complex behaviors. Mutations in flies can be induced experimentally by feeding flies with compounds that induce genetic change (chemical mutagenesis) or by inducing movement of stretches of mobile DNA to disrupt the genome (P-element transposon mutagenesis). By screening mutated flies for their ability to learn and remember olfactory cues relative to normal flies, at least forty genes have been identified which impact olfactory learning in fruit flies. Many of the mutations have been further studied to determine the proteins and biochemical pathways which impact learning when mutated. Consistent with findings in other learning models such as habituation in *Aplysia*, the fly mutation studies have identified mutations in the cAMP cell-signaling pathway as critical to *Drosophila* learning (*dunce, rutabaga*). Other mutations implicate genes involved in brain development, membrane receptors, and kinase regulating pathways (see Kahsai & Zars, 2011, for a recent review). The results from Drosophila help to identify pathways which need to be intact in order for learning to occur or be encoded into memory. Taken in conjunction with results from other animal learning models, we can begin to identify pathways which may also be relevant for learning in our own children.

The ability to identify genetic components of learning in model systems and the commonalities across models (Glanzman, 2008) suggests that human learning also will involve identifiable mechanisms based on these and other genetic pathways. Studies on specific learning related traits have begun to tease out the genetic bases of skills underlying human learning, such as the identification of genes impacting dyslexia (Powers et al., 2013). In addition, advances in data analytics, "big data," has opened up researchers' ability to plumb the entire genome to identify loci associated with disease states. Recently, genome-wide association techniques were used to identify genetic variants associated with educational attainment (Rietveld et al., 2013). Our current technology-centric culture has led to an explosion of data documenting our activities at home, at work, and at school. Most school systems maintain electronic records of student performance, which, if combined with individual genomic data, could be used to untangle the genetic underpinnings of those students who struggle in school compared to students who move through school with relative ease. Taken together, the nascent field of big data analytics of human behavior is primed to link genetic and academic information, rendering the genetic bases of academic success to be available and predictive of the performance of individual students who undergo genetic testing.

IMPACT OF GENETIC AWARENESS ON THE CLASSROOM

Presuming that we could in fact predict achievement based on genetic testing, what would that classroom of the future look like? The potential impacts of genetic testing are varied in complexity and intensity depending largely upon how this information will be used by administrators, politicians, teachers, and parents. We envision two areas of impact in the classroom. First, we imagine identification of genetic deficits which are or could be recognized under the disability categories laid out by the Individuals with Disabilities Education Improvement Act (IDEA). Second, we envision identification of genetic variants underlying types of intelligences leading to a preferred learning style of a student and hence what teaching approaches would be most beneficial for that student.

In the first arena, in which the testing results meet defined and recognized disabilities, the ability to test for the academic deficit could lead to earlier identification. Having an infant diagnosed within any disability category at an earlier age provides a greater chance of the child overcoming their deficits in future classroom settings (Muschkin, Ladd, & Dodge, 2015). Currently, receiving the necessary help often is not possible upon entering school at age five. Rather, the child is subject to screening tests and classroom modifications before a child is diagnosed as meeting an actionable learning deficit and provided an Individual Education Plan (IEP). In effect, the student tends to be failing or at least falling behind his or her peers significantly prior to receiving help. Therefore, by the time the student has been identified and is receiving services, they may already be years into their academic careers. Early genetic testing could accelerate the identification process.

It is important to remain cognizant of the imperative skills and content being taught in the first four years of schooling. These primary years are critical in teaching how to learn, rather than specific content. Without these first few years of basic learning skills – spelling, writing, phonics, grammar, addition, and subtraction – students cannot be expected to perform at the same level as their typically developing and learning peers. Thus, they enter into a cycle of falling further and further behind in each subsequent year of schooling. In this instance, having a genetic test that results in the meaningful knowledge of an infant's educational disability (or, rather, a likelihood of manifesting it) may drastically affect their performance five years later, when they enter school. With this knowledge, there is a greater chance and opportunity for parents and early childhood learning professionals to help remediate and combat precursor deficits leading to an educational disability with which the child is expected to struggle even prior to entering the formal classroom. Furthermore, providing such data to administrators and teachers allows for schoolwide and classroom supports to be established on day one of the child's entry to formal education as opposed to the waiting so often experienced by students and families today.

This testing could, for a small period of time, spread special education resources thinner, with the influx of disabled students, requiring a commensurate influx of resources. Whether this be through books, professionals, or money, the schools would need additional support to provide all of the students with an education that meets their needs as specified under IDEA and the Rehabilitation Act of 1973, most notably Section 504 containing the Free and Appropriate Public Education for Students with Disabilities. However, if students are receiving quality early childhood care to overcome precursor deficits, schools might reduce their special education spending in the long term due to the early identification of such deficits and provision of compensating services. Ideally, identification of deficits through genetics would provide a definitive determination of risk, allowing the correct intervention to be implemented early. In so doing, the student would sooner learn coping skills and less intervention would be needed over the course of the child's primary schooling. In the end, schools would have the opportunity to reduce special education spending as there would be fewer students in need of long-term accommodations (Munschkin, Ladd, & Dodge, 2015).

IMPACT OF EARLY IDENTIFICATION OF LEARNING STYLES

Taking the other area of possible impact, classroom practices would be drastically altered if genetic testing can tie so closely to learning that it provides information on different types of intelligence and, therefore, learning styles. For example, children may be identified as having high bodily kinesthetic or linguistic or intrapersonal, among other, intelligences. While a person's learning style is distinct from the profile of their intelligence(s), there is a connection between a child's cognitive strengths and weaknesses and how that child prefers to learn (Gardner, 2011; Strauss, 2013). If we can judge a child's dominant intelligences, a door opens to change the classroom environment to optimize learning by providing instruction that matches a specific student's learning style. Doing so would modify the responsibilities of our teachers who would become accountable for addressing learning styles rather than expecting the student to develop his or her own accommodating skills.

Should there become a demand by parents, administrators, or politicians for individualized accommodations, classrooms may be split beyond general education and special education, creating a need for specialized training and a whole new field of professionals versed in educating based on learning styles and intelligences. Knowing the specific intelligences of the students has the potential to make the work of the teacher easier and more effective. Instead of teaching to all modes of learning, the teacher can distinctly focus on the majority of intelligences in the classroom and potentially minimize or even ignore styles that are not a part of that specific class' demographic.

An additional ramification of addressing learning styles would be the potential to develop specialized schools and classrooms based on dominant learning style. In such an arrangement, students would be surrounded by all similar or like-minded peers, leading to exceptional scholastic improvement but potentially also hindering social and emotional development. Students would no longer be expected to work in different styles and on different projects than what matches their preferences, and in so doing, they might fail to be prepared for the diversity of real-world working environments.

LIMITATIONS ON APPLYING RESEARCH FINDINGS
TO REAL-WORLD BEHAVIOR

There are clearly a number of benefits that would be gained from a future in which information on the genetic influences on individual academic achievement are available. The animal studies described earlier demonstrate that there are individual genes which influence ability to learn and it is tantalizing to speculate on how genomics could subsequently advance education. There are, however, some significant caveats that limit the feasibility of applying behavioral genetics of learning to children in the classroom.

Most notable is the difficulty of determining the impact of a given genetic variant among the heterogeneity of school children. Genetic impacts in model organisms such as *Drosophila* can be identified because mutations are compared against a uniform background created through years of inbreeding leading to a largely genetically homogenous population. This creates a baseline learning performance against which a single gene mutation altering behavior can be identified. That is, the impact of a given gene on learning ability is measurable in flies because the remainder of the genome is held constant, allowing the impact of a single gene to be measurable. Unlike *Drosophila* and identical twins, human children are genetically unique. As such, the impact of any individual gene is mitigated by the mix of other learning related genes in the individual's genome, minimizing our ability to isolate and predict the impact of any given genetic variant except in the extreme.

Furthering the complexity is the role of the environment. Impacts on learning from environmental variables such as environmental enrichment and sleep deprivation are measureable in model learning organisms (van Praag, 2000). For example, in mice, recent studies suggest that environmental stimuli can modulate and overcome impacts of chemical inhibition of neural remodeling associated with memory formation (Bednarek & Caroni, 2011). In addition, sleep deprivation and starvation both impair learning performance in *Drosophila* and these learning impacts are modulated by the fly's genotype (Donlea et al., 2012). Environmental influences will also modulate the impact of any genetic variant found to be a source of individual differences in learning in school children, limiting the predictive nature of any genetic information.

Lastly, we need to consider that many of the genes that impact learning will only be impacting learning as a secondary effect. Developmental defects that influence the neuronal circuitry underlying learning will likewise impact student achievement. For example, the *Drosophila* mutant strain known as *latheo* was identified in a screen for memory mutants. In this case, flies derived from p-element transposon mutagenesis were tested for olfactory memory three hours after training in a classical conditioning paradigm and those that performed poorly, such as *latheo*, were further investigated (Boynton and Tully, 1992). Using *Drosophila* genetic techniques, the original *latheo* mutant strain was induced to produce additional distinct mutations in the underlying gene. Analysis of the various *latheo* mutations indicated that some alleles had cell proliferation abnormalities in the area of the brain responsible for *Drosophila* olfactory processing known as the mushroom bodies making it difficult to distinguish whether the learning deficit was based in an anatomical defect in the *latheo* flies or in their learning ability itself. Ultimately *latheo* was cloned and DNA sequenced which indicated that the gene encodes a subunit of the origin recognition complex, a protein necessary for DNA replication. These results suggest a primary role in the biological function of DNA replication rather than in the behavior function of learning (Pinto et al., 1999). Findings such as that of *latheo* underscore the difficulty in distinguishing between environmental and developmental effects on complex behaviors such as academic achievement. The ability to learn is predicated not only on the ability to take in and retain information in the classroom but also on the presence of the requisite neuronal architecture upon which the learned information may be encoded. Learning requires that a myriad of developmental pathways be completed successfully in order for learning to even be feasible. Subtle differences in development will act synergistically with those genes involved in the learning process itself and these interactive effects will make prognostication of the measurable impact of a given student's genetic profile on his or her academic performance difficult at best.

ETHICAL ISSUES OF GENOMIC-BASED EDUCATION

The field of education stands on the edge of a new era where genomics can guide instruction along the lines of how healthcare has begun to embrace personalized medicine. The bioethics community, by definition straddling the fields of health care and ethics, ensured that the move to personalized medicine was prefaced with robust discussion on the ethical and cultural implications (Juengst Flatt, & Settersten, 2012). While the ethical underpinnings of behavioral genetics are beginning to be considered (Parens, 2004), there is no community of molecular genetics, genomics, ethics and education experts working to ensure a robust discussion occurs regarding the utilization of molecular genetics and genomics to inform the education process. As linkages between molecular-genetic and genomic profiles and their academic potential are gleaned, we can expect a growth in commercial entities willing to provide testing to well intentioned parents who will lobby for educational environments that can best address their child's individual learning competencies and deficiencies (Seife, 2013). The ability of consumers to acquire their or their child's individual DNA sequence through DTCGT will likely push hard at the boundaries of available educational resources and must be tempered by the uncertainty of predicting academic achievement in an uncontrolled social and genetic environment. As a society we must grapple with what, if any, limits are appropriate in our response. Below we raise three ethical concerns that must be part of any discussion on implementing genomically informed education, that of justice, individual autonomy, and privacy.

Justice

Justice refers to the equitable distribution of burdens and benefits. For the purpose of this discussion, justice issues arise in the ability to obtain genetic testing, to obtain services in response to genetic testing results, and to ensure that provisioning services in response to testing for one child does not detract from the resources available for those students whose genetic profiles do not merit an educational intervention.

Genetic testing has now become affordable for many people. Currently genetic testing for ancestry information can be obtained for as little as $100 and a whole genome sequence is approaching $1,000. At that price range, there are likely many parents who will seek educational testing, should such testing become available. However, there are also many parents who would find such prices out of reach. These would be the same families whose children are likely to be enrolled in poorer performing schools. Financial constraints on access to testing would, therefore, further the gap in educational performance across socio-economic classes. It is plausible that this gap would lead to federal or state supported testing so as to make testing available to all at a significant tax burden.

We must also consider that justice emphasizes equity for every student to obtain a quality education. This requires a very delicate balance between providing for the individual as well as for the needs of the greater classroom community. For schools that intend to implement changes based on genetic testing, it is undoubtable that funding will factor into the quality of any programs or alterations made to meet legislative, regulatory, or community-demanded mandates to develop and maintain services in accordance with genomic profiles. The resources necessary go beyond teacher training. We envision requirements for texts, specialists, entire new classrooms or schools, and classroom tools for both student and teacher use. Already we suffer from a lack of basic necessities in our schools for students with and without disabilities (www.fundphillyschools.org/the-facts). The education crisis has left many schools suffering from a lack of resources in the form of high student–teacher ratios, lack of up-to-date teaching materials as well as the loss of ancillary professionals, such as school nurses, and even basic learning necessities. If our schools can barely meet current student needs, there is a serious question of where more funding for genomic accommodations will come from, and the fear that these costs will dilute the already minimal reserves schools have.

On the other hand, genetic testing that results in early detection of disabilities may actually provide greater resources to schools arising from there being fewer students in need of special education accommodations over the length of their education. By identifying and addressing precursor deficits early based on genomic prognostication, students may quickly be able to assimilate into classrooms rather than requiring long-term accommodations following late identification of addressable defects. Such a result could actually provide extra monetary resources to increase the number of supplementary professionals available to assist students, provide programs in the arts, and give classrooms new and better learning and teaching materials. In this way, genetic testing could be a savior for equitable teaching practices and result in a greater quality of education for students nationwide.

Alternatively, in the case where testing is not universally provided, only parents who can afford to pay for DTCGT will have access to genetic testing data. These parents would be expected to lobby for academic accommodation for their child in the classroom forcing schools and teachers to place students on a hierarchy built around parental pull and monetary resources. This could force educators away from meeting the needs of students with other, non-genetically defined deficits, redirecting the teacher, a resource in his or her own right, and his attention away from other students to focus on a handful of children whose parents hold any form of power over the school as a whole. Unlike educators, parents are concerned with only half of the educational justice problem: an equitable, quality education for their one student. In light of this, they may lose sight of the numerous other students in need. While it is generally unfair to pit one person's struggle against another's in an attempt to order their significance or difficulty, it is a factor that must be considered by educators. Teachers as a whole try to meet the needs of all their students; however, sometimes they must consider the needs of the greater class or of one student's IEP over one individual's preferences. This consideration is an everyday part of teacher planning, which would need to be conveyed to parents and guardians prior to any guarantees for special treatment due to genetic testing beyond what is currently required by the governmental and school policy.

The reality of families and school districts today suggests that the added burden of responding to genetic test results most likely will resemble the current situation seen in special education: money plays an inordinate role in testing and resources. Depending on whether or not such testing is provided to all by government resources, not all families may be able to afford the price of receiving such impactful information during their child's infancy or early childhood. Instead, they may be stuck in the current system, giving their children a disadvantage, not only against their typically developed classmates, but also against students whose families could afford DTCGT years before their academic career started.

Lastly, consider the fiscal gap between the families who can additionally afford, through either their own time or enrollment in early childhood learning programs, to have their child taught strategies and modifications while in pre-kindergarten centers. For many families of lower socioeconomic status, it will be beyond their reach and fiscally imprudent to place the child in a preschool learning program that can address any anticipated learning deficits. In these cases, there is no guarantee that the child will receive the type or amount of support they will need to catch up before school starts. Therefore, even if a child is provided such genetic testing upon entering school, there remains a potential gap between those who have already acted on the results and provided their child with professional support and those who cannot.

Autonomy

Traditional American culture is founded on a belief that hard work will be rewarded with success, i.e., the "Puritan ethic" that infuses our culture. The idea of one having a genetic predisposition to develop certain behavioral traits is antithetical to the cultural ethos that with enough drive and effort we create our own destinies and are autonomous beings. Genetic testing has the potential to restrain autonomy through its impact on students' and society's expectations being influenced by perceive genomic potential. For example, children whose genetic/genomic profile suggests learning difficulties may self-limit their educational attainment through lack of effort, perceiving themselves as being incapable of doing better and thus not worth expending additional effort. Likewise, parents may not empower their child to work toward higher achievement. This acceptance of genetic predetermination may also be subtly reinforced by teachers who, in their effort to meet performance mandates across the classroom as a whole, may focus extra efforts on those students who show promise at raising the classroom's average with the least anticipated effort. Without a concurrent increase in genomic literacy, the ability to make sense of genetic information and its limitations has the potential to set false expectations. In the case of cash strapped school systems whose funding is tied to performance and of households struggling to meet basic needs, those students who are deemed to be unlikely to excel may not be provided the effort and attention to overcome their learning deficits, creating a self-fulfilling prophecy.

In response, one can imagine parents attempting to circumvent genetic stigma by avoiding genetic testing all together. Given the uncertainty associated with determining the expected learning behavioral phenotype it is reasonable for parents to skip testing and allow the education system to provide services based on their child's actual observed behavior. In this way parents could hope to encourage students to reach further than might be expected by their genetic profile. If education systems move to incorporate genomics into the classroom it is unclear how "untested" students would be handled and if that might become a stigma unto itself.

Privacy Issues

Privacy refers to the ability to control access to personal information and was famously promoted as a right by Justice Brandeis in Olmstead v. United States, (277 U.S. 438 (1928)). Since that time, there has been a growing awareness of the need to protect individual privacy and the subsequent legislative responses. Most pertinent to this discussion are the Family Educational Rights and Privacy Act (FERPA 34 CFR 99), the Health Insurance Portability and Accountability Act (HIPAA 45 CFR 160 and 164), and the Genetic Information Nondiscrimination Act of 2008 (GINA). These regulations were promulgated in response to public concern about the privacy of academic data (FERPA), healthcare (HIPAA), and genetic information (GINA). They are similar in that they limit how identified personal information can be used or disclosed by those responsible for the information, specifically educational institutions (FERPA), health care providers, health plans and health care clearinghouses (HIPAA), and health plans and employers (GINA).

Parents or individuals who chose to obtain their genomic sequence for educational purposes can control access to the information and weigh the risks associated with disclosing the information to the relevant school system. As discussed above, parents might take this approach to avoid stigma or from lack of willingness to entrust this sensitive information to school systems. To obtain the benefit of genetic-based education on a broader community level, however, access to genetic information and associated individualized learning programs would need to be available not just to those who can afford to procure genetic testing but to all school age children. In such case, the schools or local governments would likely be called upon to sponsor testing and would have direct access to the results and/

or responsibility for maintaining the genetic information in the students' academic record. The adequacy of current standards for protection of education records is a matter of public debate and political attention (see www.whitehouse.gov/the-press-office/2015/01/12/fact-sheet-safeguarding-american-consumers-families). In an era with frequent data breaches at large corporations it is hard to imagine that local school systems would have the capacity to securely maintain sensitive student genetic information or to implement appropriate safeguards to ensure limited access and use.

Privacy concerns with regard to genetic information are multi-faceted and complex. Access to full genome sequencing by school systems would not only provide information on a student's genetic/genomic profile related to learning but also present results related to any other identifiable genetic/genomic risk factors being made available to the school as the data steward. Even with a more limited set of data released to the schools, the pleiotropic effects of "behavioral" genes will mean that providing schools only limited "learning" gene information, the genotype may implicate other conditions which may or may not be sensitive or considered sensitive by the student and his or her family. Furthermore, genetic test results speak not only to the student being tested but also reflect the genomic make up of the student's blood relatives. As such, information regarding the student's anticipated academic achievement will be suggestive of their sibling's and parent's, not to mention the family member's susceptibility to any associated traits. Current data stewardship practices in academia are not geared for genetic information. For example, FERPA places control of academic information, which would include genetic information provided for educational purposes, with the student and, in the case of minor students, their parent or guardian. In the context of genetic information in a student's record, the consent of an adult student would be adequate to allow the release of potentially sensitive information implicating their close family members under current FERPA standards. Even without consent, FERPA allows release of information for a number of purposes such as auditing and evaluation which could then place the data outside the protections of the regulations. As schools begin to be the custodians of genetic/genomic information, the regulatory schema to protect its privacy will need to be updated and the public made aware of the ramifications of the broader access to sensitive data that will exist in schools. The ability of a breach of genetic information to impact not just the student but their family members as well would necessitate a more stringent regulatory schema and associated compliance program than is currently mandated under FERPA, likely creating additional fiscal burden on school systems.

CONCLUSIONS

It is arguable that with enough additional support to overcome our genetic makeup along with enhancements to our environment, all children could reach a level of high academic achievement. We nonetheless urge caution. We expect that except in the more extreme cases, genetic and genomic information will not have sufficient impact to be cost effective. The small town where the authors live spends roughly 69 percent of its budget on education costs whereas the larger cities nearby spend only 40–50 percent of their budgets on education costs. Yet statewide rankings do not show a clear correlation in performance in our local schools relative to the larger city schools nearby. If enhancement of specialized services were directly correlated with performance, one would expect that for each additional dollar spent on education there would be a corresponding increase in student performance. This is not the case. Costs associated with providing tailored instruction based on a student's genetic profile are likely to significantly increase education costs. We expect that only a limited correlation would be found with attempts to achieve enhancement in student performance through providing specialized educational enhancements based on a student's genotype as the observed impact of a student's genomic learning profile will be difficult to interpret in the context of other environmental and developmental variables. In fact we know that teacher quality has a modulating effect on genetic influence in the case of early reading (Taylor et al., 2010). We hold that students will be better served by improvements in child nutrition and health programs, which will raise student performance of at-risk children irrespective of their genetic/genomic profile.

The number of genes influencing learning along with the unique mix of genetic alleles in any given child, some facilitating and some mitigating their learning potential, modulated by environmental influences, makes any single genotype unpredictable in its influence on student achievement in all but the most extreme cases. The impact of a given child's genetic and environmental readiness to learn would be best addressed based not on their genetic sequence alone but rather based on the actual output of their genetic and environmental circumstances – e.g., their educational performance. We advocate better testing and resources be made available to improve student performance based on the student's responses to academic rather than genomic assessments.

This is not to say that our understanding of the genetic/genomic bases of educational achievement is meritless. Our ability to understand the processes will inform educational practice and ways to maximize learning potential for all children. In the same way that knowing the impact of sleep deprivation on learning allows us to recommend better sleeping habits to improve performance, through genetics/genomics we may be better able tailor our learning environments to maximize academic achievement for all students irrespective of their individual genomic profiles.

REFERENCES

Bednarek, E., & Caroni, P. (2011). B-Adducin is required for stable assembly of new synapses and improved memory upon environmental enrichment. *Neuron*, *69*(6), 1132–1146.

Benzer, S. (1967). Behavioral mutants of *Drosophila* isolated by countercurrent distribution. *Proceedings of the National Academy of Sciences*, *58*, 1112–1119.

Boynton, S., & Tully, T. (1992). Latheo, a new gene involved in associative learning and memory in *Drosophila melanogaster*, identified from p element mutagenesis. *Genetics*, *131*(3), 655–672.

Collins, F. S., & Varmus, H. (2015). A new initiative on precision medicine. *New England Journal of Medicine*, *372*, 793–795.

Donlea, J., Leahy, A., Thimgan, M. S., Suzuki, Y., Hughson, B. N., Sokolowski, M. B., et al. (2012). Foraging alters resilience/vulnerability to sleep disruption and starvation in Drosophila. *Proceedings of the National Academy of Sciences*, *109*(7), 2613–2618.

Gardner, H. (2011). *Frames of mind: The theory of multiple intelligences*. New York, NY: Basic Books.

Feero, W. G. (2014). Clinical application of whole-genome sequencing: Proceed with care. *Journal of the American Medical Association*, *311*(10), 1017–1019.

Glanzman, D. L. (2008). New tricks for an old slug: The critical role of postsynaptic mechanisms in learning and memory in *Aplysia*. *Progress in Brain Research*, *169*, 277–292.

Juengst, E. T., Flatt, M. A., & Settersten, R. A. (2012). Personalized genomic medicine and the rhetoric of empowerment. *Hastings Center Report*, *42*(5), 34–40.

Kahsai, L., & Zars, T. (2011). Learning and memory in *Drosophila*: Behavior, genetics and neural systems. *International Review of Neurobiology*, *99*, 139–167.

Manoli, D. S., Foss, M., Villella, A., Taylor, B. J., Hall, J. C., & Baker, B. S. (2005). Male-specific fruitless specifies the neural substrates of *Drosophia* courtship behavior. *Nature*, *436*(7049), 395–400.

McGuire, S. E., Deshazer, M., & Davis, R. L. (2005). Thirty years of olfactory learning and memory research in Drosophila melanogaster. *Progress in Neurobiology*, *76*(5), 328–347.

Meyer, L. M. N., Bomfim, G. C., & El-Hani, C. N. (2013). How to understand the gene in the twenty-first century? *Science and Education*, *22*, 345–374.

Morgan, T. H. (1988). *The genetics of Drosophila*. New York: Garland Publishing.

Muschkin, C. G., Ladd, H. F., & Dodge, K. A. (2015). Impact of North Carolina's early childhood initiatives on special education placements in third grade. *Educational Evaluations and Policy Analysis*, *37*(4), 478–500.

Nagoshi, E., Sugino, K., Kula, E., Okazaki, E., Tachibana, T., Nelson, S., & Rosbash, M. (2010). Dissecting differential gene expression within the circadian neuronal circuit of Drosophila. *Nature Neuroscience, 13*(1), 60–68.

Parens, E. (2004). Genetic differences and human identities: On why talking about behavioral genetics is important and difficult. *Hastings Center Report, 34*(1), s4–35.

Pinto, S., Quintana, D. G., Smith, P., Mihalek, R. M., Hou, Z. H., Boynton, S., et al. (1999). Latheo encodes a subunit of the origin recognition complex and disrupt neuronal proliferation and adult olfactory memory when mutant. *Neuron, 23*, 45–54.

Podesta, J., Pritzker, P., Moniz, E. J., Holdren, J., & Zients, J. (2014). Big data: Seizing opportunities, preserving values. Retrieved from www.whitehouse.gov/sites/default/files/docs/big_data_privacy_report_may_1_2014.pdf

Powers, N. R., Eicher, J. D., Butter, F., Kong, Y., Miller, L. L., Ring, S. M., et al. (2013). Alleles of a polymorphic etv6 binding in dcdc2 confer risk of reading and language impairment. *American Journal of Human Genetics, 93*(1), 19–28.

Rietveld, C. A., Medland, S. E., Derringer, J., Yang, J., Esko, T., Martin, N. W., et al. (2013). Gwas of 126559 individuals identifies genetic variants associated with educational attainment. *Science, 340*, 1467–1471.

Seife, C. (2013). 23andMe is terrifying, but not for the reasons the FDA thinks. *Scientific American, SA Forum*, November 27, 2013 https://www.scientificamerican.com/article/23andme-is-terrifying-but-not-for-the-reasons-the-fda-thinks/

Strauss, V. (2013). Howard Gardner: "Multiple intelligences" are not "learning styles." Retrieved from www.washingtonpost.com/blogs/answer-sheet/wp/2013/10/16/howard-gardner-multiple-intelligences-are-not-learning-\styles/?tid=auto_complete

Taylor, J., Roehrig, A. D., Soden Hensler, B., Conner, C. M., & Schatschneider, C. (2010). Teacher quality moderates the genetic effects on early reading. *Science, 328*, 512–514.

Tully, T., & Quinn, W. G. (1985) Classical conditioning and retention in normal and mutant Drosophila melanogaster. *Journal of Comparative Physiology A, 157*, 263–277.

van Praag, H., Kempermann, G., & Gage, F. H. (2000). Neuronal consequences of environmental enrichment. *Nature Reviews Neuroscience, 1*(3), 191–198.

15

Teaching and Genetic/Genomic Variation: An Educator's Perspective

JUDI RANDI

Since antiquity, astute teachers have noticed how students differ. The Roman educator Quintilian was especially aware of nature's contribution to individual student differences: "The gifts of nature are infinite in their variety, and mind differs from mind almost as much as body from body" (Quintilian, trans. Butler, 1920). While some differences are easily discernable, such as physical appearance, the ways in which minds differ are less obvious and often highly dependent on the situation, such as the level of interest or effort a student shows toward a particular task or topic at any given time. Are these differences of mind nature's gifts, as Quintilian suggests? Educators hope not. The idea that nature determines how students think as much as how they look is appropriately resisted by educators whose work focuses on developing minds, by offering "more experiences of a certain type than nature might offer" to facilitate learning (Cronbach, 1955, p. 79).

Rather than ask if differences of mind are nature's gifts, adaptive teachers ask what kinds of experiences can education provide that enhance what nature has to offer. Fortunately, some geneticists have also moved away from asking what gifts nature has to offer to questions about how individuals' experiences contribute to their very nature. Research on the human genome is now confirming the powerful role of experience in shaping who we are, for now and the future.

In his best-selling book *Inheritance*, Sharon Moalem (2014) explains how our lives change our genes as much as genes change our lives. It is no surprise that Moalem opens his book by evoking memories of middle-school experiences to illustrate how our experiences leave an indelible mark

The author would like to acknowledge Lyn Corno, Hanna Dumont, and Eva Sapi for their comments on earlier versions of this chapter.

in our genome, or set of genes. The question of interest is no longer what is *inherited*, but what constitutes *inheritance*. The notion of genetic make-up as comprised only of the genes we inherited is no longer valid. Each experience an individual has changes that individual's genetic make-up, turning some genes on and others off, and even modifying "inherited" DNA. Moreover, these changed genes can be passed down to future generations (Champagne & Mashoodh, 2009). Thus, as Moalem explains, inheritance is not merely a matter of the genes we inherited as much as a matter of how our experiences change our genetic inheritance, transforming both what we inherited and what we pass down. Given the time that most individuals spend in school, educational experiences serve a significant role, not only in changing individuals' genomes but ultimately in shaping and reshaping the human genome as a whole. Designing these experiences is a task for adaptive teachers.

The purpose of this chapter is to provide an educator's perspective on the recent developments in the field of genetics, facilitate understanding about the genome as a source of individual differences, and explore the implications of this research for teachers and students in their classrooms. The chapter also provides some background about how genes and the environment interact so that the relevant genetic information on students, when it becomes available, can empower teaching and learning, rather than lead to the self-fulfilling prophecies educators fear.

THE GENOME AS A SOURCE OF INDIVIDUAL DIFFERENCES

It may be difficult to imagine that scientific discoveries – the sorts of discoveries that arise from laboratory experiments in test tubes and under microscopes – can actually contribute to the knowledge base for teaching and learning. Since the completion of the Human Genome Project, an international effort to map and sequence the human genome or complete set of human genes, scientists have been working to identify variations in the human genome that make us all different. Genomic refers to an organism's entire genetic make-up, whereas genetic refers to specific, individual genes (Center for Genomics and Public Health, n.d.). Genomics, a relatively new area of scientific research, includes the study of how the genome interacts with the environment to account for differences among individuals with similar genetic make-up. This kind of information about gene–environment interaction can be a powerful tool for educators. For example, almost two decades ago, from the perspective of educational psychology, Csikszentmihalyi and Schmidt (1997) encouraged educators to use knowledge of adolescent development from a biological perspective to create optimal experiences for youth. For example, they suggest that educators with an understanding of youth's natural propensity for resilience might capitalize on adolescents' potential for turning stressful situations into positive experiences. Without such an understanding, educators might attempt to reduce conflict and stress, which may in turn, lead to maladaptive behaviors, rather than help these resilient young people develop competence in dealing with stressful situations. For centuries, educational psychology has provided the knowledge base for teaching and learning; yet, teachers have traditionally been hesitant to base their practice on established research (Randi, 2007). Is it reasonable to expect that this new research from molecular biology on the human genome would be any more enthusiastically embraced by teachers than research that has been offered to them in the past?

Genetic/genomic information may be available to educators in the not too distant future (Plomin, 1998). Knowledge of what this genetic/genomic information means *and what it does not mean*, what it can *and what it cannot predict* ought to be important for educators. Teachers' reluctance to accept established educational research principles that apply to their practice seems to arise in part from the ways research is typically disseminated to teachers, such as with scripted programs of instruction that discourage teachers from making their own decisions about teaching individuals in their classrooms (Smagorinsky, Lakley, & Johnson, 2002).

Will the availability of genetic information prompt the development of scripted interventions matched to students' genetic make-up? Or will genetic information be available to teachers as a decision-making tool? In any case, as Bouregy, Grigorenko, Tan, and Latham (this volume) advise, it may be prudent for teachers to develop understandings about the "new" genetics so they are able to distinguish fact from fallacy and to critically evaluate what is likely to be disseminated to them, so that they might avoid misconceptions that may inappropriately influence their beliefs and practices. Sternberg and Grigorenko (1999) provided several examples of teacher misconceptions about how genes and the environment interact. Specifically, some educators may believe that if a disorder is genetic, no intervention will have an effect; others may believe that information about individual students' genetic traits may lead to labeling, discrimination, and self-fulfilling prophecies (see also Hodapp & Fidler, 1999). In the future, certain forms of genetic/genomic information may be available to teachers, and, if used appropriately, can add to the pool of evidence teachers use to make informed decisions about teaching individuals in their care, including selecting appropriate interventions early on, for particular genetically influenced disorders, such as Down's syndrome (see Asbury, Rimfield, & Kraphold, this volume).

Genomic Variations

Completed in 2003, the Human Genome Project was an international effort to produce a map and a sequence of genes on the chromosomes of humans (US Department of Energy Office of Science, 2003). The human genome (the complete set of human genes) has been compared to a book that has twenty-three chapters, each representing one of the twenty-three pairs of chromosomes in the human genome (Ridley, 1999). To continue Ridley's analogy, each individual's genome has the same chapters and the same paragraphs, in the same order. Those sequenced chapters are common to all humans and thus make all humans alike at the same time that they make us different from chimpanzees and other species. Our capacity for language, for example, is distinctly human.

Just as humans are distinguished from other species by their genome, individuals are distinguished from each other by variations within their genomes. Using Ridley's book analogy, these variations occur in "words" on the pages of the human genome, such as different spellings of words, typographical errors, or transposed letter sequences. Some of these variations are present from conception, while others occur through interactions with the environment. For example, infants' language development is strongly influenced by the language(s) they hear and learn to speak; the environment shapes a child's brain (McLeod & Bliele, 2003). The innate ability to discriminate speech sounds is altered by experiences in the cultural and linguistic environment. Infants can discriminate the full range speech sounds but subsequently lose the ability to discriminate certain sounds that they do not hear in their native language (Molfese et al., 2005). We can observe some of these differences, such as language, physical appearance, or personality differences with the naked eye as measurement. Yet, there are other genetic variations that cannot be so easily observed, such as genes that may lead to a disease, or may predispose an individual to hyperactivity, or a language disorder.

The complete human genome that has been constructed is actually based on the DNA of only a few individuals (Venter et al., 2001). Scientists are now studying DNA from a broader array of individuals to identify variations within the human genome. New technologies for analyzing DNA are making it possible for scientists to study what causes genetic variability among individuals. One recent study published in *Science Express* found that even in healthy individuals, there was a great deal of variation in their DNA sequence as well as environmentally influenced variations in the proteins of cells, suggesting that a common gene (such as a gene for a particular disorder) may vary in its structure from individual to individual and that the human genome may be more variable and flexible than previously thought (Snyder et al., 2010).

So what does this mean for educators? Which of the individual differences that teachers observe in students might be wired in their genes? Which of these differences are the result of experiences that have shaped students as learners? The answer to these questions most likely lies somewhere in between, at the interaction of genes and the environment.

Molecular biologists are beginning to discover some differences between students' cognitive abilities that appear to be of genetic origin. The Human Genome Project has been pinpointing particular regions in the human genome that are responsible for human characteristics, such as language, and new technologies are making it possible to study how individuals' genetic make-ups differ. This does not mean, however, that how well a student performs in school is determined entirely by genetic make-up (Thomas et al., 2015). Molecular biologists study the biological mechanisms of inheritance and gene expression. Individual differences are explained, at least in part, by what happens outside the genotype or gene itself. Factors in the environment, including biological or chemical reactions, as well as differences among families, cultures, and classrooms, can change how genes are expressed or appear. Recent research is providing strong evidence that the interaction between genes and environment is critical in determining what an individual is like. Moreover, the human genome changes across an individual's lifespan (see Grigorenko & Dozier, 2013). Whether or not cognitive abilities and other factors that affect learning are of genetic origin or acquired, there is increasing evidence that education (an environmental input) can contribute to the developmental changes that occur across an individual's lifespan, as well as to gene expression. Findings from the new science of epigenetics should be particularly salient to teachers. Epigenetics is concerned with environmentally influenced changes in gene expression that are passed down from generation to generation (Grigorenko & Dozier, 2013). What does this mean for teachers? Potentially, the different school and classroom environments children experience, over time, can influence gene expression, not only for children growing up today, but for future generations.

Genes and the Environment

Historically, philosophers, scientists, psychologists, and others have debated which influences human behavior more, genes or the environment. Behavioral geneticists, who seek to understand what accounts for individual differences, are interested in studying the interactions between genes and the environment, studying heritability or the proportion of genetic (versus environmental) influences on gene expression within a particular population (see Tan, this volume, for a discussion of heritability). In the past, classic scientists (see, e.g., Galton, 1876; Terman, 1925) have argued that many talents are innate, giving rise to educators' fears about determinism and doubts about self-efficacy: they may ask "Can teaching make a difference?" The new science of epigenetics is raising awareness that the nature–nurture issue is not an "either–or" issue, nor are there simple explanations about which is more influential on an individual's development. Rather, human development is much more complex, and must be understood within the context of gene–environment interactions. Most important, given the findings of the new science of epigenetics, studies from the scientific community are confirming that classroom environments and the instruction teachers provide matter a great deal.

It is important to understand that environment is broadly defined by geneticists as any non-genetic factor that may influence gene expression, including biological events and chemicals that can turn genes on and off (Plomin & Asbury, 2005). In laboratory studies, researchers are demonstrating the importance of experience and other non-genetic factors in regulating gene activity (Champagne & Mashoodh, 2009). Some scientists are studying the biological mechanisms that influence the expression of genes, or epigenetics. Epigenetic factors influence gene expression (the appearance of the gene as an observable trait, or a phenotype), but do not change the genotype, the genetic make-up of the cell itself. Even individuals who share the same genotypes can differ widely from each other (Holliday, 2006). For educators, this research supports conventional thinking among psychologists and social scientists that education, experience, and other forms of support (i.e., non-genetic factors in the environment) matter when it comes to gene expression. On the other hand, educators ought not ignore what is in the genes, because as Plomin and Asbury (2005) emphasize, developmental change depends on the interaction between nature and nurture, not one or the other.

How interactions between the environment and the individual's unique genome influence behavior and development may be especially salient in classrooms where different individuals share the same environment. A similar shared environment occurs in families where siblings are raised by the same parents, live in the same household, and share similar parenting experiences. Yet, siblings are often very different, with different interests, personalities, and behaviors. Parents express frustration that they raised all their children in the same way, yet one is successful and productive while the other is unruly and troublesome. Plomin, Asbury, and Dunn (2001) have studied environmental influences on children growing up in the same family who differ, despite similar genetic make-up. Their study suggests that siblings may react differently to shared environments and that peer influence and other experiences outside the family, may be important sources of non-shared environment that account for these familial individual differences.

One can extrapolate these findings to the shared classroom environment. The classroom accounts for only part of the educational environment. For example, educators and psychologists have pointed to peer influence and other factors outside the classroom that may impact learning. At the same time, within the same classroom or group setting, individual learners may react to a common or shared learning experience in different ways. This is why adaptive teachers provide varied learning experiences tailored to individual differences. Similarly, genetic research is beginning to support the idea that the same environment may be beneficial for some students and detrimental to others because individuals react differently to the environment (and instruction) they share. Future research on shared and non-shared environments may find that the shared classroom environment is more important than previously thought. What may be critically important is not that the environment is conducive to learning in general or on average, but that the environment is sufficiently flexible to be conducive to *different* learners at low, middle, and high ends of the distribution.

In short, findings on genome variations may one day be able to provide information about structuring learning environments to reach different types of learners as well as explain why some individuals thrive and others struggle in the same environment. For thoughtfully adaptive teachers who have always resisted "one size fits all" scripts, empirical research that demonstrates how genetic variations account for the individual differences they have noted since antiquity should be especially affirming. This research may also explain why individual differences are often especially salient in group settings, such as traditional classroom learning environments.

Recent developments in the field of genetics are coming to a common understanding of the role of gene–environment interactions in contributing to individual differences. Researchers are hopeful that studies of the mechanisms that undergird learning and development will lead to an understanding of why particular educational methods are effective and for whom (Thomas et al., 2015). It stands to reason that educators ought to be part of this conversation. Without an understanding of how to use genetic information to structure environments in ways that will, for example, minimize heritability, rather than exaggerate differences, teachers may find that nature overcomes nurture, rather than interacts appropriately with what nurture has to offer.

Reconciling Differences

The "either/or" nature–nurture debate may be coming to a close, reaching a common ground somewhere in the middle. There remains, however, considerable disagreement about what is of genetic origin and what is not. Some of the disagreement stems from the different methods scientists use to predict heritability. Heritability is defined as a population statistic or "the proportion of phenotypic variability (individual differences) in a trait or behavior explained by genetic variation, with the rest of the variation assumed to be of environmental influence" (Thomas et al., 2015, p. 2). Through different methods, molecular biologists and behavioral geneticists have worked to identify the particular genes responsible for the heritability of complex traits and disorders (Plomin, 2013). In laboratory studies, molecular biologists study the actual biological pathways that mediate gene expression. Behavioral geneticists explain heritability by determining the statistical probability that a particular gene associated with a behavior predicts a phenotypic outcome, or the expression of the behavior (Meaney, 2010). Studies in molecular biology typically derive lower estimates of heritability than the estimates determined in statistical methods in behavioral studies. This discrepancy has been termed the "missing heritability" problem (Grigorenko & Dozier, 2013). Regardless of differing methods and conclusions, genetic research is offering new insights about the causal mechanisms underpinning learning. Genetic research can complement the work of educational psychologists and teachers by contributing to understandings about why current and future educational methods work and for whom (Thomas et al., 2015).

For decades, behavioral geneticists have used statistical methods to estimate the degree of heritability of certain traits by studying shared characteristics of twins. Twin studies are a valuable source of data for behavioral geneticists because identical twins share 100 percent of their genes, whereas non-identical twins, like the rest of the population, vary in their genetic make-up from individual to individual. Both identical and non-identical twins, however, share the same family environment, thus minimizing the environmental influence on shared traits (see, e.g., Plomin, Asbury, & Asbury, 2001). Plomin, Asbury, and Dunn (2001) pointed out that statistical methods employed in twin studies are not sufficient to explain why children in the same family, even twins, are different. Later, Plomin and Daniels (2011) studied differences between children in the same family, including siblings whose genetic make-ups are similar and adopted children living in the same family (see also Plomin, 2011). They concluded that different children, whether related or not, have different reactions to the same family environments; how similar environments are experienced or perceived by children is a primary driver of their behavioral development. Thus the same environmental influences may make children in the same family more different than similar. This is a lesson for classroom teachers: providing all children the same instruction and the same learning environment, may actually make them more different. Some students will thrive in the environment provided; others will flounder in the same environment. Will treating each student differently, by addressing each student's learning needs and interests, help all students adapt to the same instructional environment, making them appear more similar than different? Writing about teaching and learner variation from the perspective of educational psychology, Randi and Corno (2005) have suggested that successfully adaptive teachers actually make students more alike than different, by adapting their instruction to individual student differences, simultaneously adapting learners to the instructional context, and thus facilitating their own teaching (see also Randi, in press).

Some scholars have criticized the methods used by behavioral geneticists, discounting that any behavior in response to environments can be considered genetic in origin (Charney, 2008). Charney points out the absurdity of studies concluding that beliefs, such as political ideologies, are genetically influenced. He does, however, acknowledge that intelligence may be at least partially heritable. Charney (2012) emphasized that the complex interactions between individuals and their environments make it especially difficult to predict what an individual will be like or how that individual will respond to the environment, based on genotype alone. Charney explained individuals' ability to adapt to their environments as "adaptive phenotypic plasticity" or the ability of phenotypes to change in response to the environment individuals inherited (i.e., share with their parents). Thus, genetic phenotypes or "traits" are more dynamic than previously thought. These advances in genetic research parallel recent research in social-emotional learning that demonstrates how children can be taught to adapt to environmental stress, by controlling emotions, thoughts, and behaviors in different situations (Dufall, 2016). The idea that instruction in social-emotional learning might actually be promoting changes in gene expression is powerful.

The reification of genetic make-up as a dynamic process may explain some of the paradoxes that have surfaced in debates about whether or not cognitive abilities are inherited. Thomas and colleagues (2015) describe two developmental phenomena, one concerning height, which is generally presumed to be of genetic origin, and the other concerning intelligence, the heritability of which has been the subject of much controversy (see, e.g., Asbury & Plomin, 2014; Flynn, 2007). To demonstrate the paradox, Thomas and colleagues (2015) cited studies that have demonstrated that the average height of men, although considered to be of genetic origin, has increased over the past 150 years, which can be attributed to environmental factors such as better nutrition and health care. Intelligence scores have also increased over the last five decades (Flynn, 1999). Thomas et al. (2010) suggested that practice and experience with abstract reasoning of the kind used to measure general intelligence on IQ tests might account for the increase ("the Flynn Effect"). Other scholars attributed this rise to environmental influences, such as exposure to stimulating environments created by television and the Internet (Nettlebeck & Wilson, 2013). Whether or not intelligence has a genetic origin, like height, the point here is that environment matters and intelligence, even on a population level, is demonstrably dynamic rather than fixed.

Academic Achievement and Individual Differences

It is difficult to imagine that academic achievement can ever be attributed to one's genes. Education and other forms of training have been found to increase cognitive abilities and expertise in different domains. Moreover, definitions of academic achievement vary from test score measurement to school completion with much measurement variation in-between. (For a review of the research on cognitive abilities measurement and other cognitive factors, see Kyllonen, 2016.) And yet, some researchers claim that how children perform in school can be explained by the differences in their genome. For example, Krapohl et al. (2014) suggested that there are many genetic factors that can influence achievement, and these authors include among genetic "traits" intelligence, self-efficacy beliefs, personality, and behavior problems, along with other factors that influence how easy and enjoyable they find learning. For a geneticist, a "trait" is the expression of a gene, or an observable behavior that appears to be of genetic origin (i.e., located in an identified region of the genome). In the case of intelligence, which Ridley (1999) calls "the most impenetrable and least easy of all the brambles in the genetic forest" (p. 76), geneticists have located intelligence on chromosome 6. Chorney and colleagues (1998) found that a sample of teenagers with high IQ scores (>160) shared a particular genetic sequence in a gene on chromosome 6 that was different from the sequence in other people. These researchers cautioned, however, that the genetic contribution to general cognitive ability (g) is difficult to establish because g is likely associated with several genes that have small effect sizes (quantitative trait loci). Moreover, cognitive ability involves environmental as well as genetic sources of variance. They concluded that there are no "genes for genius" as cognitive ability involves much more than genes.

In the new genetic paradigm, it is important to understand that the expression of traits is considered to be dynamic and situational, influenced by factors in the environment at any given time. Moreover, environmental factors are thought to influence expression patterns in the genome itself. Because the environment actually influences changes in genetic expression patterns, it is logical to hypothesize that environmental factors and opportunities (i.e., education) might influence alterations in gene expression, accounting for individual differences in cognitive ability. As Asbury, Rimfield, and Krapohl (this volume) point out, there is still much work to be done in order to understand the complex interactions between environmental influences and academic achievement, including understanding when and how these associations may be mediated by genes.

It is in this genetic context that Krapohl et al. (2014) conducted their study of educational achievement. Nevertheless, this research is still controversial. For example, some would question labeling self-efficacy and other beliefs as "traits" based on studies that correlated self-efficacy and other factors with academic achievement (e.g., Diseth, 2011). Krapohl et al. (2014) reported that identical twins they studied not only shared genes but also educational outcomes, such as similar grades on exams. In addition to exam grades, Krapohl et al. correlated scores on measures of intelligence, self-efficacy, and personality with achievement, and used data from twins to infer which of these so-called "traits" might be inherited. They concluded that intelligence accounted for more of the heritability of educational achievement than other behavioral "traits" they studied, especially in the core subject areas of English, math, and science. Krapohl et al. (2010) explained that phenotypic (expressed gene) correlations between such traits and educational achievements can be mediated genetically or environmentally, suggesting that environmentally mediated phenotypic traits can be shaped by educational interventions. Regardless of whether or not these inferences about heritability can be considered valid, data from molecular biology studies have demonstrated that genetic phenotype traits are dynamic and responsive to environmental influences. Krapohl and her colleagues hypothesize that, if all children were provided sufficient instruction to succeed at academic tasks, they would be more alike than different in their achievement – in effect, that factors in the environment could erase genetic influences. On the other hand, without environmental effects (i.e., education), genetics would be solely responsible for differences in children's achievement and those differences would be more pronounced. The point here is that environments can be structured in such a way as to reduce the potential impact of genetic differences. Researchers concerned with the educational implications of genetic research hope that, some day, genetic information can be used to identify which individuals will benefit most from particular interventions (Thomas et al., 2015). This genomic science view on tailored instruction is not unlike the theory of adaptive teaching, which posits that adaptive teachers create a common teaching ground, adjusting instruction to like groups of learners and learners to particular forms of instruction to minimize individual differences (Randi & Corno, 2005).

Research on the heritability of cognitive traits is, to date, inconclusive. Since the original study that associated IQ with a particular region in the genome (Chorney et al., 1998), more recent evidence suggests that aspects of intellectual ability are of genetic origin and associated with a particular region in the human genome (Benyamin et al., 2014). In addition to intelligence, certain talents may be genetically influenced. For example, exceptional athletic competence may have a genetic component but the contributions of nature are difficult to disentangle from the contributions of nurture. So-called "giftedness" in a particular area is a good example of how gene–environments interactions contribute to the range of individual differences. Dai and Colemen (2005) discussed giftedness as a norm-referenced concept that depends on just how rare one's talent is. For example, a five-year-old pianist who can play Chopin etudes might be considered "gifted," yet for an adult trained musician, such a feat would not be considered exceptional competence. The point here is that musical talent can be nurtured and innate talents can manifest with varying degrees of competence, depending on how they are nurtured. Another theory (Jensen,

1998), however, is that exceptional talent comes about as the result of a rare combination of inherited genes, as the luck of the draw so to speak.

"Creativity" is another, often vaguely defined, quality with aspects that may be heritable (Simonton, 2008; Tan & Grigorenko, 2013). It is unlikely that a single gene for creativity or constructs like this will be identified, however; there are numerous cognitive and dispositional factors that contribute to creativity and genetic influences operate in complex ways over the course of development (Simonton, 2012). For example, an innate musical ability might facilitate and speed up musical talent acquisition, as may be the case for a young musical prodigy. According to Simonton, "nature enhances nurture" by self-selection of experience (Simonton, 2012, p. 220). Put another way, the interaction of nature and nurture reinforces a propensity that can account for a wide range of individual differences, including the range of developmental differences occurring among students in a typical contemporary classroom (see Stanford Aptitude Seminar, 2002). More generally, all students can learn, but not all students will learn in precisely the same way, and at the same rate. Understanding individual differences as a result of natural qualities that may be "speeding up" learning for some more than others may counter the possibility that children who lag behind their peers will be labeled as low achievers.

Given that some research on the role of genes in academic achievement is controversial, adopting educational practices derived from this research without a basic understanding of genetics, runs the risk of structuring learning environments that will accentuate, rather than minimize differences. Leaving nature to its own devices, or capitalizing on what may be merely perceived as innate talents, will only serve to create a wider gap between successful and unsuccessful learners. In their controversial book, *G Is for Genes*, Asbury and Plomin (2013) argued that genes play a role in predisposing individuals in particular ways, such as a propensity for math or music. They advocated that potentially discoverable genetic predispositions in children should be nurtured and developed. These authors support the idea that intelligence, although it appears to be genetically influenced, should be considered malleable rather than fixed. Asbury and Plomin made a strong case for personalized learning environments and instruction tailored to individual student differences. The idea that intelligence is not fixed and that instruction should be tailored to individual differences is likely to be welcomed by adaptive teachers. On the other hand, some of the practices advocated by Asbury and Plomin are controversial, such as using genetic information to label or categorize children. They also advocated paring down the curriculum to basic skills and offering students additional options based on their genetic predispositions for things such as musical talent. They recommended leveraging technology (e.g., personalized tutoring software) to create personalized learning situations. Other researchers (Thomas et al., 2015) have pointed out that aligning environments with special talents, such as developing musical talent to exclusion of other general skills, can exaggerate gene expression, making differences more pronounced. Asbury and Plomin's book is an eminent example of the kinds of recommendations coming out of genetic research that can influence policy, and through it, practice – for better, or worse. These may also be the kinds of educational "innovations" that leave teachers skeptical of what research can offer.

GENETIC/GENOMIC RESEARCH AND PRACTICAL APPLICATIONS

Among the public, there has been much interest in health studies that identify genes that predict the likelihood of disease and the role environment plays in gene expression. Of particular interest is the science of epigenetics, a promising line of research that investigates biological mechanisms that influence the expression of genes. The expectation is that this line of research into the etiology of diseases such as cancer, can lead to individualized and targeted interventions with fewer side effects (Collins & Mansoura, 2001). In the future, there may be ways to design interventions not only for diseases but for mental and behavioral disorders as well. Thomas et al. (2015) consider this one of the most achievable near term goals for the practical application of genetic research.

Mental Disorders

In the area of special education, etiology-based interventions are becoming possible. In 1999, Hodapp and Fidler predicted that the genetic profiles of children with developmental disabilities could be valuable in designing environmental interventions. Recall that geneticists broadly define "environmental intervention" as any influence outside of the genes. In the case of developmental delay or disability, the authors explained that severe symptoms may be avoided by early genetic screening and feeding affected infants a particular diet that regulates gene expression. In contrast, without such early genetic screening and intervention, traditional education interventions, such as behavior modification, would be necessary to remediate behavior. Research on the genetic basis of autism and other developmental disorders seems especially promising (Eapen, 2011).

More than a decade later, another special education researcher (Reilly, 2012) provided further examples of how knowing the etiology of a student's special education needs could inform the work of educators in the design of cognitive and behavioral interventions. Although children with intellectual disabilities share many of the same deficits in neuropsychological functioning, there are particular patterns of abilities and deficits associated with different genetic syndromes. Reilly proposed that educators use guidelines for selecting education interventions matched with different learning profiles, such as providing phonics-based reading instruction to children whose genetic syndrome is associated with strengths in auditory processing. Reilly also emphasized that some relative strengths and weaknesses are shared by children across syndromes; and, in these cases, the same interventions can support the needs of students with different genetic syndromes, such as individual work stations for easily distracted children, whether they are affected with attention-deficit hyperactivity disorder (ADHD), or genetic syndromes, such as Fragile X or Williams syndromes, both of which are associated with high distractibility. Reilly explained that, without information about children's genetic syndromes and their profiles, teachers choose interventions based on their informal observations of children. Genetic profiles and information about a child's particular syndrome and patterns of deficits and strengths associated with that syndrome could guide educators to interventions that are more likely to lead to positive outcomes. Even with this information, however, the author emphasized that teacher judgment is still critical in that there is often considerable variability across individuals with the same genetic syndrome.

Genetic research is also beginning to address etiologies for reading disorders. There has been considerable debate in the literature about whether or not dyslexia is a generic term that encompasses a variety of related reading disorders or a separate disorder (Elliott & Grigorenko, 2014). Regardless of this debate, studies of the genetic origins of language impairment and other reading disabilities related to poor reading could facilitate earlier diagnoses and more effective interventions. One recent study conducted by researchers at Yale (Powers et al., 2013) investigated the genetic components that underlie reading disability and language impairment. They identified the genetic variants – specific parts of the gene that are responsible for the disorders, and they found that some variants of a gene regulator (which they termed READ1 or regulatory element associated with dyslexia1) are associated with reading problems. Other variants are associated with language impairment. Individuals with both variants have multiple risks that impact reading, language, and general intellectual ability. Their findings underscore the complexity of gene–gene interactions. They concluded that the magnitude of the risk for developing reading disorders depends on how these genes and others interact, as well as environmental factors. With continued research along these lines, genetic testing could predict which children are at risk and the magnitude of the risk before symptoms appear. This is especially critical for early childhood education where expulsion rates for preschool children due to behavioral concerns exceed those of elementary and secondary school students (Fox & Hemmeter, 2009). Currently, preschool educators use behavior modification techniques to address problems, often labeling inattentive and unruly children as ADHD. Fox and Hemmeter suggested that language and communicative impairments might also be a cause of similar behavioral problems. Understanding the genetic etiology of disorders that can potentially cause problems in school may not only lead to earlier interventions, but also to interventions that address the root of the problem, whether it is ADHD or a communication disorder.

Language impairment and deficits in auditory processing are not the only disorders that impact reading development. Stein (2014) theorized that impairments in visual processing of genetic origin can contribute to dyslexia and may account for some of the other symptoms associated with dyslexia, such as difficulty focusing attention, poor sequencing, and left–right confusions. Stein developed a strong argument supporting the "M cell" theory, but more research is needed to confirm this hypothesis. Stein suggested that nutrition supplements could be important for proper functioning of M cells. Stein also suggested several interventions, such as the use of colored filters. While the use of visual treatments such as colored filters and special glasses with one lens occluded were tested in double-blind studies and found to be effective in improving vision problems in his samples, the caution here is that genetic research has not yet demonstrated deficits in the M nerve cells as an etiology of dyslexia.

There are lessons to be learned. For teachers who often hear children complaining that letters are jumping around or that reading gives them headaches, the M cell theory is intuitively appealing. Even if visual deficits contribute to reading disabilities, not all children with reading deficits have visual deficits rooted in their genes. For example, the symptoms teachers note can be caused by poor readers' frustration or other non-genetic causes. Researchers in molecular biology and behavioral genetics, however, could potentially work together to draw valid conclusions about the etiology of reading deficits and identify appropriate interventions based on genetic profiles. A basic understanding of how genetic studies can explain and pinpoint the causes underlying reading disorders in individual students might then be available to assist teachers in selecting tailored interventions that go beyond more generic visual treatments, such as special glasses, or nutritional supplements for all poor readers.

There is another reason for identifying the etiologies of specific disorders. Byrne, Khlentzos, Olson, and Samuelsson (2010) discussed the impact of a hypothetical reading intervention that could raise all children's reading scores above the level that could be classified as a reading disability. As these authors explained, individual differences of genetic origin may remain between students but those differences will no longer impede academic progress, if all children can read well enough to access content in other school subjects. The caveat here is that it is unlikely that a single intervention will raise the reading scores of all students. Knowing the genetic etiology or the cause of a given reading disorder should facilitate the design of interventions tailored to individual student needs.

Genes That Influence Social Behavior

In addition to the mental disorders just described, some social behaviors have also been shown to have genetic origins. Like cognitive abilities, social behaviors are environmentally mediated. Prosocial behavior (behavior that benefits others, such as volunteering or sharing toys with friends) may have some genetic origins, but environmental factors have also been found to contribute to individual variation in prosocial behavior, including parent modeling and peer influence (Knafo & Plomin, 2006). Prosocial behavior is an example of a personal quality that is positively influenced by intervention; intervention studies have been conducted even before the proliferation of genetic studies (see, e.g., McMahon & Washburn, 2003; Solomon, Watson, Delucchi, Scraps, & Battistich, 1988). A better understanding of which, if any, genetic origins of prosocial behavior can be confirmed and how these genes interact with the environment could once again lead to earlier identification of children at risk and targeted interventions.

Prosocial behavior does seem influenced, at least in part, by genetics. Genetics also plays a role in influencing antisocial behaviors. Historically, educators have relied on behavior modification techniques to improve students' behavior. Behavioral modification as a general intervention, such as the use of rewards and punishments to shape behavior, has a long history in educational psychology. Some teachers claim, however, that these techniques are variably effective, especially in cases of severe behavior problems. In instances like this, it may be that genetic research will yield knowledge of the inherited traits that can predispose individuals to aggressive and dissocial behaviors. Identification of the genes contributing to the etiology of these behaviors may then produce new forms of interventions, including remediation of biological hazards and chemical imbalances that may have contributed to gene expression. Behavioral geneticists have demonstrated that childhood conduct disorder is genetically influenced and associated with alcohol dependency later in life (True et al., 1999). More recently, molecular biologists, Dick and colleagues (2004) also concluded that conduct disorder is genetically influenced. They conducted a genome-wide study to identify genes contributing to conduct disorder and found that those genes may be shared with those that contribute to alcohol dependence. They also found another genetic region contributing to conduct disorder but not associated with alcohol dependence. These scientists called for further studies to investigate the linkages between alcohol dependence on conduct disorder. If studies such as these can lead to the development of interventions, the new science of genomics may counteract deeply entrenched fears of biological determinism.

One particularly promising line of research that potentially informs behavioral intervention programs is the emerging field of social genomics. Researchers in this field study the relationship between genes and social behavior. They argue that at times, genes influence social behavior, but at other times, social behavior may influence gene expression. For example, in the health field, Cole (2009) cited dozens of situations in which an individual's life style (e.g., diet, stress) can trigger or and turn off genes that may lead to disease. Another study in behavioral genetics investigated the influence of the school and peer environment as a contributor to gene expression (Boardman et al., 2008). These researchers concluded that smoking is highest within schools in which the most popular students are smokers. It is not difficult to imagine that research may one day be able to predict how school and classroom environments can be structured to turn off genes that lead to undesirable behavior and turn on genes that lead to more positive behavior.

The gene–environment interaction is complex and more research is needed to investigate the millions of individual variations in the genome. Studies of shared and non-shared environments have shown that different students react to the same environment in different ways (Plomin, Asbury, & Dunn, 2001). It is not simply a matter of inherited genes but how individuals' experiences (even in the case of twins) shape individuals' manifested "traits" and how individuals shape their environments. Advances in molecular biology may broaden the range of interventions available to include cognitive and behavioral interventions, using biological and chemical interventions that have an effect on gene expression. This research may also lead to recommendations for teaching students to draw upon techniques for the social regulation of genes that may have the same positive effects for learning that have been shown in disease prevention.

In medical science, physicians and pharmacologists are working together to analyze individual DNA samples and predict which pharmaceutical treatments will be effective for particular individuals. In education, neuroscientists likewise have begun to identify the etiology of mental and behavioral disorders, increasing the likelihood of more effective interventions. Similar findings about the etiology of learning disabilities may lead to more tailored and more effective interventions and environments. With these innovative types of interventions on the horizon and the increasing availability of genetic information, attention may shift away from school improvement to improved instruction for individual students, but these new practices will undoubtedly also raise questions about the role of genetics in education.

IMPLICATIONS FOR PRACTICE

Imagine a group of teachers gathered around a conference table, reviewing reams of student data, and deliberating about instructional decisions, student groupings, and appropriate interventions. Teachers' required participation in such data team meetings is not new; data-driven decision making has been a common educational practice for some time now (see, e.g., Mandinach & Honey, 2008). Typically, the information available to teachers includes such data as students' scores on norm-referenced achievement tests and/or state-wide criterion-referenced standardized tests as well as the results of classroom-based measures of reading comprehension and oral reading fluency, and teacher-developed unit tests in math, language arts, and other subject areas. Teachers may also review report card data, which in some cases, include anecdotal comments on individual students, such as their effort, interests, self-confidence, and behavior. Often, the results of IQ tests and other specialized tests administered by school psychologists to identify students in need of special services are available to teachers in these data team meetings. The goal of these data team meetings is to identify instructional approaches tailored to the needs of a given small group or groups of students, as well as individual students. For example, teachers often use the data to group students homogeneously for instruction tailored to the strengths or weaknesses.

What if, in addition to achievement data, genetic profiles of students become part of the data set available to teachers at these data team meetings? Even without this additional information to review, teachers complain that data team meetings take them away from the classroom. The emphasis on data use for school improvement and comparisons between groups taught by different teachers has made teachers skeptical about data (Mandinach, Parton, Gummer, & Anderson, 2015). The availability of genetic profiles adds to existing concerns about data use, including the ethical and appropriate use of genetic data families agree to share with schools. Already, there are concerns about the ethical and appropriate use of non-genetic assessment data by educators. Mandinach and her colleagues provide guidelines for ethical and responsible data use, including protecting student privacy and confidentiality of data, as well as guidelines that promote good relationships with parents. These guidelines may be even more critical if genetic information becomes part of the shared data set. Moreover, additional guidelines may need to be established for the responsible use of genetic information.

The Human Genome Project anticipated ethical and policy issues from the outset, so scientists have been identifying and addressing these issues (The Human Genome Project, 2008). For example, issues center on the privacy and confidentiality of genetic information and fairness of use of the information by schools and other agencies. Researchers at the US National Human Genome Research Institute (Collins, Green, Guttmacher, & Gruer, 2003) called for policy on the use of genomic information in medical and non-medical settings. For example, they questioned whether or not genetic information on predisposition to hyperactivity should be available to schools or if such behavioral information should be admissible in a court of law. If information on behavior that is genetically influenced might affect jury decisions, it could potentially influence teachers' expectations about students as well. In education, genetic information used to track and label students raises many ethical issues. It is quite possible that some parents will make genetic information available and others will not. Does this disadvantage a student who might benefit from tailored intervention? Also of concern is the psychological impact and potential discrimination due to an individual's genetic make-up. For educators, this means that genetic information should be one among many tools to improve an individual student's learning or behavior, not a reason to expect less or more of that student. If teachers can tend to have lower expectations for students they perceive to be low achievers (Rosenthal & Jacobson, 1968), will *knowing* that a student's genes predispose the student for a disability influence their expectations even more? These are serious issues, which, fortunately, are beginning to be discussed in anticipation of the availability of genetic data (see, e.g., Bouregy & Bouregy, this volume).

Apart from ethical considerations, it is critical for educators to acquire "genetic/genomic literacy" as Tan, Grigorenko, Bouregy, and Latham (this volume) advise. Unfortunately, the impetus to make genetic information available to educators may lead to unreasonable recommendations and policies about how teachers should acquire knowledge about genomics, such as additional coursework in genetics and rigorous assessments of genetic knowledge as requirements for teacher licensure. It seems to make more sense to integrate discussions about the *implications* of genomic research into educational psychology courses, where teachers can hone their skills in using genetic information to select educational methods matched to individual students' learning needs. Already some textbooks in educational psychology and human growth and development are including recent developments in genomics and epigenetics (see, e.g., Woolfolk & Perry, 2015), but the field is advancing so quickly that providing the most up-to-date information through textbooks will be challenging at best. Alternate routes to teacher licensure, which provide more practical experiences and little formal instruction in the knowledge base, pose another challenge to ensuring new teachers will enter with the knowledge base necessary to understand the implications of genomic science.

Nonetheless, there are basic concepts of the new science of genomics that are especially critical for educators to understand, including an understanding that genes *do not* control destiny and that environments (i.e., non-genetic factors) *can be* structured in ways that minimize genetic influence. Thomas and colleagues (2015) explain individual differences as a rank order in a class, not an arithmetic average – that is, as where an individual stands relative to others in the class, not as how well the whole class is doing. If the environment changes for all students, then all students can potentially move up, even if the rank order (on some genetic variable) stays the same. If the educational environment (the instructional scaffolds, interventions, and other forms of support) is not provided, then heredity explains the differences. The implication is that changing the environment can influence the expression of genes. Geneticists Thomas and colleagues (2015) explained, "environments can be matched with genetic differences in such a way as to reduce the effect of those differences" (p. 4). That is, educators can reduce the differences by strategically designing environmental conditions for those individuals whose genes put them at risk for developing atypically. Strategically adaptive teachers, even without this genetic information, have been designing instruction tailored to individual differences and helping learners adapt themselves to the learning environment (Corno & Snow, 1986). One would hope that genetic/genomic information will facilitate, rather than constrain the work of adaptive teachers, by providing them another source of data on which to base their instructional decisions, rather than a genetic prescription.

In short, the findings from genetic/genomic research are likely to become part of the educational landscape, and will no doubt begin to influence educational practice. If teachers are not sufficiently familiar with genomic science, they may misinterpret the research, especially if the "research" is disseminated to them through popular media, or professional literature often three or four times removed from the actual studies, Genomic research (and students' genetic profiles when they become available), like educational research, and student achievement data ought to be provided to teachers as decision-making tools; so teachers can use the information to select appropriate interventions or design instruction tailored to the particular students in their classrooms. Genetic/genomic science is complex, and except in rare genetic disorders, it is unlikely that genetic profiles can be used to predict outcomes with confidence (Charney, 2012). An intervention targeting a particular learning disability may be appropriate for some students but not other students with the same disability. Individuals differ, not only in genetic make-up that might predispose some and not others to particular learning challenges, but also in a myriad of other ways, both of genetic and non-genetic origin. Teachers work with students in classrooms every day of the school year; they not only have access to "test results" but they also observe students, noting their particular strengths, weaknesses, interests, work habits, and a host of other factors that make individuals different from one another. Despite the rapid advances in genomic/genetic science, despite advances in technology that facilitate personalized learning, in the end, it is the teacher whose decisions, skills, and diligence matter the most for the students entrusted to her care.

Teachers, often skeptical of research, might be encouraged that genetic/genomic science may be confirming what adaptive teachers have noticed since antiquity. As Ridley (1999) explains, an individual's environment is as much a consequence of genes as it is of external factors, in that the individual seeks out and creates his own environment according to propensities. Teachers serve a critical role in creating learning environments for children – environments that contribute to their developmental trajectories. Whether children come to school endowed with innate talents waiting to be cultivated, or arrive with a variety of social experiences that prime them to experience school in different ways, genetic research underscores that children are not blank slates. Both genetic and environmental factors interact in complex ways that engrave unique codes on individuals. Unraveling not only the genetic code but also what experience encodes is a challenge best tackled by the concerted efforts of molecular biologists, behavioral geneticists, psychologists, parents, and classroom teachers – one would hope, working together.

REFERENCES

Asbury, K., & Plomin, R. (2014). *G is for genes: The impact of genetics on education and achievement* (Vol. 24). Malden, MA: John Wiley & Sons.

Benyamin, B., Pourcain, B., Davis, O. S., Davies, G., Hansell, N. K., Brion, M. J., et al. (2014). Childhood intelligence is heritable, highly polygenic and associated with FNBP1L. *Molecular Psychiatry, 19*(2), 253–258.

Boardman, J. D., Saint Onge, J. M., Haberstick, B. C., Timberlake, D. S., & Hewitt, J. K. (2008). Do schools moderate the genetic determinants of smoking? *Behavior Genetics, 38*(3), 234–246.

Byrne, B., Khlentzos, D., Olson, R. K., & Samuelsson, S. (2010). Evolutionary and genetic perspectives on educational attainment. In K. Littleton, C. Wood, & J. K. Staarman (Eds.), *International handbook of psychology in education*, (pp. 3–34). Bingly, UK: Emerald.

Center for Genomics and Public Health. (n.d.) Genomics versus genetics: What's the difference? Retrieved from http://depts.washington.edu/cgph/GenomicsGenetics.htm.

Champagne, F. A., & Mashoodh, R. (2009). Genes in context gene–environment interplay and the origins of individual differences in behavior. *Current Directions in Psychological Science, 18*(3), 127–131.

Charney, E. (2008). Politics, genetics, and "greedy reductionism." *Perspectives on Politics, 6*(2), 337–343.

(2012). Behavior genetics and postgenomics. *Behavioral and Brain Sciences, 35*(5), 331–358.

Chorney, M. J., Chorney, K., Seese, N., Owen, M. J., Daniels, J., McGuffin, P., et al. (1998). A quantitative trait locus associated with cognitive ability in children. *Psychological Science, 9*(3), 159–166.

Cole, S. W. (2009). Social regulation of human gene expression. *Current Directions in Psychological Science, 18*(3), 132–137.

Collins, F. S., & Mansoura, M. K. (2001). The Human Genome Project. Revealing the shared inheritance of all humankind. *Cancer, 91*(1 Suppl.), 221–225.

Collins, F. S., Green, E. D., Guttmacher, A. E., & Guyer, M. S. (2003). A vision for the future of genomics research. *Nature, 422*(6934), 835–847.

Corno, L., & Snow, R. E. (1986). Adapting teaching to individual differences in learners. In M. C. Wittrock (Ed.), *Third handbook of research on teaching* (pp. 605–629). Washington, DC: American Educational Research Association.

Cronbach, L. J. (1955). *Text materials in modern education: A comprehensive theory and platform for research.* Chicago: University of Illinois Press.

Csikszentmihalyi, M., & Schmidt, J. (1997). Stress and resilience in adolescence: An evolutionary perspective. In K. Borman & B. Schneider (Eds.), *The adolescent years: Social influences and educational challenges,* Yearbook of the National Society for the Study of Education, *99*(5), 1–17.

Dai, D. Y., & Coleman, L. J. (2005). Introduction to the special issue on nature, nurture, and the development of exceptional competence. *Journal for the Education of the Gifted, 28*(3–4), 254–269.

Dick, D. M., Li, T. K., Edenberg, H. J., Hesselbrock, V., Kramer, J., Kuperman, S., et al. (2004). A genome-wide screen for genes influencing conduct disorder. *Molecular Psychiatry, 9*(1), 81–86.

Diseth, Å. (2011). Self-efficacy, goal orientations and learning strategies as mediators between preceding and subsequent academic achievement. *Learning and Individual Differences, 21*(2), 191–195.

Dufell, J. C. (2016). Global greatness: How social-emotional learning helps children success in school, in the workplace, and in life. Committee for Children. Retrieved from www.cfchildren.org

Eapen, V. (2011). Genetic basis of autism: Is there a way forward? *Current Opinions in Psychiatry, 24*(3), 226–236.

Elliott, J. G., & Grigorenko, E. L. (2014). The neurobiological bases of reading and reading disability. In J. G. Elliott & E. L. Grigorenko (Eds.) *The dyslexia debate* (pp. 88–122). New York, NY: Cambridge University Press.

Fairbanks, C. M., Duffy, G. G., Faircloth, B. S., He, Y., Levin, B., Rohr, J., & Stein, C. (2009). Beyond knowledge: Exploring why some teachers are more thoughtfully adaptive than others. *Journal of Teacher Education, 61*(1–2), 161–171.

Flynn, J. R. (1999). Searching for justice: The discovery of IQ gains over time. *American Psychologist, 54*(1), 5–20.

——— (2007). *What is intelligence? Beyond the Flynn effect.* New York: Cambridge University Press.

Fox, L., & Hemmeter, M. L. (2009). A programwide model for supporting social emotional development and addressing challenging behavior in early childhood settings. In W. Sailor, G. Dunlap, G. Sugai, & R. Horner (Eds.), *Handbook of positive behavior support* (pp. 177–202). New York, NY: Springer.

Galton, F. (1876). The history of twins, as a criterion of the relative powers of nature and nurture. *Journal of the Anthropological Institute of Great Britain and Ireland, 5,* 391–406.

Grigorenko, E. L., & Dozier, M. (2013). Introduction to the special section on genomics. *Child Development, 84*(1), 6–16.

Hodapp, R. M., & Fidler, D. J. (1999). Special education and genetics connections for the 21st century. *Journal of Special Education, 33*(3), 130–137.

Holliday, R. (2006). Epigenetics: A historical overview. *Epigenetics, 1*(2), 76–80.

Jensen, A. R. (1998). *The g factor: The science of mental ability.* Westport, CT: Praeger.

Knafo, A., & Plomin, R. (2006). Prosocial behavior from early to middle childhood: Genetic and environmental influences on stability and change. *Developmental Psychology, 42*(5), 771.

Krapohl, E., Rimfeld, K., Shakeshaft, N. G., Trzaskowski, M., McMillan, A., Pingault, J. B., et al. (2014). The high heritability of educational achievement reflects many genetically influenced traits, not just intelligence. *Proceedings of the National Academy of Sciences, 111*(42), 15273–15278.

Kyllonen, P. C. (2016). Human cognitive abilities: Their organization, development, and use. In L. Corno & E. Anderman (Eds.), *Handbook of educational psychology* (3rd ed.). London: Routledge.

Mandinach, E. B., & Honey, M. (Eds.). (2008). *Data-driven school improvement: Linking data and learning. Technology, education–connections (TEC) series.* New York, NY: Teachers College Press.

Mandinach, E. B., Parton, B. M., Gummer, E. S., & Anderson, R. (2015). Ethical and appropriate data use requires data literacy. *Phi Delta Kappan, 96*(5), 25–28.

McLeod, S., & Bleile, K. (2003). *Neurological and developmental foundations of speech acquisition.* Chicago, IL: American Speech Language Hearing Association Convent.

McMahon, S. D., & Washburn, J. J. (2003). Violence prevention: An evaluation of program effects with urban African American students. *Journal of Primary Prevention, 24*(1), 43–62.

Meaney, M. J. (2010). Epigenetics and the biological definition of gene x environment interactions. *Child Development, 81*(1), 41–79.

Moalem, S. (2014). *Inheritance: How our genes change our lives -- and our lives change our genes.* New York, NY: Grand Central Publishing.

Molfese, D., Key, A. F., Maguire, M., Dove, G., & Molfese, V. (2005). Event-related evoked potentials (ERPs) in speech perception. In D. B. Pisoni & R. E. Remez (Eds.),*The handbook of speech perception* (pp. 99–121). Malden, MA: John Wiley & Sons.

Nettelbeck, T., & Wilson, C. (2013). Intelligence and IQ. In K. Wheldall (Ed.), *Developments in educational psychology.* New York, NY: Routledge.

Plomin, R. (1998). Using DNA in health psychology. *Health Psychology, 17*(1), 53–55.

(2011). Commentary: Why are children in the same family so different? Nonshared environment three decades later. *International Journal of Epidemiology, 40*(3), 582–592.

(2013). Child development and molecular genetics: 13 years later. *Child Development, 84*, 104–120.

Plomin, R., & Asbury, K. (2005). Nature and nurture: Genetic and environmental influences on behavior. *Annals of the American Academy of Political and Social Science, 600*(1), 86–98.

Plomin, R., & Daniels, D. (2011). Why are children in the same family so different from one another? *International Journal of Epidemiology, 40* (3), 563–582.

Plomin, R., Asbury, K., & Dunn, J. (2001). Why are children in the same family so different? Nonshared environment a decade later. *Canadian Journal of Psychiatry, 46*(3), 225–233.

Powers, N. R., Eicher, J. D., Butter, F., Kong, Y., Miller, L. L., Ring, S. M., et al. (2013). Alleles of a polymorphic ETV6 binding site in DCDC2 confer risk of reading and language impairment. *American Journal of Human Genetics, 93*(1), 19–28.

Quintilian *Institutes of Oratory* (trans. H. E. Butler) (1920). London: William Heinermann.

Randi, J., & Corno, L. (2005). Teaching and learner variation. Pedagogy: Teaching for learning. *British Journal of Educational Psychology: Monograph Series II, 3*, 47–69.

Randi, J. (Ed.). (2007). Research in the service of practice. *Theory into Practice, 46* (4).

Randi, J. (2017). Teaching and learning hand in hand: Adaptive teaching and self-regulated learning. *Teachers College Record, 119*(13). Retrieved from http://www.tcrecord on April 21, 2017.

Reilly, C. (2012). Behavioural phenotypes and special educational needs: Is aetiology important in the classroom? *Journal of Intellectual Disability Research, 56*(10), 929–946.

Ridley, M. (1999). *Genome: The autobiography of a species in 23 chapters.* London, UK: Fourth Estate.

Rosenthal, R., & Jacobson, L. (1968). Pygmalion in the classroom. *Urban Review, 3*(1), 16–20.

Simonton, D. K. (2008). Scientific talent, training, and performance: Intellect, personality, and genetic endowment. *Review of General Psychology, 12*(1), 28.

(2012). Teaching creativity current findings, trends, and controversies in the psychology of creativity. *Teaching of Psychology, 39*(3), 217–222.

Smagorinsky, P., Lakly, A., & Johnson, T. S. (2002). Acquiescence, accommodation, and resistance in learning to teach within a prescribed curriculum. *English Education,* 187–213.

Snow, R. (1989). Aptitude-treatment interaction as a framework for research on individual differences in learning. In P. L. Ackerman, R. J. Sternberg, & R. Glaser (Eds.), *Learning and individual differences: Advances in theory and research* (pp. 13–59). New York: W. H. Freeman.

Snyder, M., Grubert, F., Heffelfinger, C., Hariharan, M., Asabere, A., Waszak, S. M., ... Snyder, M. (2010). Variation in transcription factor binding among humans. *Science,* 232–235. doi:10.1126/science.1183621

Solomon, D., Watson, M. S., Delucchi, K. L., Schaps, E., & Battistich, V. (1988). Enhancing children's prosocial behavior in the classroom. *American Educational Research Journal, 25*(4), 527–554.

Stanford Aptitude Seminar. (2002). *Remaking the concept of aptitude: Extending the legacy of Richard E. Snow.* Mahwah, NJ: Lawrence Erlbaum.

Stein, J. (2014). Dyslexia: The role of vision and visual attention. *Current Developmental Disorders Reports, 1*(4), 267–280.

Sternberg, R., & Grigorenko, E. (1999). Myths in psychology and education regarding the gene-environment debate. *Teachers College Record, 100*(3), 536–553.

Tan, M., & Grigorenko, E. L. (2013). All in the family: Is creative writing familial and heritable? *Learning and Individual Differences, 28*, 177–180.

Terman, L. M. (1925). *Genetic studies of genius: Mental and physical traits of a thousand gifted children.* Palo Alto, CA: Stanford Universtiy Press.

Thomas, M. S., Kovas, Y., Meaburn, E., & Tolmie, A. (2015) What can the study of genetics offer to educators? Working Paper #12077. *Mind, Brain, and Education.*

True, W. R., Heath, A. C., Scherrer, J. F., Xian, H., Lin, N., Eisen, S. A., ... Tsuang, M. T. (1999). Interrelationship of genetic and environmental influences on conduct disorder and alcohol and marijuana dependence symptoms. *American Journal of Medical Genetics, 88*(4), 391–397.

US Department of Energy Office of Science. (2003). *Genomics and its impact on science and society: The Human Genome Project and beyond.* Washington, DC: US Department of Energy. Available: http://www.ornl.gov/sci/techresources/ Human_Genome/publicat/primer2001/PrimerColor.pdf

Venter, J. C., Adams, M. D., Myers, E. W., Li, P. W., Mural, R. J., Sutton, G. G., ... Beasley, E. (2001). The sequence of the human genome. *Science, 291*(5507), 1304–1351.

Walker, S. O., Petrill, S. A., Spinath, F. M., & Plomin, R. (2004). Nature, nurture and academic achievement: A twin study of teacher assessments of 7-year-olds. *British Journal of Educational Psychology, 74*(3), 323–342.

Woolfolk, A., & Perry, N.E. (2015). *Child and adolescent development* (2nd ed.). Upper Saddle River, NJ: Pearson.

16

Will the Next Einstein Get Left in the Petri Dish? Be Careful What You Wish for in the Designer Baby Era

CAROLYN D. COWEN

Be careful what you set your heart upon – for it will surely be yours.
– James Baldwin

What is the impact of the Human Genome Project on parents and families?

How has mapping the human genome catalyzed reproductive-genetic technologies and practices and, perhaps, opened the door for genetic manipulation and selection procedures to modify behavioral and cognitive traits? How might these modifications begin altering the human genome and shaping our destiny? This chapter explores these big hypothetical questions. But we start by tackling this more manageable question, how might mapping the human genome, and the associated developments before and after, affect parents and families? Let us go straight to the heart of the matter, where a parent's hopes and dreams begin: having a child.

Imagine that you wish desperately to bring a genetically related baby into the world, but you have been unsuccessful thus far. This brings you to a fertility clinic for in vitro fertilization (IVF), which produces two viable embryos for implantation. The clinic's menu of services includes pre-implantation genetic diagnosis (PGD), which tests embryos for various genetic conditions. PGD also can identify sex. Now, imagine five different scenarios in which you are faced with a yes-or-no question. How do you answer each?

> Question 1: Imagine that you just discovered that you are at great risk for bearing a child with Tay-Sachs, a deadly disease that typically kills a child by four years of age. *Will you have your embryos tested with the intention of discarding any with genetic abnormalities? Yes or no?*
>
> Question 2: Imagine that you have been told that you have significant risk factors for having a child with Down syndrome, which is characterized by moderate to severe intellectual impairment and higher-than-normal risk for various health conditions. Social and environmental variables

can alleviate the degree of disability associated with Down syndrome, but there is no cure. *Will you have your embryos tested with the intention of discarding any with genetic abnormalities? Yes or no?*

Question 3: Imagine that members of your family have been afflicted with the genetic disorder Gerstmann-Straussler-Scheinker (GSS), a neurological condition that leads to a lingering death, but does not strike until middle adulthood. Quality of life is not affected until the onset of symptoms, though fear of the Damocles sword of GSS is highly stressful. *Will you have your embryos tested with the intention of discarding any with genetic abnormalities? Yes or no?*

Question 4: Imagine that breast cancer has stalked your family for generations. Embryos with the cancer-associated genes might develop breast cancer as adults. But they might not. Those who do would enjoy four or five decades of good health. *Will you have your embryos tested with the intention of discarding any with genetic abnormalities? Yes or no?*

Question 5: Finally, imagine a very different scenario. You already have two children of one sex, but your heart is set on a family of both sexes. In addition to genetic disease screening, the fertility clinic offers family balancing – meaning that it can screen and select for the desired gender. *Will you have your embryos tested with the intention of discarding (or cryo-freezing[1]) any that are not the gender you want? Yes or no?*

How did you answer those five questions? Did you draw a line at some point? This thought exercise – pure science fiction just a few decades ago – follows a progression of increasingly difficult ethical considerations. Where each of us falls along this ethical continuum depends on our beliefs and values and, probably, any history we may have had with these diseases or scenarios.

Some of us might answer yes to the first two questions, but draw the line at the third because quality of life is good through much of adulthood. Others may feel that having a child with Down syndrome is not entirely bad (see Krahn, 2011, for a discussion) and will say no to question two. Some might draw the line at question four because the gene is associated only with risk and perhaps a cure will be available by the time the child becomes an adult. Many will balk at question five because there is no medical condition. Some will answer no to all five questions and, for that matter, would not even consider IVF for various deeply held convictions. Others will answer yes to all five questions with no qualms, or at least not enough to change any answers.

---

[1] Cyropreservation: Commonly used for embryonic storage, cells are preserved by cooling to sub-zero temperatures. Thus far, duration of storage has not been found to have any negative effect on implantation, pregnancy, live birth rate, birth defects, or developmental abnormalities.

These once far-fetched scenarios and questions not only illuminate how advances in reproductive-genetic technologies beget tricky ethical considerations, but also how breakthrough advances transition to clinical practice and then become culturally normalized. (For example, the first "test tube baby," Louise Brown, was a big deal in 1978. Now? Hardly.) This normalizing, in turn, helps propagate new advances that further shift cultural norms.

Might this dynamic, progressive process someday yield genetic manipulation and selection practices to alter behavioral and cognitive traits and prevent conditions such as dyslexia, attention-deficit hyperactive disorder (ADHD), delayed language, autism spectrum disorders (ASD), and mood and anxiety disorders? In the quest to have perfect babies, might such practices eventually have a homogenizing effect, one that begins decreasing neurodevelopmental variation? Is such variation as vital to the human species as biodiversity is to life in general? This chapter speculates on such possibilities, explores the forces that might propagate these possibilities, and proposes steppingstones to initiate considered, coherent planning.

TODAY'S HEADLINES ARE YESTERDAY'S SCIENCE FICTION

However you answered the five questions above, they are not academic. Hopeful parents weigh these and similar thorny questions in fertility clinics across the United States every day thanks to a confluence of advances in genomics, genetics, and reproductive technologies. The wish to avoid a multiple-birth pregnancy leads prospective parents into the same thicket of tricky ethical considerations. If all embryos are healthy, parents and clinicians face a *Sophie's Choice*:[2] Which ones get implanted? What happens to those that are not? Pre-natal tests, such as amniocentesis,[3] stir up even more ethical and emotional turmoil since pregnancy is well underway. This chapter, however, explores the various ethical questions and biological-social considerations arising from the newer upstream reproductive-genetic technologies, AKA *reprogenetics*,[4] particularly as they and their ongoing advances might affect future neurodevelopmental variation.

---

[2] *Sophie's Choice*: A 1982 film about a mother who, upon arrival at Auschwitz, is forced to choose which of her two children would be gassed and which would be sent to the labor camp.

[3] Amniocentesis: Administered in the 16th–22nd week of gestation to detect chromosomal (e.g., Down syndrome) and genetic abnormalities in women at increased risk.

[4] Reprogenetics: Coined by Lee M. Silver, professor of molecular biology, Princeton, in his 1998 book *Remaking Eden*.

Until recently, such ethical questions and biosocial considerations were the domain of science fiction. Now, even a cursory online search turns up headlines and stories about parents confronting scenarios like those in questions 1–5. (Indeed, actual news stories inspired those scenarios.) These parents are the vanguard, venturing into frontiers once unimaginable. A brief review of today's practices and the driving forces in reprogenetics offers a glimpse over the horizon of these frontiers and insights into how current trajectories in human genetic manipulation and selection might set the stage for practices that eventually affect neurodevelopmental variation and, perhaps, shape our destiny. Let us take a peek at today's practices and forces.

Unthinkable Scenarios. Are PGD and other reproductive-genetic technologies already outpacing consideration of biosocial implications? Even now, these technologies have rendered formerly unthinkable scenarios thinkable – such as a deaf couple seeking to give birth only to a deaf child or parents of a sick child conceiving a savior sibling to provide the ill child with a compatible organ or cell transplant. True stories. In fact, they are old news (see Baruch, Kaufman, & Hudson, 2006; Fahmy, 2011; Hadley, 2003). Where does one draw the line? Sex selection, controversial but practiced, segues pretty seamlessly to aesthetic trait selection (e.g., eye or hair color), which, yes, is possible,[5] though so controversial, it is not advertised. So far.

Many of the ethical dilemmas and dramas depicted in films such as *My Sister's Keeper*[6] or even in *Gattaca*[7] no longer are the stuff of Hollywood imaginings. They are today's headlines. As this chapter was being written, new headlines announced development of a three-person IVF technique expected to become available in the United Kingdom to treat infertility and to prevent transmission of diseases caused by genetic abnormalities in the mitochondria[8] of a woman's egg. The US Food and Drug Administration (FDA) also is considering this procedure,[9] which entails implanting the nucleus of a woman's egg cell into another woman's egg cell that has healthy

---

[5] See the Center for Genetics and Society website: www.geneticsandsociety.org, "Genetic Selection." Also see Sidhu, 2012.

[6] *My Sister's Keeper*, a 2009 film about a "savior sister," conceived as a genetic match for a sibling suffering from leukemia.

[7] *Gattaca*, a futuristic 1997 film depicting a society practicing genetic engineering to perfect the human species.

[8] Mitochondria – power plants within cells surrounded by two membranes and with their own genome.

[9] Prior to this volume going to press, a committee of scientists and bioethicists convened by the FDA concluded that it was ethically permissible to proceed with mitochondrial

mitochondria but has had its nucleus removed. Offspring would carry the DNA of *three parents*, though the contribution from the donated mitochondria would be small.

Driving Forces. Largely free from state or federal regulation, the science and application of IVF and PGD – the primary tools of today's genetic manipulation and selection practices – have quietly mushroomed in US fertility clinics since being introduced in the early 1990s. This growth rests on the bedrock of reproductive freedom and the quest to liberate generations of families from the shackles of lethal or debilitating genes. Minimal regulation, disease prevention goals, reproductive freedom, and the drive to produce a genetically related child converge powerfully to propel reprogenetics.

This is not to say that reprogenetics has not sparked plenty of debate, much of it echoing themes debated in human embryonic stem cell research and centering on the fate of embryos left over from IVF procedures. The Roman Catholic Church, for example, believes that PGD entails destruction of human life. Bioethicists and others from the scientific, medical, and social science communities voice concerns about "playing God" and unintended biosocial consequences. Concerns include promoting eugenics (attempting to perfect the human species by eliminating genes deemed "unfit"); widening socioeconomic gaps (families that can afford it would become the genetic elites, or "genobility"); subordinating children's well-being for parental goals (as in the case of a deaf couple hoping to use PGD in an effort to produce a deaf child); and tampering with the human germline[10] by changing the genome[11] of descendants (a concern raised vis-à-vis three-parent IVF).

However, unlike stem cell research, PGD in the United States is not subject to restrictions and gets little oversight.[12] This means fertility clinics and would-be parents are the primary drivers of both the advancement

---

replacement techniques (MRT), albeit with a number of specified caveats. However, federal regulation tacked on to the 2016 budget prohibited the agency from permitting procedures that involve heritable genetic modification of human embryos. As this chapter was nearing completion, it was unclear exactly when or how (or if?) MRT will go forward in the United States.

[10] Germline: The sequence of germ cells that have genetic material that may be passed to a child, the lineage of cells spanning many generations.

[11] Genome: The genetic material of an organism.

[12] The FDA regulates only the safety and efficacy of such technologies. Coordinated oversight or guidelines eventually might come from a coalition of medical societies such as the American Society for Reproductive Medicine (ASRM) and the American Congress of Obstetricians and Gynecologists (ACOG). Now, though, they do not agree. For example,

and expansion of reproductive tools such as PGD. Soon, though, market forces may start factoring into the equation. In a *Science* article, "Stirring the Simmering 'Designer Baby' Pot," Thomas Murray (2014), a bioethicist at Hastings Center (a non-profit bioethics research center), described the entrance of the private company, 23andMe, into the gene selection marketplace.

> In September 2013, the company was granted a patent, "Gamete[13] donor selection based on genetic calculations," that allows would-be parents "to select a donor and view other possible phenotype[14] of the hypothetical child resulting from the recipient's and the donor's gametes, such as alcohol flush reaction, lactose tolerance, muscle performance, and any other appropriate phenotype ..." ... What makes the fuss stirred up by the 23andMe patent distinct from preimplantation genetic diagnosis (genetic profiling of embryos prior to implantation) and prenatal screening (procedures during pregnancy to detect health problems in the fetus) is its disconnection from the abortion debate. There is no existing embryo or fetus here. At stake is not whether to have a child, but how much discretion parents should exercise in determining what sort of child they have. (pp. 1208–1210)

23andMe[15] may be a harbinger of what lies ahead as technologies advance, consumer demand for personal genetic information grows, and subsequent market opportunities emerge. Coupled with fertility clinic and parental forces, an entrepreneurial marketplace will only accelerate the undermining of biosocial pillars. These driving forces and related ethical debates promise to intensify as genomics, genetics, and reproductive technologies evolve, as new services become accessible, and as parents seek to

---

ASRM defers usually to a parent's wishes on sex selection, while ACOG advocates prohibiting such selection.

[13] Gamete: A mature reproductive cell, such as a sperm or egg, able to unite with another of the opposite sex to form a new organism.

[14] Phenotype: The composite of an organism's observable traits. A phenotype results from the expression of an organism's genes, influence of environmental factors, and interactions between the two.

[15] In November 2013, the FDA ordered 23andMe to stop marketing its Saliva Collection Kits and Personal Genome Service over concerns about potential for making consequential health decisions based on false positive or negative assessment of genetic risk (see Gray, 2013). The company continued to sell its personal genome test without health-related results. In October 2015, the company announced it would include a revised health-related component with FDA approval. 23andMe also has expanded its business to include brokered access to its database of more than one million people's DNA. This genetic data, coupled with phenotypic traits, is of interest to researchers and already is contributing to insights and published studies (see Farr, 2016).

gain genetic advantage for their offspring. Yes, cost – about $20,000 for PGD – is a major deterrent, but it is not an insurmountable barrier, especially for families of means (again, raising the specter of genetic elites). Indeed, PGD for non-medical reasons (i.e., family balancing) is becoming a multimillion-dollar industry (Sidhu, 2012).

Largely unrestrained by regulation in the United States and despite high costs, an amalgam of powerful forces might be carrying us incrementally but inexorably into the *Brave New World*[16] of genetic engineering without much national conversation about ethical or social policy issues or even much public awareness.[17] True, about 99 percent of US babies are still made the good old-fashioned way, but with the milestone of sequencing the human genome well behind us and in light of ongoing advances in reprogenetics, perhaps it is not too far-fetched to imagine that future sociocultural anthropologists will regard our time as the era that ushered humanity to the threshold of the so-called "Designer Baby Era."

UP NEXT: BEHAVIORAL TRAIT SELECTION?

Let us extend our thought exercise into this *Brave New World* of genetic engineering and imagine another scenario taking place sometime in the future.

> Question 6: Imagine that gene screening, selection, and modification have progressed to the point where it is possible to determine behavioral and cognitive traits. Imagine further that you have a positive family history for dyslexia and that you are painfully aware of the academic struggles that typically accompany this reading disability. *Given the opportunity to implant only embryos without genes that predispose for dyslexia or to treat gametes or embryos with a proven gene modification that directs the developing brain to achieve a baby without predisposition for dyslexia, would you avail yourself of either opportunity? Yes or no?*

Unlike the previous five questions, this question is hypothetical. Using PGD to select behavioral traits or deploying other emerging gene-modifying techniques[18] to tweak fetal or gamete DNA to produce desired

---

[16] *Brave New World*, a 1932 novel by Aldous Huxley set in 2540 London, which predicted developments in reproductive technology, sleep-learning, psychological manipulation, and classical conditioning, all of which converge to change society.

[17] See Murray's article, "Stirring the Simmering 'Designer Baby' Pot" (2014), for deeper discussion about the lack of legislation, regulation, and professional guidelines.

[18] For example, CRISPR-Cas9 (Clustered Regularly Interspaced Palindromic Repeats) – a gene-editing technique enabling scientists to make precise changes to DNA

temperament, intelligence, or athleticism far exceeds today's capabilities. These traits are influenced by cocktails of genes interacting dynamically not only with each other, but also with their environments. Such complex polygenetic–environmental interplay is not fully understood, putting this level of genetically engineered human enhancement squarely in the domain of science fiction. Will this complex interplay keep any such genetic engineering beyond our grasp forever?

Not according to Ronald Green – professor emeritus for the Study of Ethics and Human Values at Dartmouth College and founding director at the National Human Genome Research Institute at the National Institutes of Health. In his book *Babies by Design* (2007), Green made several controversial predictions about the future of gene selection and modification, including the possibility that someday we may develop the ability to target dyslexia[19] and other cognitive profiles genetically.

> We will begin with gene selection aimed at reducing the likelihood that a child will be born with a genetic disease, and eventually include changes designed to permanently eliminate serious disorders like cystic fibrosis and sickle cell disease from a family line. Beyond this, gene modification will encompass the first hesitant steps to improve the genetic endowment of our children so they can flourish in new ways. This may include increased natural resistance to diseases like AIDS and cancer or to problems like diabetes or obesity. Somewhere down the line, we will see the emergence of what I call "cosmetico-genomics," as parents strive to give their children more attractive physical features, including normal height,

sequences – may open pathways to manipulate human DNA to correct genetic flaws in human embryos (e.g., mutations causing blindness) or to cure autism and schizophrenia in children and adults along with other conditions linked to genetic vulnerabilities affecting neurodevelopment. First shown to be an effective genome-engineering/editing tool in human cell culture in 2012, CRISPR-based methods have the potential to manipulate *multiple* genes (neurodevelopmental disorders typically involve multiple genes). As this chapter was nearing completion, researchers from the University of California and the Broad Institute still were fighting for control of the CRISPR patent. Meanwhile, CRISPR-related research barrels ahead. Clinical trials with humans are expected in the US in the near future and already have been pioneered in China. Nevertheless, the National Academy of Sciences and the National Academy of Medicine convened a yearlong public consensus study on human gene editing and has published its report, Human Genome Editing: Science, Ethics, and Governance (2017). (For more on CRISPR history, controversies, promise, and concerns, see Pollack, 2015; also see Kozubek, 2016.)

[19] As discussed by Olumide, Gilger, Talavage, Hynd, and McAteer (2012, p. 618), a relatively small number of "key genes" appear to put individuals at risk for reading disability (RD). "In fact, some 10 genes or gene locations have been tentatively identified as contributors to RD risk by the Human Genome Organization (HUGO), although these 'genes' have yet to be thoroughly identified or characterized" (Schumacher, Hoffman, Schmal, Schulte-Korne, & Nothen, 2007).

good teeth, clear complexions, and pleasing faces. In the more distant future, we may see cognitive and neurological enhancements, ranging from reduced susceptibility to *dyslexia*, [italics added] learning disorders, and depression to improved memory and enhanced IQ. (p. 7)

Dyslexia's Possible Upsides. How did you answer question six? Your answer probably depends on how much you know about dyslexia,[20] whether you have had any personal experiences with this biologically based but educationally malleable neurodevelopmental condition, the nature of those experiences, and your stage of life.

For example, adults with dyslexia sometimes say that during their school years, the humiliation of chronic academic failure was so traumatic and the struggle to learn to read was so awful, they would have traded in their "dyslexic brains" in a flash. But as adults (especially if they have overcome dyslexia's challenges, found workarounds, discovered strengths, or experienced success), many people with dyslexia view their learning difference as an advantage that not only imparts tenacity and compensatory strengths, but also, sometimes, exceptional talents and affinities. Dyslexia champion, Sir Richard Branson (founder of the Virgin Group and knighted for "services to entrepreneurship") writes,

My teachers thought I was lazy and dumb, and I couldn't keep up or fit in. Out in the real world, my dyslexia became my massive advantage: It helped me to think creatively and laterally, and see solutions where others saw problems (2017).

Branson[21] maintains that dyslexia is just a different way of thinking, not a disadvantage. In a survey of entrepreneurs conducted by Cass Business School (Logan, 2009), 35 percent of the participants identified

[20] Definition adopted by the International Dyslexia Association and the National Institutes of Health in 2002: "Dyslexia is a specific learning disability that is neurological in origin. It is characterized by difficulties with accurate and/or fluent word recognition and by poor spelling and decoding abilities. These difficulties typically result from a deficit in the phonological component of language that is often unexpected in relation to other cognitive abilities and the provision of effective classroom instruction. Secondary consequences may include problems in reading comprehension and reduced reading experience that can impede the growth of vocabulary and background knowledge."
[21] In 2017, Branson helped launch Made by Dyslexia, a UK-based global charity dedicated to developing campaigns to explain "dyslexic thinking." As a promotional device, Branson debuted Made by Dyslexia at "the world's first dyslexic sperm bank," which does not actually accept donors but serves to underscore the view that dyslexia is a positive trait. The London Sperm Bank (a real sperm bank) made headlines in late 2016 and early 2017 for declining donations from people with dyslexia. Also see endnote 42 below.

themselves as having dyslexia. This less-than-rigorous study[22] is commonly cited as evidence of a "dyslexic advantage"[23] in entrepreneurial contexts.[24] InventiveLabs,[25] a US-based social enterprise, is betting on this proposition. Their pitch competition (cosponsored by Microsoft) requires competitors to have at least one person with a learning difference (e.g., dyslexia, ADHD, ASD) on their pitch teams. This is not some kind of affirmative action requirement to level the playing field for people with learning challenges. It is a strategic decision grounded in the hypothesis that this population often posses traits highly prized in entrepreneurial contexts.

The genetic, neurobiological, and cognitive bases of dyslexia[26] and its educational treatments[27] are well studied. Dyslexia's downsides are clear. Do we have evidence of upsides? Empirical data supporting a dyslexia–talent

---

[22] Mark Seidenberg points out in *Language at the Speed of Sight* (2017) that the study is flawed – it is based on self-reported information from an informal survey with only a seven percent return rate. In detailing its problems, Seidenberg undercuts much of Malcolm Gladwell's "desirable-difficulty" thesis – the premise of *David and Goliath* (2013), in which Gladwell cites dyslexia as an example desirable difficulty and maintains that people with dyslexia are overrepresented among high achieving entrepreneurs. Seidenberg argues that while people with dyslexia can be highly successful and this message is helpful to convey, the probability of becoming a "superachiever" due to dyslexia is unlikely: "If dyslexia were represented five times as often among superachievers, that would be very interesting, but it would have a negligible effect on the probability of becoming a member of that crowd. It would be analogous to holding several tickets in the Powerball lottery instead of one" (2017, p. 185).

[23] A book (Eide and Eide, 2011) by the same name focuses almost entirely on dyslexia as a strength.

[24] A *Fortune Magazine* cover story (Morris, 2002), "The Dyslexic CEO" featured "four dead-end kids" (Richard Branson, Charles Schwab, John Chambers, and David Boise) who, despite difficulty learning to read, went on to become hugely successful. This, too, often is cited to illustrate a "dyslexic advantage," especially in entrepreneurial contexts.

[25] Rick Fiery and Tom Bergeron cofounded InventiveLabs in 2014 to foster talent in young entrepreneurs. Think business incubator meets gap year meets structured college alternative and you get the idea. InventiveLabs' programs are small, "high-touch," and infused with a strong dose of acceptance, mentorship, guidance, and support (see Cowen, 2016 and see InventiveLabs website: www.inventivelabs.org).

[26] No discussion about dyslexia is complete without noting that debate continues about its definition and theoretical construct, as underscored by the book, *The Dyslexia Debate* (Elliot and Grigorenko, 2014).

[27] Most reading difficulties can be resolved or diminished when reading is taught by highly knowledgeable and skilled teachers. Effective reading instruction (AKA "Structured Literacy") benefits all children, but it is vital for children with dyslexia. The International Dyslexia Association (IDA) delineates the components of Structured Literacy in their "Knowledge and Practice Standards for Teachers of Reading." See IDA's website: www .dyslexiaida.org.

link remains scant, but intriguing. A handful of studies suggest that people with dyslexia might have strengths in visual-spatial domains or might process spatial information differently (more globally?) than normal readers (e.g., Geiger & Letvin, 1987; Gilger, Talavage, & Olulade, 2013; Olulade, Gilger, Talavage, Hynd, & McAteer, 2012; Schneps, Rose, & Fischer, 2007; von Károlyi, Winner, & Sherman, 2003). A recent functional magnetic resonance imaging study conducted at Haskins Laboratories not only replicated the von Károlyi et al. study, but also found a pattern of functioning suggesting that (a) figures are processed more automatically in individuals with dyslexia, and (b) print is processed more automatically in typically developing peers – providing "evidence for a tradeoff[28] and of the potential for efficient brain organization for the domain of visuospatial processing" (Diehl et al., 2014).

We must be careful not to over-interpret the results of these studies. Ken Pugh, senior author of the Haskins study, cautions that a richer scientific foundation is necessary to determine (a) whether these strengths are a consequence of less reading experience; and (b) if they translate into a significant real-world benefit (Cowen, 2014). Furthermore, any discussion about a possible upside to dyslexia must include the sobering reminder that for every celebrity or millionaire dyslexia success story, there are thousands who struggle with the harsh social consequences of school failure and the print illiteracy or marginal print literacy[29] that typically accompany dyslexia. *To be clear, this handful of studies does not yet constitute the empirical evidence[30] needed to firmly establish a dyslexia–talent link.[31]*

---

[28] Norman Geschwind, too, speculated on this tradeoff: "We are thus brought to the apparently paradoxical notion that the very same anomalies on the left side of the brain that have led to the disability of dyslexia in certain literate societies also determine superiority in the same brain" (p. 23, 1982).

[29] Reports consistently underscore disproportionately low levels of literacy among prisoners and adjudicated and incarcerated juveniles (e.g., see Coley & Barton, 2006; National Center for Learning Disabilities, 2014).

[30] The Dyslexia Foundation convened a panel of practitioners and researchers to focus on complex nonverbal/spatial skills and career choices in dyslexia. They concluded that there was insufficient data to support a conclusion that dyslexia correlates with specialized skills or career choices, which is not to say that a body of evidence points in the opposite direction - only that more research is needed. The panel recognized that the dyslexia brain may be unique beyond well documented neural differences related to reading and called for more research on (a) how people with dyslexia reason spatially and (b) whether such abilities are expressed in academic, personal, and career behaviors (Gilger, 2017a, 2017b).

[31] Carl Sagan's popular and controversial quote – "absence of evidence is not evidence of absence" – helps raise another important caution. Lack of sufficient evidence demonstrating a dyslexia-talent link *due to insufficient scientific study* is not the same thing as scientific evidence demonstrating no dyslexia-talent link.

With those cautions and caveats underscored, it is worth noting that these studies are consistent with anecdotal reports and clinical observations[32] dating back to the earliest days of dyslexia's discovery[33] that support the notion that dyslexia has upsides,[34] particularly in visual-spatial domains. These empirical studies also are consistent with speculations by the late Norman Geschwind, a pioneering behavioral neurologist,[35] who spoke and wrote often about what he called the "pathology of superiority." Several Geschwind quotes below take us though his thinking on talent in dyslexia, the role of the social environment in determining disability, and the possibility that dyslexia reflects a diversity mechanism.

> Many dyslexics have superior talents in certain areas of non-verbal skill, such as art, architecture, engineering, and athletics. The immediate naïve presumption is that success in these fields is simply the result of compensatory achievement in non-verbal fields on the part of those who do not succeed in readily acquiring reading. I believe that this explanation must convey at best a very small fraction of the truth. (Geschwind, 1982, p. 22)

> The overwhelming majority of humans who ever have lived have been illiterate ... Most of us come from families that four generations ago did not possess the ability to read. If certain changes on the left side of the brain lead to superiority of other regions, particularly on the right side of the brain, then there would be little disadvantage to the carrier of such changes in an illiterate society; their talents would make them highly successful citizens. It is not surprising that this type of brain organization should occur with such high frequency. (Geschwind, 1982, pp. 22–23)

> The knowledge of every aspect of dyslexia will be enriched by seeing it in its broadest biological and sociological settings. We must understand its

---

[32] Anecdotal reports and clinical observations have merit, offering insights that can lead to hypothesis development, empirical research, and improved practices. Anecdotal reports and clinical observations also have limitations, e.g., the *illusory correlation* – the phenomenon of perceiving relationships among variables when no relationships exist. Humans are prone to assuming that certain groups and traits co-occur and often overestimate the strength of any relationships. See Chapman (1967), who coined the term *illusory correlation*.

[33] In his seminal report, W. Pringle Morgan (1896) described the case of fourteen-year-old Percy F., who could not read and wrote his name as "Precy," but could multiply 749 by 887 correctly and *quickly*.

[34] A rich body of literature champions a talent-dyslexia link, e.g., *The Many Faces of Dyslexia* (Rawson, 1988), "Gifts, Talents, and the Dyslexias" (Vail, 1990), *In the Mind's Eye* (West, 1991).

[35] Geschwind (father of modern behavioral neurology) made important contributions to understanding the neurological basis of dyslexia and related language disorders and brought a new emphasis to research on the relation of structure to function in the brain.

relationships to high talent and the societal setting in which it becomes a disability. (Geschwind, 1984, p. 327)

The pattern of cortical development may well reflect a mechanism that is advantageous to the population as a whole, since it leads to a great diversity ... and therefore patterns of talent. (Geschwind & Galaburda, 1987, p. 143)

Let us return to our sixth question. If you could implant embryos without the genes that predispose for dyslexia or if you could modify genes in gametes or embryos to design a baby without susceptibility for dyslexia, would you? As someone who has dyslexia in nearly every branch of my family (assortative mating[36] at work!), who has children with dyslexia, and who was diagnosed with dyslexia in first grade, the prospect of weeding dyslexia out of the gene pool worries me. So does this second quote from Green (2007):

Eventually, dyslexia may come to be seen as a genetic disorder that is routinely screened for in prenatal testing and responded to either by embryo selection or by targeted gene modification. (p. 79)

Square Pegs in Today's Schools. If Green's vision turns out to be prophetic, people like me might not exist in the future. Unsettling to contemplate on a personal level, yes, but my concern goes deeper. Could there be unintended consequences in eliminating or reducing future generations of people with dyslexia or other learning differences from humanity's gene pool? Maybe this would subtract from prison populations and reduce special education costs. Then again, maybe this would subtract people humankind might miss and need.

In her book *Understanding Dyslexia and Other Learning Disabilities* (2013), Linda Siegel profiles a number of notables who struggled as children in school and who, very likely, would have been placed in special education had they been in school today.

The stories of Agatha Christie, Susan Hampshire, Greg Louganis, William Butler Yeats, Pablo Picasso, Hans Christian Anderson, and Winston Churchill when they were children are typical of cases found in today's clinics for children with learning disabilities. (p. 257)

What if Geschwind was on to something? What if these children exist because their brain organization – square pegs in today's one-size-fits-all

---

[36] Assortative mating: Frequent mating among individuals with similar genotypes and phenotypes. See Gilger (1991) for a study on assortative mating among those with reading disability.

schools – and the diverse thinking these brains generate are advantageous to our species? Are there potential risks in genetically eliminating or altering humanity's square pegs to fit into the round holes of a transient socio-cultural context, one that is merely a brief way station in the evolution of our species and, with any luck, our journey into the future?

Granted, today's industrial era–designed and agrarian era–influenced schools are notoriously slow to apply the technologies that are revolutionizing society globally. But today's outmoded learning factory model is a mere blink of humanity's 200,000-year-old eye.[37] And, given the exponential pace of technological advancement and social change, it is unlikely that twenty years from now school will look much like it does today, even if school today looks a lot like it did 200 years ago. The angst and turmoil of today's reform efforts notwithstanding, education's past probably will not be its prologue.

So, let us imagine that someday we do develop the technical capability to weed out or modify the genes that predispose for the brain organization[38] that makes learning to read so difficult in today's education system. In fact, let us imagine that in addition to diseases and physical traits, parents can select from a list of detectable neurodevelopmental "disorders" to screen for the genes that predispose for dyslexia; ADHD; ASD; learning risks, such as delayed language; and mood and anxiety disorders. Why shouldn't parents seek to screen out these risks and gain every possible genetic advantage for their offspring in school, one of society's most consequential first tests?

Several factors complicate any such aspiration. For starters, some of these conditions are more challenging than others and their multiplicity of traits, environmentally malleable to varying degrees,[39] fall along continuums of severity.[40] In other words, not every person with one of these

---

[37] Just prior to publication of this chapter, new discoveries suggest that the origin of our species may even date back 300,000 years. See Callaway (2017).

[38] Imaging research of prereaders with a family history of dyslexia showed in two studies – one published in the *Proceedings of the National Academy of Sciences* (Raschle, Zuk, and Gaab, January 23, 2012), the other published in the journal *Cerebral Cortex* (Langer et al., 2015) – that brain activity associated with developmental dyslexia is present as early as infancy. Also see Gaab (February, 2017) for discussion of additional more recent studies (Kraft, 2016; Wang, In press) with similar findings.

[39] Several neuroimaging studies (e.g., Shaywitz et al., 2004; Simos et al., 2002; Temple et al., 2003) indicate that with intensive evidence-based educational intervention, many children with dyslexia show some "normalization" of left-hemisphere reading circuits and improved reading skills. However, see Bishop, (2013) for interesting discussion about limitations of imaging studies for assessing effective instruction.

[40] Dyslexia, like other neurodevelopmental conditions (e.g., attention deficit hyper-activity disorder, autism spectrum disorder), is less of a distinct syndrome with clear diagnostic biomarkers and obvious demarcations and more of a *developmental*

conditions will be like another person with the same condition. Not only that, the same person can present with different challenges at various developmental stages. Each condition, each context, each life stage, and each person is different.

Some individuals will have a mild version of a particular condition, or will inhabit environments that do not penalize their weaknesses,[41] or will receive the early intervention needed to succeed, or will find that their struggles impart the fortitude for dealing with life's challenges. Some will discover relative strengths – perhaps even exceptional talents – that enrich their lives and the lives of others (e.g., Temple Grandin, noted doctor of animal sciences and author with ASD) or that influence events affecting countless lives (e.g., Winston Churchill, who struggled initially to read and write and was plagued throughout his life with dyscalculia;[42] see Siegel, 2013). Others might not have any of this good fortune, might be defeated by early failure, and might become burdens to society. Many more people with such neurodevelopmental variation will fall somewhere between these extremes.

The point is: Having dyslexia or any of these neurodevelopmental "disorders" does not consign someone automatically to a doomed life. How things turn out depend on a host of variables, including environmental context, a transient dynamic that sometimes determines whether a particular neurobiological-cognitive profile translates into a disability or an advantage.

Nuances and Unknowns vs. Precise Interventions. How might parents in fertility clinics across the nation weigh these nuances and unknowns against any targeted genetic interventions of the future? Could parents in their current zeitgeist foresee a sociocultural context decades hence that might render a particular brain organization an advantage or disadvantage? Could they or anyone know which dynamic gene–environment interactions

---

multidimensional-spectrum condition. The various hallmark characteristics of dyslexia fall along continuums of severity that are not fixed – they can change developmentally across a lifespan and improve with environmental influence, e.g., effective teaching. (Perhaps dyslexia might be best conceptualized as *developmental dyslexia spectrum disorder.*) Therein lies one of the great challenges in dyslexia's construct, thus, in how individuals with dyslexia are identified and, consequently, for whom treatment and resources are allocated and accommodations are granted.

[41] For example, thanks to spell-check, a so-called residual spelling problem is not as handicapping for an adult with dyslexia today as it was just a few decades ago. I can attest to this. Even with my fairly extensive orthographic knowledge, I benefit from spell-check - which enables me to avoid disrupting "writing flow."

[42] Dyscalculia: A math learning disability.

might help produce a Winston Churchill? Could anyone anticipate what technologies and educational reforms might evolve to mitigate which learning challenges? Could anyone divine what events requiring a unique cognitive profile might thrust someone forward to play a pivotal role for humankind? These unknowns are, well, unknowable.

We can, however, hazard reasonable predictions based on a few knowns. For example, as long as the driving forces in US reprogenetics continue to be fertility clinics, the marketplace, and parents, they and the motivations that propel them are likely to play a big role in powering us into whatever this *Brave New World* might hold. Policies at the United Kingdom's largest sperm bank banning donations from individuals with dyslexia[43] illustrate just how powerful, pervasive, and ill-considered those motivations might be. Thus, any effort to anticipate and shape this future must include: (a) a deep appreciation of these driving forces and motivations and (b) efforts to facilitate understanding among decision makers (mostly hopeful parents) of the various biosocial considerations.

TALE OF TWO TERMS: NEURODIVERSITY AND CEREBRODIVERSITY

Two terms – emerging independently and nearly simultaneously from different corners of the special education field – offer portals into a conceptual framework that might help scaffold the understanding decision makers would need as advances in genetic selection and manipulation might be rocketing us into the Designer Baby Era. *Neurodiversity* (coined by a mother from the ASD community) and *cerebrodiversity* (coined by a neuroscientist-educator working in the dyslexia domain) address overlapping themes and together help illuminate important considerations.

Neurodiversity. In his book, *Neurodiversity* (2010), Thomas Armstrong traces the origins of the term *neurodiversity*.

> The first use of the word "neurodiversity" in print was in an article by journalist Harvey Blume published in *The Atlantic*, in September 1998. Blume wrote, "Neurodiversity may be seen as every bit as crucial for the human race as biodiversity is for life in general. Who can say what

---

[43] Until accusations of eugenics were leveled at the London Sperm Bank and became headlines in late December 2015 and early January 2016, the clinic turned away donors with dyslexia and similar conditions (ADD, ADHD, Asperger's). As this book was going to press, the UK's Human Fertilisation and Embryology Authority was reviewing the sperm bank's policies and practices (see Weaver, 2015).

form of wiring will prove best at any given moment? Cybernetics and computer culture, for example, may favor a somewhat autistic cast of mind." The actual coining of the term is attributed to Judy Singer – a self-described parent of an "aspie" (person with Asperger's syndrome). (p. 7)

Armstrong goes on to explore neurodiversity in the context of seven conditions (ADHD, ASD, dyslexia, mood disorders, anxiety disorders, intellectual disabilities, and schizophrenia), which he suggests may "represent alternative forms of natural human difference," rather than mental disorders. He delineates eight principles of neurodiversity:

1. The human brain works more like an ecosystem than a machine.
2. Human beings and human brains exist along continuums of competence.
3. Human competence is defined by the values of the culture to which you belong.
4. Whether you are regarded as disabled or gifted depends largely on when and where you were born.
5. Success in life is based on adapting one's brain to the needs of the surrounding environment.
6. Success in life also depends on modifying your surrounding environment to fit the needs of your unique brain (niche construction).
7. Niche construction includes career and lifestyle choices, assistive technologies, human resources, and other life-enhancing strategies tailored to the specific needs of a neurodiverse individual.
8. Positive niche construction directly modifies the brain, which in turn enhances its ability to adapt to the environment. (pp. 9–23)

Throughout the book, Armstrong emphasizes the "strengths, talents, abilities, intelligences," and "extraordinary gifts" of "neurodiverse individuals," but stresses that he does not mean to romanticize mental illness or to trivialize the damage these conditions can inflict, a point he underscores by sharing his own struggles with depression and an anxiety disorder. His thesis, "backed by substantial research in brain science, evolutionary psychology, anthropology, and other fields" (p. 25), is that diversity of minds in humans is good.

> It provides civilization with a multiplicity of possibilities, a variety of styles of living, a number of unique perspectives on life that enrich our world rather than impoverish it, as would happen if we only had a narrow spectrum of human beings represented on the planet. As autism advocate Temple Grandin points out, "Aware adults with autism and

their parents are often angry about autism. They may ask why nature or God created such horrible conditions as autism, manic depression, and schizophrenia. However, if the genes that cause these conditions were eliminated, there might be a terrible price to pay. It is possible that persons with bits and pieces of these traits are more creative, or possibly even genius ... If science eliminated these genes, maybe the whole world would be taken over by accountants." (p. 218)

At the end of his book, Armstrong touches on genetic engineering and raises the specter of eugenics.

Soon it may be possible to detect genes for ADHD, dyslexia, schizophrenia, depression, or other forms of neurodiversity in utero, or even to engage in genetic engineering to eliminate them entirely from the Human Genome. Such practices would echo and perhaps exceed those of the eugenics movement of the first part of the twentieth century. (pp. 218–219)

Cerebrodiversity. Gordon Sherman – executive director of Newgrange School, Ann Robinowitz Education Center, and Laurel School; and former director of the Dyslexia Research Laboratory at Beth Israel Deaconess Medical Center – coined the term *cerebrodiversity* in a series of talks for teachers in 2002.

Sherman sees dyslexia as a byproduct of cerebrodiversity, which he defines as humanity's collective neural heterogeneity and describes as an important adaptive advantage that enabled our species to leverage individual strengths for collective success. Sherman's cerebrodiversity framework builds on the speculations of Geschwind, with whom Sherman worked for the first four of his thirty-year career in neuroscience, before heading up two schools and an outreach center for individuals with learning differences. In 2010, Sherman and I co-authored an article on cerebrodiversity for the International Dyslexia Association's *Perspectives*, from which these excerpts are taken.

Geschwind's thinking lays the groundwork for a model of developmental dyslexia that transcends today's disability paradigm and enables us to see such learning differences as byproducts of a complex mechanism – a dynamic gene-brain-environment interplay that has enabled our species to adapt and succeed for over 200,000 years. Forged in interstellar and planetary forces eons in the making and sculpted by evolution's agents (diversity, environmental change, and adaptability), the human brain develops according to a dynamic interaction between a genetic blueprint

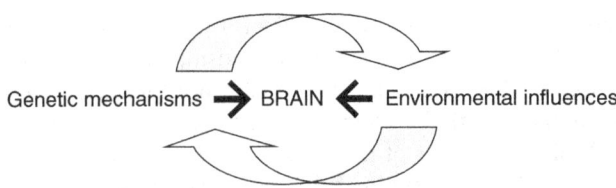

FIGURE 16.1 Learning differences are byproducts of a complex mechanism – a dynamic gene-brain-environment interplay. The human brain develops according to a dynamic interaction between a genetic blueprint and environmental experience (e.g., social, natural, even uterine), which is facilitated by neuroplasticity (the brain's ability to alter and form new neuronal connections and to reorganize).

bequeathed by natural selection and environmental experience (e.g., social, natural, even uterine).[44] A simple representation looks like [Figure 16.1].

This brain development is facilitated by *neuroplasticity*, the brain's ability to alter and form new neuronal connections and to reorganize. These genetic mechanisms, environmental influences, and their neural interplay yield considerable cerebrodiversity – tiny neural differences (anatomical, cellular, and connectional) as unique to each of us as our fingerprints. These neural differences produce subtle cognitive strengths and weaknesses that differ from person to person.

This evolution perspective helps us recognize the neural profiles and spectrums of dyslexia as byproducts of cerebrodiversity and highlights how gene-brain-environment interaction yields neural strengths and weaknesses, which, depending on environmental demand, can translate into socially defined talents and disabilities. For example, the dyslexia brain organization was not a disability before the advent of the printing press some 600 years ago. Before then, books were not ubiquitous and print literacy was not fundamental to functioning in society. (Sherman and Cowen, 2010, p. 16)

Sherman's evolution-cerebrodiversity explanation of dyslexia is speculative, but draws heavily from established knowledge bases in neuroscience, evolution, and human history to posit an explanation for the paradoxical strengths and weaknesses (and co-morbidities[45]) often reported in dyslexia and other neurodevelopmental conditions. He hypothesizes that there is an

---

[44] See Chapters 14 and 15 in this volume for further discussion of this dynamic gene–environment interaction and the complex educational implications.

[45] Comorbid: Presence of additional disorders or diseases co-occurring with a primary disease or disorder. Another Geschwind quote may offer insight into dyslexia's high incidence of comorbidities (e.g., ADHD). "The learning disabled are not merely byproducts of diversity. They are probably a more variable population than those without special disability" (Geschwind, 1984, p. 327).

over-representation of corresponding talents in dyslexia (e.g., in visual-spatial domains) and that these result from the same cerebrodiversity mechanisms that produce dyslexia's hallmark difficulties learning to read, write, and spell.[46]

Sherman's evolution lens sharpens perspective on the transience of the environmental context defining today's learning assets and liabilities. As long as print literacy remains fundamental to society, some will struggle with dyslexia. But others will be disabled by tomorrow's demands. To illustrate how quickly the social environment can change, Sherman often asks in talks about cerebrodiversity how many people in the audience are discovering that they have an acute case of *dystechnia*.[47] Hands go up.

The lessons of humanity's 200,000-plus-year history suggest that our future will continue to depend, at least in part, on our species' ability to exploit individual strengths for collective success. It is difficult to predict what talents or strengths will be needed to solve tomorrow's challenges or to discover future opportunities. Cerebrodiversity, an adaptive advantage and the ultimate human resource, ensures that all kinds of thinkers will be available, unless we find some way to deplete this resource.

Cautionary Messages. The neurodiversity and cerebrodiversity frameworks[48] have much in common. Both are speculative, but draw from established domains of knowledge. Both are forged from extensive first-hand experience studying and treating people with significant neurodevelopmental variation. Both posit that brain diversity may be a selective biosocial advantage in human evolution. Both remind us that the environment not only helps shape the brain,[49] but also, to some degree, defines disabilities

---

[46] Sherman co-authored two of the papers (Diehl et al., 2014; von Károlyi, Winner, & Sherman, 2003) cited earlier in this chapter studying a relationship between dyslexia and visual-spatial strengths.

[47] Indeed, researchers at Cornell and UNC-Chapel Hill (Tufekci & Brashears, 2014) say they have identified a new condition – *cyberasociality* – the inability or unwillingness to relate to others via social media, which, according to the authors, is far from trivial in a digitally mediated society.

[48] Gilger and Hynd (2008) propose a similar neurodevelopmental variation framework – *Atypical Brain Development* – for understanding the "great variation and covariation that exists within and across children," particularly vis-à-vis the "twice exceptional" child (a specific learning disability concomitant with a cognitive gift or talent).

[49] Even the environmental experience of learning to read shapes the brain, which recruits neural circuitry bequeathed by a genetic blueprint that evolved in our preliterate ancestors to accomplish non-reading tasks. Although our brain structures have changed little since those preliteracy times, becoming literate alters the brain by changing how we *connect* those neural structures. So, while the human brain invented reading, learning to read changes the human brain. It is an enigma all the more fascinating because the neural operations involved in fluent reading are subconscious. Exactly how that neural machinery operates is not apparent to the fluent reader. So the enigma is compounded

and talents. Both caution that the environment is subject to change. Both suggest that many individuals with neurodevelopmental differences and disorders often seem to have tradeoff strengths relative to their weaknesses and, perhaps, even talents[50] that can enrich all of our lives. Finally, as repro-genetics offers parents an expanding array of possibilities that someday may include cognitive-behavioral trait selection or modification, both frameworks embody a cautionary message: *Be careful what you wish for.*

Perhaps the most powerful way to convey this cautionary message is through a short poem written by a nephew of mine when he was just seven years old.

THE BRAIN THINKING

If everybody was the same
And someone had a thought
And, it was wrong,
Then everybody would be wrong
So …
Everybody needs to be different.

– Abbott Cowen

INTO THE GREAT UNKNOWN WITHOUT
A FLIGHT PLAN

Question 7: Now, our final question. Imagine that gene selection and modification have advanced tenfold. Not only is it possible to prevent a host of diseases, but also to determine behavioral and cognitive abilities, temperament, athleticism, and aesthetic features. Imagine further that you have two healthy male embryos. One appears to be a good "low-risk

---

- the brain invented reading, reading changes the brain, but the brain doesn't perceive how it reads. And this, as Seidenberg (2017) points out, helps explain the proliferation of wrong-headed ideas about effective reading instruction. Translated into practice, these misguided ideas have harmed generations of children. The difficulty in learning to read and the opacity of fluent reading processes are beautifully captured in this John Steinbeck quote (published posthumously), "Some people there are who, being grown, forget the horrible task of learning to read. It is perhaps the greatest single effort the human undertakes, and he must do it as a child" (1976, p. xi). For a deep dive into the wonders and science of the reading brain and the important instructional implications, see Dehaene (2009), Seidenberg (2017) and Wolf (2007).

[50] In a similar vein, Gilger, Talavage, and Olulade (2013) speculate that "twice exceptionality" may derive from a shared etiology.

specimen," with no genetic abnormalities. The other appears to be more risky, with probabilities for high intelligence and musical aptitude, but also for short stature and clumsiness. There also may be increased risk for delayed language development and, perhaps, a predisposition for rebelliousness and temper tantrums and for characteristics consistent with mild Asperger's syndrome or ASD (e.g., perhaps a tendency for social awkwardness and perseveration).[51] Imagine that a *Sophie's Choice* is necessary and only one child can be carried to term safely. *Which will you choose, the first or the second?*

If you were suspicious of a set-up in this last question and recognized Einstein-like traits in the second profile, you were right. But before we go further, this important caveat cannot be overemphasized: The story that Einstein had dyslexia is turning out to be more fiction than fact (Siegel, 2013), illustrating the hazards of post-mortem diagnoses of famous people, a popular practice over the years. There is evidence that Einstein was delayed in learning to talk and some scholars have observed that he shared some of the characteristic peaks and valleys associated with Asperger's (see Ioan, 2006; Isaacson, 2007; Katz, 1995; Muir, 2003). *But this is not to say that Einstein was on the autism spectrum.* The most we can definitively say is that he appeared to have mild manifestations of traits we now associate with this condition. And, again, context is everything! Asperger's or ASD–like traits in one circumstance might be considered in another just the eccentricities of a brilliant mind.

With those caveats firmly in mind, the point illustrated by the final question is this: Any quest for whatever is deemed good in any given zeitgeist might be misguided. We may never understand the powerfully nuanced interplay between the genetic and environmental variables that might align to set the stage for an Einstein to shine brightly or might misalign to marginalize or disable him (or her). Without the benefit of prescience, or perhaps without recognition of oneself or of a successful loved one in the second profile, how many aspiring parents would risk all their hopes and dreams on the second choice? This is another hypothetical question, of course, fraught with endless what-ifs, but you get the idea.

---

[51] Asperger's syndrome: An autism spectrum disorder (ASD) characterized by difficulties in social interaction and nonverbal communication, along with restricted or repetitive behaviors or interests. Physical clumsiness and odd use of language often are reported. Asperger's was removed from the fifth edition of the *Diagnostic and Statistical Manual of Mental Disorders* (American Psychiatric Association, 2013) and replaced with a diagnosis of ASD on a scale of severity.

The scenarios in questions six and seven are fiction. For now, at least. They are posed to stretch our thinking as, perhaps, we edge closer to the threshold of the Designer Baby Era. Already, hopeful parents face mind-boggling opportunities and decisions in fertility clinics across the country. Options and choices unthinkable just a few decades ago now are common-place. Genetic engineering technologies will continue to advance and pro-liferate. The upsides are myriad and miraculous. But are there downsides too? How prepared would we be as parents, a society, and a species to step across this threshold? How well do we understand the possible unintended consequences? How wisely would we wield these advancing gene selection and modification technologies?

No Plan. Since the Human Genome Project was declared complete in April 2003, researchers have continued identifying protein-coding genes and their functions, unraveling the complexities of epigenetic[52] mechanisms and how they affect individual differences, discovering disease-causing genes, and developing breakthrough treatments to save and improve lives. The pace of these advances stands in stark relief against sluggish efforts to monitor, regulate, guide, or study the growing practices of genetic manipu-lation and selection. This dichotomy suggests that we will encroach ever deeper into biosocial frontiers. As new manipulation and selection tech-nologies are introduced into clinical practice, inevitably, they will become progressively routine and culturally normalized. For example, we haven't seen splashy headlines about "test tube babies" in decades.

True, dramatic headlines about designer baby breakthroughs punctu-ate the news cycle periodically, esoteric debate among bioethicists appears on the public radar for a few moments, and politicizing along the lines of stem cell research sometimes heats up. But there is no coherence here. No plan. In the absence of leadership and coordination at policy levels,[53] there is no thoughtful national dialog or planning as we may be getting closer to "crossing the Rubicon." This puts parents center stage as the primary deci-sion makers. But this stage has no spotlight, no audience. Decisions are being made quietly, privately, and with little thought about or understand-ing of possible unintended consequences. Like climate change, it may be years before any impact will be noticed.

---

[52] Epigenetics: Heritable changes in gene expression not caused by changes in the underly-ing DNA sequence.

[53] Compare the 2004 report from The President's Council on Bioethics, "Reproduction and Responsibility: The Regulation of New Biotechnologies," and Murray's 2014 essay in *Science* to see how little movement there has been at the policy level.

Is brain diversity as vital to the human species as biodiversity is to life in general? No one knows. Perhaps such diversity is an unstoppable force. Maybe our stunning advances in reprogenetics will cause hardly a ripple in the vast and dynamic ocean of biosocial forces that brought *Homo sapiens* from the African savannah to this moment in our story. Perhaps the odds of identifying, never mind manipulating, the complex genes associated with temperament, intelligence, and various neurodevelopmental conditions are too slim to worry about.

On the other hand, maybe human-engineered selection will not only be possible, but will also be a big improvement on natural selection. Perhaps we can and should weed out "unfit genes" to improve the human species and can and should eradicate inherited and incidental genetic diseases and disorders to eliminate a lot of suffering. Perhaps technology-assisted genetic screening is just an extension of the mate screening that humans have always practiced. Maybe genetic engineering of ourselves is as much the evolutionary course of a big-brained species as making an obsidian spear tip, or fire, or any of the tools or technologies that brought us from the Stone Age to the Information Age and now, perhaps, to the dawn of the Designer Baby Era.

Or, maybe, seduced by the siren song of the perfect baby, we will founder on the rocks of our own hubris. Again, no one knows.

Four Steppingstones to a Plan. In any case, we seem to be launching headlong into the great unknown of genetic engineering without a flight plan. Is it too late to engage in some sort of national dialog that might stimulate and inform a plan – such as the conversation conducted by the United Kingdom over a decade ago, which harnessed polling, focus groups, and the Internet (Murray, 2014)? Can we have this conversation without inflaming passions and bogging down in polarizing politicization? As Murray suggested in his *Science* article, planning and coordinating a forum for this national dialog would fall under the aegis of the US Presidential Commission for the Study of Bioethical Issues.[54] Facilitating this national conversation would be an important, albeit tricky, step. Conducting the conversation at a high level might be trickier still.

A second step might be for this same commission to convene consensus-building summits among the various professional groups[55]

---

[54] Beyond a 2012 report ("Privacy and Progress in Whole Genome Sequencing"), there does not seem to be action along the lines Murray recommends.

[55] Groups such as the American College of Obstetrics and Gynecology's Committee on Ethics, the Ethics Committee of the American Society for Reproductive Medicine, the American Academy of Pediatrics, the American College of Medical Genetics.

already providing self-regulation guidelines, many of which are at odds. Finding consensus among these groups would not be easy. They would need to navigate the ethical and biosocial intricacies of competing tar-baby[56] issues such as reproductive freedom, right-to-life, safeguarding children, genetic disease prevention and intervention, equal access, and protecting the integrity of the human genome. Conflict abounds in each of these issues. Combined, they might be combustible. These groups also would struggle not only to align the various perspectives each brings to these issues, but also to sort out turf each has staked out. But, certainly, there is common ground to be found that could help inform a plan. And, each group is grappling with these complex issues already, so there is some momentum.

A third step might be to launch a website under the auspices of a joint committee of these professional and stakeholder parent groups. Even without consensus, just reporting the various pros, cons, and viewpoints and curating information about emerging technological advances, scientific breakthroughs, and biosocial implications via a web-based platform could be enormously helpful to would-be parents. Barring any unforeseen legislation or regulation, these parents will remain on the frontlines, making the consequential decisions. Naturally, they will view these decisions as personal and private; but they also might appreciate and benefit from a comprehensive, neutral, web-based resource to inform their decision making. Perhaps the professional and stakeholder groups could support the management of this website, to which they might link their own websites and refer prospective parents. This step, one that could help build professional and stakeholder consensus to inform a plan, might also become an objective or deliverable (i.e., a website) in such a plan. In other words, the plan and the website could evolve dynamically.

A fourth step might be to reach out to the stakeholder groups and grass-roots networks coalescing around particular neurodevelopmental conditions, many of which have a strong genetic basis (e.g., dyslexia, ADHD, and ASD). These communities, which include families with these conditions, would be riveted by the prospect of genetic selection and modification technologies targeting the genes that predispose for these neurodevelopmental conditions in future generations. Some prospective parents from these communities might wish to liberate their family lines from the burdens and challenges of these conditions. Others – probably those with milder

---

[56] Tar-baby: The doll made of tar and turpentine used to trap Br'er Rabbit. Modern usage: Any sticky situation that is aggravated by additional contact.

manifestations of dyslexia, ADHD, or ASD – might not, especially if they view such conditions as a gift. Either way, these stakeholder communities would be highly motivated to understand and educate their members and others about the emerging ethical and biosocial complexities. Thus, partnering with these communities not only might inform a plan, but it might also accomplish an objective in such a plan – i.e., leveraging a distribution network for disseminating information and raising awareness, particularly among prospective stakeholder parents.

THE END: PERFECT BABY

In this chapter, we considered seven questions of increasing ethical difficulty to help stretch our thinking and to get into a parent's mindset vis-à-vis impact of the Human Genome Project. Parents are the central characters in our story – acting as both protagonist, the main character with whom we must empathize, and as antagonist, responsible for the major events around which our story revolves. As a subplot, we explored learning differences, particularly dyslexia, to (a) illustrate complexities, such as the role and transience of the environment (particularly the educational environment), in determining beneficial or disabling traits of potential offspring and (b) to introduce the possibility that neurodevelopmental conditions, such as dyslexia, reflect the adaptive advantage of a diversity mechanism at work. The plot thickened as we delved into the neurodevelopmental variation framework (neurodiversity and cerebrodiversity) to help scaffold the deep understanding we may need to consider questions like: (a) Is brain diversity as vital to *Homo sapiens* as biodiversity is to life in general, and (b) can we ever know enough about possible unintended consequences to be able to wisely engineer behavioral, cognitive, and temperament traits?

All this set the stage for the following four proposed steppingstones to help generate a plan as we explore the many unknowns of genetic engineering:

1. Initiate national dialog facilitated by the US Presidential Commission for the Study of Bioethical Issues (or some similar group).
2. Convene consensus-building summits among various professional groups already providing self-regulation guidelines (sponsored by this same commission or similar group).
3. Create a jointly sponsored website curating various pros/cons, perspectives, viewpoints, and information for parents making the consequential decisions.

4. Initiate outreach efforts to stakeholder groups and grassroots networks for bidirectional input.

Finally, we come full circle – back to our opening scene, where a would-be parent's hopes and dreams begin: having a child. These hopes and dreams, the inciting force, lie at the heart of all the complexities discussed in this chapter and must be understood deeply and respected highly in any plans or policies, be they top-down and regulatory, bottom-up and laissez-faire, or some combination thereof. The wish to bring a healthy baby into the world and to give this child every possible advantage might be encoded biosocially into much of humankind as indelibly as our DNA. Yes, the marketplace and fertility clinics are powerful forces in their own right. So, too, is science. But these are more means than ends. A perfect baby is the end most parents seek.

Without other steering, guiding, or countering forces, this parental aspiration is likely to override nearly everything else. Indeed, any such forces might be impotent in the face of this quest, which could be rocketing us into the Designer Baby Era, where, for better and worse, we might begin altering our genome and reshaping our destiny. Something to ponder.

ACKNOWLEDGMENTS

My heartfelt thanks go to Jeff Gilger and Linda Siegel, who reviewed this manuscript and provided insightful comments and suggestions. Thanks, too, to the editors of this volume for their very helpful critiques. Naturally, any errors, incomplete information, or poorly framed discussion are mine. Finally, thanks to Abbott Cowen for his brilliant little poem.

REFERENCES

American Psychiatric Association. (2013). *Diagnostic and statistical manual of mental disorders* (5th ed.). Washington, DC: American Psychiatric Association.

Armstrong, T. (2010). *Neurodiversity: Discovering the extraordinary gifts of autism, ADHD, dyslexia, and other brain differences.* Philadelphia, PA: Da Capo Press.

Baruch, S., Kaufman, D. J., & Hudson, K. (2006). Genetic testing of embryos: Practices and perspectives of U.S. IVF clinics. *Fertility and Sterility, 89*(5), 1053–1058. doi:10.1016/j.fertnstert.2007.05.048

Bishop, D.V. M. (2013). Neuroscientific studies of intervention for language impairment in children: Interpretive and methodological problems. *Journal of Child Psychology and Psychiatry, 54,* 247–259.

Branson, R. (2017, April). Richard Branson: Dyslexia is merely another way of thinking. *The Sunday Times.* Retrieved June 11, 2017 from www.the-times.co.uk/edition/news-review/richard-branson-dyslexia-is-merely-another-way-of-thinking-8tlmgsndw

Callaway, E. (2017, June). Oldest *Homo sapiens* fossil claim rewrites our species' history. *Nature*. Retrieved June 19, 2017 from http://www.nature.com/news/oldest-homo-sapiens-fossil-claim-rewrites-our-species-history-1.22114

Center for Genetics and Society. (n.d.). About genetic selection. Retrieved April 20, 2014 from www.geneticsandsociety.org/section.php?id=82

Chapman, L. (1967). Illusory correlation in observational report. *Journal of Verbal Learning and Verbal Behavior*, 6(1) 151–155. doi:10.1016/S0022-5371(67)80066-5

Coley, R. J., & Barton, P. E. (2006). *Locked up and locked out: An educational perspective on the U.S. prison population*. Princeton, NJ: Educational Testing Service.

Cortiella, C., & Horowitz, S. H. (2014). *The state of learning disabilities: Facts, trends, and emerging issues*. New York: National Center for Learning Disabilities.

Cowen, C. D. (2014, January). Dyslexia and visuospatial processing strengths: New research sheds light. *International Dyslexia Association Examiner*. Retrieved April 26, 2014 from https://dyslexiaida.org/dyslexia-and-visuospatial-processing/

Cowen, C. D. (2016, December). Who are the next great entrepreneurs? *International Dyslexia Association Examiner*. Retrieved June 11, 2017 from https://dyslexiaida.org/who-are-the-next-great-entrepreneurs/

Diehl, J. J., Frost, S. J., Sherman, G. F., Mencl, W. E., Kurian, A., Molfese, P., et al. (2014). Neural correlates of language and non-language visuospatial processing in adolescents with reading disability. *NeuroImage*. doi:10.1016/j.neuroimage.2014.07.029

Dehaene, S. (2009). *Reading in the brain: The new science and evolution of a human invention*. New York: Viking.

Eide, B. L., & Eide, F. F. (2011). *The dyslexic advantage: Unlocking the hidden potential of the dyslexic brain*. New York: Hudson Street Press.

Elliot, J. G., & Grigorenko, E. L. (2014). *The dyslexia debate*. Cambridge: Cambridge University Press. doi:10.1017/CBO9781139017824

Fahmy, M. S. (2011). On the supposed moral harm of selecting for deafness. *Bioethics*, 25(3), 128–136. doi:10.1111/j.1467-8519.2009.01752.x

Farr, C. (2016). 23 and Me's consumer DNA data goldmine is starting to pay off. Retrieved August 21, 2016 from www.fastcompany.com/3062731/most-innovative-companies/23andmes-consumer-dna-data-gold-mine-is-starting-to-pay-off

Gaab, N. (2017). It's a myth that young children cannot be screened for dyslexia. *International Dyslexia Association Examiner*. Retrieved June 18, 2017 from https://dyslexiaida.org/its-a-myth-that-young-children-cannot-be-screened-for-dyslexia/

Geiger, G., & Lettvin, J. Y. (1987). Peripheral vision in persons with dyslexia. *New England Journal of Medicine*, 316, 1238–1243. doi:10.1056/NEJM198705143162003

Gilger, J. W. (1991). Differential assortative mating found for academic and demographic variables as a function of time of assessment. *Behavior Genetics*, 21(2), 131–150. doi:10.1007/BF01066332

Gilger, J. W., & Hynd, G. W. (2008). Neurodevelopmental variation as a framework for thinking about the twice exceptional. *Roeper Review*, 30, 214–228. doi:10.1080/02783190802363893

Gilger, J. W., Talavage, T. M., & Olulade, O. A. (2013). An fMRI study of nonverbally gifted reading disabled adults: Has deficit compensation effected gifted potential? *Frontiers in Human Neuroscience.*

Gilger, J. W. (2017). Beyond a reading disability: Comments on the need to examine the full spectrum of abilities/disabilities of the atypical dyslexic brain. *Annals of Dyslexia, 67,* 1–5. doi 10.1007/s11881-017-0142-x

(2017). Why study nonverbal skills in dyslexia? *International Dyslexia Association Examiner.* Retrieved June 13, 2017 from https://dyslexiaida.org/why-study-nonverbal-skills-in-dyslexia/

Geschwind, N. (1982). Why Orton was right. *Annals of Dyslexia, 32,* 13–30. doi:10.1007/BF02647951

(1984). Brain of a learning-disabled individual. *Annals of Dyslexia, 34,* 319–327. doi:10.1007/BF02663629

Geschwind, N., & Galaburda, A.M. (1987). *Cerebral lateralization: Biological mechanisms, associations, and pathology.* Cambridge, MA: MIT Press.

Gladwell, M. (2013). *David and Goliath: Underdogs, misfits, and the art of battling giants.* Boston: Little, Brown & Co.

Gray, T. (2013, November 25). FDA to 23andme founder Anne Wojciki: Stop marketing $99 DNA test or face penalties. Retrieved April 29, 2014 from www.fastcompany.com/3022208/fda-tells-23andme-founder-anne-wojcicki-to-stop-marketing-99-genetic-test-or-face-penalties

Green, R. N. (2007). *Babies by design: The ethics of genetic choice.* New Haven, CT: Yale University Press.

Hadely, C. (2003). Building healthy families. *European Molecular Biology Organization Reports, 4*(11), 1017–1019.

Ioan, J. (2006). *Asperger's syndrome and high achievement: Some very remarkable people.* London: Jessica Kingsley.

Isaacson, W. (2007). *Einstein: His life and universe.* New York: Simon & Schuster.

Katz, I. (1995). *In a world of his own: A storybook about Albert Einstein.* New York: Cimino Publishing.

Kozubek, J. (2016). *Modern Prometheus: Editing the human genome with CRISPR-CAS9.* Cambridge: Cambridge University Press.

Kraft, I., et al. (2016). Predicting early signs of dyslexia at a preliterate age by combining behavioral assessment with structural MRI. *Neuroimage, 143,* 378–386.

Krahn, M. T. (2011). Regulating preimplantation genetic diagnosis: The case of Down's syndrome. *Oxford University Press Medical Law Review, 19*(2), 157–191.

Langer, N., Peysakhovich, B., Zuk, J., Drotta, M., Sliva, D. D., Smith, S., et al. (2015). White matter alterations in infants at risk for developmental dyslexia. *Cerebral Cortex.* doi: 10.1093/cercor/bhv281

Logan, J. (2009). Dyslexia entrepreneurs: The incidence, their coping strategies, and their business skills. *Dyslexia, 15*(2), 328–346. doi:10.1002/dys.388

Murray, T. H. (2014, March). Stirring the simmering "designer baby" pot. *Science, 343,* 1208–1210. doi:10.1126/science.1248080

Morgan, W. P. (1896). A case of congenital word blindness. *British Medical Journal, 2,* 1378. doi:10.1136/bmj.2.1871.1378

Morris, B. (2002, May 13). Overcoming dyslexia. *Fortune Magazine.*

Muir, H. (2003, April). Einstein and Newton showed signs of autism. *New Scientist.* Retrieved April 30, 2014 from www.newscientist.com/article/dn3676# .U2KZGyhDg5s

National Academy of Sciences, Engineering, and Medicine. (2017). *Human genome editing: Science, ethics, and governance.* Washington, DC: The National Academies of Press. doi:https://doi.org/10.17226/24623

Olulade, O. A., Gilger, J. W., Talavage, T. M., Hynd, G. W., & McAteer, C. I. (2012). Beyond phonological processing deficits in adult dyslexics: Atypical fMRI activation patterns for spatial problem solving. *Developmental Neuropsychology,* 37, 617–635. doi:10.1080/87565641.2012.702826

Pollack, A. (2015, May 11). Jennifer Doudna, a pioneer who helped simplify genome editing. *New York Times.* Retrieved January 21, 2015, from www.nytimes.com/ 2015/05/12/science/jennifer-doudna-crispr-cas9-genetic-engineering.html?_r=0

President's Council on Bioethics. (2004). *Reproduction and responsibility: The regulation of new biotechnologies.* Retrieved March 18, 2014 from https://bioethicsarchive.georgetown.edu/pcbe/reports/reproductionandresponsibility/

President's Council on Bioethics. (2012). *Privacy and progress in whole genome sequencing.* Retrieved June 18, 2017 from http://bioethics.gov/sites/default/ files/PrivacyProgress508_1.pdf President's Council on Bioethics. (2012). Privacy and progress in whole genome sequencing. Retrieved June 18, 2017 from http://bioethics.gov/sites/default/files/PrivacyProgress508_1.pdf

Raschle, N. M., Zuk, J., & Gaab, N. (2012). Functional characteristics of developmental in left-hemispheric posterior brain regions predate reading onset. *Proceedings of the National Academy of Sciences,* 109(6), 2156–61. doi:10.1073/ pnas.1107721109

Rawson, M. B. (1988). *The many faces of dyslexia.* Baltimore, MD: International Dyslexia Association.

Schneps, M. R., Rose, T. L., & Fischer, K. W. (2007). Visual learning and the brain: Implications for dyslexia. *Mind, Brain, and Education,* 1(3), 128–129. doi:10.1111/j.1751-228X.2007.00013.x

Schumacher, J., Hoffman, P., Schmal, C., Schulte-Korne, G., & Nothen, M. M. (2007). Genetics of dyslexia: The evolving landscape. *Journal of Medical Genetics,* 44, 289–297.

Seidenberg, M. (2017). *Language at the speed of sight: How we read, why so many can't, and what can be done about it.* New York: Basic Books.

Seigel, L. (2013). *Understanding dyslexia and other learning disabilities.* Vancouver: Pacific Educational Press.

Shaywitz, B. A., Shaywitz, S. E., Blachman, B., Pugh, K. R., Fulbright, R., Skudlarski, P., et al. (2004). Development of left occipito-temporal systems for skilled reading following a phonologically-based intervention in children. *Biological Psychiatry,* 55, 926–933. doi:10.1016/j.biopsych.2003.12.019

Sidhu, J. (2012, September). How to buy a daughter: Choosing the sex of your baby has become a multi-million dollar industry. *Slate Medical Examiner/ Health and Medicine Explained.* Retrieved April 28, 2014 from www.slate.com/ articles/health_and_science/medical_examiner/2012/09/sex_selection_in_ babies_through_pgd_americans_are_paying_to_have_daughters_rather_ than_sons_.html

Silver, L. M. (1998). *Remaking Eden: How genetic engineering and cloning will transform the American family.* New York: Avon.

Simos, P. G., Fletcher, J. M., Bergman, E., Breier, J. I., Foorman, B. R., Castillo, E. M., et al. (2002). Dyslexia-specific brain activation becomes normal following successful remedial training. *Neurology, 58,* 1203–1213. doi:10.1212/WNL.58.8.1203

Sherman, G. F., & Cowen, C. D. (2010). Evolutionary lens offers alternative perspective on dyslexia. *Perspectives, 36*(1), 16–17.

Steinbeck, J. E. (1976). *The acts of King Arthur and his noble knights.* New York: Farrar, Straus and Giroux.

Temple, E., Deutsch, G. K., Poldrack, R. A., Miller, S. L., Tallal, P., Merzenich, M. M., et al. (2003). Neural deficits in children with dyslexia ameliorated by behavioral remediation: Evidence from functional MRI. *Proceedings of the National Academy of Sciences, 100,* 2860–2865.

Tufekci, Z., & Brashears, M. E. (2014). Are we all equally at home socializing online? Cyberasociality and evidence for an unequal distribution of disdain for digitally-mediated sociality. *Information, Communication & Society: Special Issue: Communication and Information Technologies Section,* 17(4), 486–502.

Vail, P. (1990). Gifts, talents, and the dyslexias: Wellsprings, springboards, and finding Foley's rocks. *Annals of Dyslexia, 40,* 3–7.

von Károlyi, C., Gray, W., & Sherman, G. F. (2003). Dyslexia linked to talent: Global visual-spatial ability. *Brain and Language, 85,* 427–431. doi:10.1016/S0093-934X(03)00052-X

Wang, Y., et al. (In press.) Development of tract-specific white matter pathways during early reading development in at-risk children and controls. *Cerebral Cortex.* doi:10.1093/cercor/bhw095

Weaver, M. (2015). Largest sperm bank turns away dyslexic donors. *The Guardian.* Retrieved December 29, 2015, from www.theguardian.com/society/2015/dec/29/largest-uk-sperm-bank-turns-away-dyslexic-onors?utm_source=esp&utm_medium=Email&utm_campaign=GU+Today+main+NEW+H&utm_term=146715&subid=15169943&CMP=EMCNEWEML6619I2

West, T. G. (1997). *In the mind's eye: Visual thinkers, gifted people with dyslexia, and other learning difficulties, computer images, and the ironies of creativity.* New York: Prometheus Books.

Wolf, M. (2007). *Proust and the squid: The story and science of the reading brain.* New York: Harper.

Conclusion: How Might School Systems Use Genetic Data?

STEPHEN R. LATHAM

New findings on the genetic roots of complex traits, and particularly of learning ability and disability, are being produced literally every day. It is impossible to imagine that this influx of new information will not have some effect on the structure of public schooling, and particularly on the ways in which public schools deal with children with learning disabilities. In some areas, greater understanding of the genetic bases of learning will lead educators to want to make use of genetic test results to guide their choices about the educational approaches that will work best for particular students. In other areas, advances in the genetics of learning will actually counsel against educators' making use of genetic data, because the contribution of genetics to learning ability will be understood to be so affected by environmental factors as to reveal genetic data to be less useful than data about actual student performance. In any case, even the simplest and most straightforward moves toward educational use of genetic data will require enormous amounts of preparation, will come at great expense, and will give rise to difficult questions of ethics and policy. In what follows, I shall summarize some of the possibilities and difficulties related to each of three main ways in which schools might use student genetic data: (1) to confirm an already existing diagnosis of a known genetic condition such as Down syndrome, and then perhaps to use information about the genetic etiology of that condition to forecast and prepare for as-yet-latent developmental and learning problems commonly experienced by students with that condition; (2) to make early discovery of previously undetected learning disabilities such as dyslexia in order to facilitate educational interventions before disabilities make themselves manifest; and (3) to fully map out the genetic roots of a student's learning abilities and disabilities in order to create "personalized education" for the student.

USING GENETIC INFORMATION TO CONFIRM
KNOWN CONDITIONS AND GUIDE EARLY
INTERVENTION

It might seem that the simplest case for the use of genetic testing to guide educational planning would involve the relatively small number of well-known and easily detected conditions such as Williams syndrome, Down syndrome, Prader-Willi syndrome, or Fragile X (see Chapter 3 of this volume). Children diagnosed with these and other commonly recognized syndromes have known risks for learning disability, both with regard to cognitive factors and to non-cognitive behavioral issues which affect school performance (on non-cognitive genetic traits that affect learning, see Chapter 6 of this volume). Children with these diagnoses are already introduced to public school systems, and those systems have already been required by federal law (Individuals with Disabilities Education Act, IDEA) to work out the correct approaches to their education. It seems fairly plain that genetic information could help school systems in their efforts to serve these students. Genetic information could confirm uncertain diagnoses based on behavior and appearance, and (as pointed out in Chapter 8 of this volume) could also help school systems predict as-yet-undetected learning issues associated with any given genetic etiology for a learning disorder.

It is perhaps worth noticing one way in which the increased availability of genetic screening is already having – and, barring major legal changes in the United States, will increasingly have – an important effect on American school systems: via parental selection away from children with OGOD conditions. Many studies have confirmed that a large percentage of parents choose to terminate pregnancy when they learn that their developing embryo has Down syndrome. Different studies have made varying estimates of termination rates, but population studies between 1995 and 2011 found an average termination rate of 67 percent, and hospital-based studies in that same period found a rate of 85 percent (Natoli, Ackerman, McDermott, & Edwards, 2012). There is some indication that termination rates for Down syndrome are going down; but the ready availability of new non-invasive pre-natal testing may increase testing rates and, therefore, also termination rates. Termination of pregnancy has reduced the population of people living with Down syndrome by an estimated 30 percent (de Graaf, Buckley & Skotko, 2015), and is likely having similar effects on other easy-to-diagnose OGOD conditions. And in the not too distant future, according to scholars such as Stanford Law School's Henry T. Greely, easier methods of pre-implantation genetic diagnosis ("PGD") will make it possible for large numbers of parents to avoid becoming pregnant with genetically disabled children in the first place (Greely, 2016). There are two upshots, here, for the interaction of disabled children with school systems. On the one hand, costs of special education programs will be reduced as genetic screening, pregnancy termination, and PGD reduce the numbers of children presenting to school systems with OGOD conditions. On the other hand, the comparative rarity of such children may make it more difficult for schools to establish routines and policies for dealing with them, and may increase stigmatization both of the children and of their parents.

But let us consider, now, the possible use by schools of genetic data related to children with OGOD conditions who actually turn up, needing education tailored to their special needs. A best-case (though still rough) account of what might happen could be this: A child's parents and their pediatrician decide on the basis of various behaviors and physical traits that the child has a genetic disorder. This disorder has been shown in the literature to be associated with certain genetic abnormalities. A genetic test is, therefore, done on the child, to confirm the diagnosis by confirming the presence of the relevant abnormalities. The results of this test enable educators to offer the child appropriate educational interventions, including, importantly, early interventions for problems that are associated with the abnormality but which may not yet have manifested in the child. (On the effect of knowing the genetic etiology of a student's special education needs on interventions, see Chapter 15 in this volume.) We might call this approach a "confirming-and-guiding" use of genetic information.

Confirming-and-guiding seems conceptually easy to justify. The science is easy, given the strong relationship of genotype to phenotype in these cases. And the downside ethical risk of mislabeling or mischaracterizing a child is fairly limited in such cases, where a known genetic disease or syndrome is already diagnosed on the basis of already detectable behaviors and physical characteristics. The utility of the genetic confirmation also seems fairly high – though of course, school systems may wonder whether the marginal certainty and predictive value of genetic information is worth the cost to the system of obtaining and processing it.

It is sobering, however, to think about the complexities of introducing even this fairly limited confirm-and-guide use of genetic testing to the public school system. Today, parents present to the school system with a diagnosis from a pediatrician (or perhaps from a specialist) indicating that their child has a certain syndrome. The school understands what this diagnosis means, and attempts to develop an educational plan for the child based on the diagnosis. What would be different if a school required, or even merely opted to accept, genetic evidence for a given syndrome, and proposed to make use of that evidence in framing an educational plan for a given student?

First, the school system would have to articulate a rule about exactly which syndromes it was willing to accept genetic information to confirm, and whether it would require such information to confirm a diagnosis in every case. Formulation of such a policy would require school systems to examine and evaluate evidence about the genetic bases of different syndromes, the reliability of available tests, and the degree to which test results are apt to map onto available and effective teaching approaches designed to address the syndromes in question. In some cases, genetic tests confirming a given diagnosis will map fairly well onto strong predictions about learning abilities and styles; these, in turn, can justify expenditures on child-specific educational programs. Of course, a given school system need not make these evaluations on its own; national expert bodies will undoubtedly inform the discussion and may recommend model policies. But local political problems are apt to be substantial here. Given that even within the universe of One-Gene-One-Disorder learning problems, there is substantial variation in learning ability (see Chapter 2, this volume) – and, therefore, substantial variation in which teaching approaches prove efficacious – parents may be worried that policies will pigeonhole their children.

Next, school systems will have to designate particular sources of genetic information about students as valid. Not just any report from any online gene-testing firm will do (on personal genomics providers, see Chapter 13 of this volume); schools will have to choose from among FDA-approved testing firms. FDA approval (as detailed in Chapter 11 of this volume) may be affected by showings of disparate impact of testing on different groups, and may also require that genetic data be used only in conjunction with educational test results and other data. Depending on budgetary and other administrative concerns, this latter requirement may mean that genetic testing may not be performed meaningfully and usefully in advance of ordinary aptitude testing. So any "early intervention" advantage from genetic testing may be lost.

Third, school systems will have to make basic decisions about justice in the allocation of scarce resources (see Chapter 14 of this volume). A school system that commits to responding to genetic information about students which is supplied by parents will need (as mentioned earlier) some standards about the range of syndromes about which it is willing to receive evidence from parents. It will also need some standards about the labs which supply, and the physicians or commercial services which interpret, the raw genetic data presented by parents. But in addition to those practical problems, it will have to face the social problem of whether it will support, financially, genetic testing of children by parents who would not otherwise be able to afford it. If school systems do not support such testing, then wealthy parents may have a systematic advantage in persuading school systems to deploy their resources for the special needs of their children. This is an obvious injustice, but not one which most school systems are well equipped to avoid.

Whole-genome sequencing costs (in 2016) upwards of $1,500; whole-exome screening nearly $1,000. This is amazingly inexpensive compared to the costs of similar sequencing only a decade ago, but it may nonetheless be prohibitively expensive for most school systems. Moreover, given the fairly small number of students who would be included in a confirming-and-guiding genetic testing program, the actual cost of testing may, in fact, be the least of the costs that schools would incur if they committed to using genetic data to guide educational plans in the case (only) of One-Gene-One-Disorder learning disabilities.

It will cost money to formulate and settle policy on which genetic data will be considered, and from what sources. But once that policy is in place, special education teachers and school administrators will have to be trained in the proper interpretation of genetic data, and in the proper range of educational interventions that are appropriate to different disorders, including the range of disabilities associated with each. (A good policy would also encourage teachers not only to address disabilities associated with different syndromes, but also not to forget the positive traits associated with certain of them, e.g., the sociability and musical talent frequently found in children with Williams syndrome. On the general topic of "neurodiversity," see Chapter 16 in this volume.) In addition, there may be technical expenditures necessary to protect genetic data in the possession of the school system. Decisions will also have to be made about who will have access to that data for public finance and audit purposes, and perhaps also for downstream research purposes. (On the ethics of genetic research in this context, see Chapter 12 in this volume.) And of course, additional policy will have to be developed – at some expense – for the creation of informed consent and child-assent procedures for genetic testing, or for the release of genetic information to the school system. Such policy would likely have to include optout systems for parents who do not want genetic data used by schools, as well as for those who want the schools to use the data, but not to release it for future research. Development of such policies may involve attempts to educate the public about the meaning and implications of genetic findings relating to learning ability. (On the general state of genomic literacy, see Chapter 10 of this volume).

All of this takes place in a context in which, at least at the moment, public school funding levels are terribly low. For the school year ending in 2014, thirty-one of the fifty US states spent less on education per student than they had in 2008, before the recession (Leachman, Albares, Masterson, & Wallace, 2016). Per capita per year expenditures for 2014 averaged around $11,000, with state expenditures varying from New York's high of about $20,000 to Utah's low of $6,500. At this writing, Oklahoma's school budgets are so tight that much of the state has moved to a four-day school week; yet in the most recent election, Oklahoma citizens rejected a 1 percent sales tax hike that would have provided additional funding for schools (Murphy, 2016). Also in 2014, federal IDEA funding covered less than 16 percent of the estimated excess costs of educating disabled children in the public schools in the manner prescribed by IDEA. The funding shortfall is absorbed by already stretched state and local school districts (McCann, 2016). Given these budgetary realities, school systems may be reluctant to afford even a fairly straightforward, uncontroversial move toward confirm-and-predict use of genetic data for a small number of students.

USE OF DATA TO DETECT RISK FOR LEARNING DISABILITIES

In the previous section I addressed the possible use by educators of genetic data gathered to confirm already existing genetic diagnoses. Another possible use of genetic data by educators is both more controversial and also possibly more beneficial to students. I refer to the use of genetic data to diagnose or to identify risk for learning disabilities (paradigmatically, dyslexia) and to address them with early educational interventions even before they are detectable by ordinary scholastic testing.

As has been discussed in Chapters 7 and 9 of this volume, dyslexia and other learning disorders are frequently not discovered until children begin to fail in school. In extreme cases they may not even be identified until a student is in high school, with a long trail of poor grades and frustration behind her. By the time parents and the school system take steps to address the problem, the affected children have already suffered serious impacts on their confidence and self-esteem, and they, as well as their teachers and peers, may already have begun to set low expectations for their academic performance. If genetic data could be used reliably to diagnose or to establish a child's high risk for a specific reading disorder before that child reaches reading age, then early interventions could be used to minimize the disheartening experience of early learning failure, and to allow students and their teachers to frame realistic educational expectations and goals from day one. Spotting a child's learning problems and providing for her learning needs from the very beginning of her education could both prevent a lot of confusion and heartache and facilitate more efficient learning.

Notably, learning disabilities like dyslexia do not have anywhere near as simple a genetic etiology as Down syndrome or Williams syndrome. Dyslexia has been shown to be associated with a number of different genes, and with non-gene parts of the genome that up- or downregulate the activity of those genes (Powers et al., 2013). The upshot is that genetic testing cannot show that a given student has or will have dyslexia. But we may not be too far from the day when genetic testing will at least be able to offer valuable predictive insight into a given child's risk for reading difficulties (and for the intellectual gifts often associated with it).

Implementation by a school system of a program that uses genetic testing to detect previously undetected risks for learning disability, and to use information about those risks to "track" students into different educational programs, is considerably more difficult and controversial than the confirm-and-predict use of genetic information already discussed. As with confirm-and-predict, risk-detecting schemes will require that costs associated with policy formation, education, data management, and consent be incurred. But new and additional problems are posed by the fact that this application of genetic data involves the school system in uncovering and acting upon previously undetected risks.

The allocative justice problems associated with risk detection by schools are more profound than those associated with confirmatory testing. Once again, a school system will have to determine whether it will pay for students to undergo genetic testing. The form of testing in question would likely be comparatively expensive whole-genome sequencing, since the up- and downregulation of genes related to dyslexia seems to be controlled by genomic sequences not examined in less expensive whole-exome screening. More importantly, since the point of risk-detecting genomic screening is to identify otherwise unidentified at-risk children who might benefit from special educational interventions, the screening would have to be offered *to everyone*. This over-$1,000-per-student expense is extremely unlikely to be workable for the average American school system, which devotes only about ten times that much to the total education of each of its students each year.

Budget constraint might, therefore, leave school systems unwilling to pay for genetic testing. Might they nonetheless be willing to use certain kinds of genetic data to guide educational interventions, if that data are supplied to them by parents? The justice problems with doing so are profound. It will only be fairly wealthy parents who have sufficient resources to spend on "just-in-case," risk-detecting genetic screening of their children; and it will only be well-educated and very well-informed parents who will think to make that expenditure and then call the school system's attention to their test results. Thus the most privileged of parents would have the ability to use their privilege to trigger a school system's allocation of public resources to their children's education. Less privileged children with learning disabilities would be condemned to the ordinary, non–gene-testing pathway toward detection of their problems – the pathway which begins with the student's experience of failure and disappointment.

The state of the science with regard to using genetic testing to diagnose, or to make reasonably accurate risk predictions about, different sorts of learning disability is far from clear. There is promising (though not yet conclusive) work on the genetic roots of dyslexia, but fairly little actionable work beyond that. This second method by which school systems might use genetic data requires, first, a fairly small set of genetic anomalies related to a given disability, so that detection is simple and reliable; and second, that the disability in question have some fairly stable characteristics, such that an educational method adequate to a range of affected students' particular problems can usefully be developed as a response to the genetic information. The main practical problem with any school system's adopting this second use of genetic data is that, while the number of target conditions (and, therefore, the number of required educational interventions) may be fairly small, the screening – since it is aimed at early detection of otherwise as-yet-undetected conditions – really ought to be offered to every student. School systems would have to (1) figure out how to afford that; or (2) come up with arguments justifying the use of data supplied by comparatively well-off parents to give their children privileged access to publicly funded special education; or (3) defend on egalitarian grounds a policy that prohibits use of potentially useful data, even when (privileged) parents have gathered it at their own expense.

BROADER GENETIC PROFILING

Most traits related to learning are not tied to specific "telltale" genetic mutations, but are the cumulative result of the small effects of many genes (see

Chapter 9 of this volume). Unlike the cases of OGOD genetic illnesses and dyslexia, there is considerable debate about whether and how to characterize combinations of such traits as "disabilities" (see Chapter 7 in this volume). Such traits are commonly normally distributed across the population. Some researchers have speculated that, as science in this area progresses, it may be possible to match educational approaches to individual students not based on specific syndromes they have, or even on the basis of well-recognized categories of learning disability, but simply on their overall genetic learning profile – the student's unique combination of genetically based reading and mathematical skills and sub-skills, memory and recall, as well as learning-relevant non-cognitive personality traits.

But there are reasons to suspect that this is a fantasy of personalized education that no public school system will ever be able achieve. This in part because of the enormous and likely prohibitive expense involved in training and staffing a school system with educators who are actually able to tailor educational approaches to the unique combinations of variable cognitive and non-cognitive learning skills possessed by each student. But it is more fundamentally a function of what we already know about the genetic science of learning.

The rough-and-ready estimate is that genetic variation among students explains about half of the variation in their academic achievement. Different studies have posited different levels of heritability for different species and sub-species of learning ability, but "roughly half" is a reasonable estimate for purposes of thinking about broad policy questions. That is in some ways a stunningly high number – no doubt many people would be surprised to learn that genetics has so very much influence on school performance. But it is also a low enough number to guarantee that, no matter how much better we get at understanding and mapping the associations of particular genetic variations with particular capacities for learning, the information about a given student's unique learning capacity that can be provided by genetic testing will never be a good substitute for the information we can gain from plain, old ordinary in-person assessment of that student's classroom performance.

This should be no surprise. If half of the variation in a population's academic performance is explained by genetic factors, the other half is explained by environmental factors that can vary tremendously, not only between families in the same community, but even for siblings in the same home. The assumption of environmental similarity in twin studies has been shown not to hold for siblings of different ages (Plomin, DeFries, Knopik, & Neiderhiser, 2016). This is both because home environments change with time as more siblings are born, parents age, and family income and security levels change, and also because genetic differences among siblings influence the ways in which their families, teachers, and communities respond to them. The developmental impact of lower socio-economic status may diminish the heritability of learning-related traits in those populations (see Chapter 9 of this volume). Tan (Chapter 1 of this volume) points out that large-scale environmental factors such as recession and natural disaster can actually diminish the heritability of learning-related traits within a given population temporarily, making genetic data less informative in some time periods and in some populations than in others. Finally, as discussed in Chapter 4 of this volume, epigenetic factors in early life can greatly alter the function of the genome. In sum: even given a goal of "personalized education" which tailors teaching approaches to the particular strengths and weaknesses of particular students, with the complex effects of gene X environment interactions and epigenetics on phenotypical trait expression mean that it is likely to remain more efficient to "personalize" on the basis of direct information gathered about student performance and learning style through educational testing and interpersonal contact, rather than to "personalize" by enlisting genetic data.

Policy formation problems in this sphere are apt to be more challenging than in either of the two areas earlier discussed. There are understandable benefits to making a genetic confirmation of a (known-to-be) disabled child's particular condition, in order to make a better job of individualizing educational plans you were planning to individualize anyway. There are also obvious benefits to discovering addressable learning disabilities early, before they have undesirable psychological impact on children. But the case for the social benefits of pinning down the genetic roots of particular children's learning difficulties (and/or gifts) – particularly when the difficulties cannot be defined as "disabilities" – is harder to make. How much money, how much teacher training, how many public consultations with genetic and special education experts, can we justify in the quest to pin down the multiple and difficult-to-characterize genetic learning traits of a given child – traits which may already have been altered by the environment, and which, while they are broadly predictive of the characteristics of populations, are only problematically (and probabilistically) predictive of a given student's performance and needs? Moreover, the science in this area is apt to be provisional for a long time, which means that policies and standards and consent forms will need to be revisited with some frequency, as our understanding of genetic interactions and gene/environment interactions and epigenetic effects changes our view of the implications of the genetic data we have in hand.

To sum up: there are a relatively few ways in which the extraordinary progress in our understanding of the genetics of learning might actually affect teaching in the public schools. It is most likely to help in our better addressing the needs of students already recognized as disabled. It is possible that it may also help us uncover and provide early intervention for certain specific learning disabilities such as dyslexia, though the problem of how to pay for and manage the population-wide screening program that would best justify such genetic screening – or the alternative problem of how to justify a program that fails to include the whole population – is vexing. For the vast majority of learning difficulties, however – those which result from the small effects of very many genes, combined with environmental and epigenetic effects – it is unlikely for many practical reasons that our progress in science will have much practical payoff in the school system. For most students' learning challenges, we would be better off relying on old-fashioned testing and observation rather than genetic testing. This is both because the old-fashioned methods capture environmental and epigenetic impacts that gene testing does not, and because it is almost undoubtedly true that scarce school system resources would be better spent on high-impact environmental changes (extra teacher time, school meals, preschool programs, etc.) than on gathering genetic data.

Regardless of whether this relatively pessimistic assessment of the utility of genomics for education is correct, it remains imperative that educators and school systems get ready to deal with the inevitable requests that they take account of and use students' genetic data. It is very likely that, very soon, some parents will obtain genetic information about their children's genetic learning traits from private firms, and will come into school meetings with that data in hand, asking for individually tailored educational plans for their children. Schools need to be ready either to make use of that information or to explain why making use of it would be inappropriate, either because of the facts in a particular case or because of concerns about systemwide fairness. Other parents will be given genetic data about their children as part of their ordinary medical care, or even as part of routine newborn screening. Those parents may well come to the school system asking for interpretations of that data, or for clarity as to how it might be used to inform their children's education. Requests for use of genetic data may also come from sources other than parents. Forward-thinking teachers or school administrators may challenge a school system to create a cutting edge educational program and to begin to implement genetic testing of students; advocates for disabled students may press school systems into using genetic data for early detection of and intervention for learning disabilities.

Wherever the requests come from, it is clear that educators will need to familiarize themselves with the many difficult scientific and ethical questions, the answers to which will determine the value of genetic testing in education. Unless such questions are thoughtfully considered before genetics moves into the classroom, we risk making a range of mistakes from missing important opportunities for children to mistakenly channeling and stigmatizing children on the basis of their genes. This is a challenging issue that crosses disciplines and will require bringing together the diverse voices of educators, geneticists, ethicists and the lay public. The sooner we can begin the conversation the better prepared we can be to make wise choices about whether and how we might best incorporate genomics into education, the better to serve our children.

REFERENCES

de Graaf, G., Buckley, F., & Skotko, B. G. (2015). Estimates of the live births, natural losses, and elective terminations with Down syndrome in the United States. *American Journal of Medical Genetics A, 167A*(4), 756–767.

Greely, H. T. (2016). In 20 to 40 years, most Americans won't have sex to reproduce. Get ready. *Vox.* Retrieved from www.vox.com/2016/9/16/12931962/future-sex-reproductive-technology-ethics-ivf/

Leachman, M., Albares, N., Masterson, K., & Wallace, M. (2016). Most states have cut school funding, and some continue cutting. Center on Budget and Policy Priorities. Retrieved from www.cbpp.org/research/state-budget-and-tax/most-states-have-cut-school-funding-and-some-continue-cutting.

Murphy, S. (2016, November 16). Oklahoma voters reject sales tax hike for public education. *Lincoln Journal Star.* Retrieved from http://journalstar.com/news/national/govt-and-politics/oklahoma-voters-reject-sales-tax-hike-for-public-education/article_0b3a7876-ffde-545f-bf39-a4f913399539.html.

McCann, C. (2016, December 1). IDEA funding. Retrieved from www.edcentral.org/edcyclopedia/individuals-with-disabilities-education-act-funding-distribution/.

Natoli, J. L., Ackerman, D. L., McDermott, S., & Edwards, J. G. (2012). Prenatal diagnosis of Down syndrome: A systematic review of termination rates (1995–2011). *Prenat Diagn, 32,* 142–153.

Plomin, R., DeFries, J. C., Knopik, V. S. A., & Neiderhiser, J. M. (2016). Top 10 replicated findings from behavioral genetics. *Perspectives on Psychological Science, 11*(1), 3–23.

Powers, N. R., Eicher, J. D., Butter, F., Kong, Y., Miller, L. L., Ring, S. M., et al. (2013). Press alleles of a polymorphic ETV6 binding site in DCDC2 confer risk of reading and language impairment. *American Journal of Human Genetics, 93*(1), 19–28.

INDEX

twin studies (*cont.*)
  on heritability of autism spectrum
    disorder, 35
  on intellectual disabilities, 110
  intelligence and, 22, 113
  Minnesota Study of Twins Reared Apart, 27
  on reading ability and disability, 164
  UK study on General Certificate of
    Secondary Education and, 71–72
Twins Early Development Study, 207, 208

U.S. Presidential Commission for the Study of
  Bioethical Issues, 375

values
  defined, 148
VCF syndrome, 191

Watson, Dr. John, 108
Western Reserve Reading and Math
  Project, 208
whole-exome sequencing (WES), 58–59
Williams syndrome, 52, 69
  educational interventions
    based on knowledge of genetic
    etiologies, 193
  personality and vulnerability in, 189–91